Biopolymer Composites Handbook

Biopolymer Composites Handbook

Editor: Lauren Dixon

MURPHY & MOORE
www.murphy-moorepublishing.com

www.murphy-moorepublishing.com

MURPHY & MOORE

Cataloging-in-Publication Data

Biopolymer composites handbook / edited by Lauren Dixon.
p. cm.
Includes bibliographical references and index.
ISBN 978-1-63987-723-2
1. Biopolymers. 2. Polymeric composites. I. Dixon, Lauren.
QP801.B69 B56 2023
572.33--dc23

Murphy & Moore Publishing
1 Rockefeller Plaza,
New York City,
NY 10020, USA

ISBN 978-1-63987-723-2

Contents

Preface

Over the recent decade, advancements and applications have progressed exponentially. This has led to the increased interest in this field and projects are being conducted to enhance knowledge. The main objective of this book is to present some of the critical challenges and provide insights into possible solutions. This book will answer the varied questions that arise in the field and also provide an increased scope for furthering studies.

Biopolymer composites are those composites, which have one biodegradable and bio-based constituent. These include natural fibers which are combined with natural or synthetic biopolymers. Natural fibers are biodegradable, renewable and carbon neutral resources. This versatility makes them a potential alternative of synthetic fibers in the biopolymer composite industry. Natural fibers are categorized into animal and plant fibers. The performance of biopolymer composites improves through chemical alteration, which is based on the processing needs, type and environmental circumstances. Some common processing methods utilized for biopolymer composite preparations include extrusion, compression molding and injection molding. The composites which are formed through synergistic combination of organic and inorganic components at the molecular or nanometer level are known as hybrid materials. These are distinguished from standard composites since their constituents differ at the macroscopic level. This book provides comprehensive insights with respect to biopolymer composites. It will prove to be immensely beneficial to students and researchers in this field.

I hope that this book, with its visionary approach, will be a valuable addition and will promote interest among readers. Each of the authors has provided their extraordinary competence in their specific fields by providing different perspectives as they come from diverse nations and regions. I thank them for their contributions.

Editor

1

Tensile and Shear Testing of Basalt Fiber Reinforced Polymer (BFRP) and Hybrid Basalt/Carbon Fiber Reinforced Polymer (HFRP) Bars

Kostiantyn Protchenko *, Fares Zayoud, Marek Urbański and Elżbieta Szmigiera

Faculty of Civil Engineering, Warsaw University of Technology, 00-637 Warsaw, Poland; fares.zayoud.stud@pw.edu.pl (F.Z.); m.urbanski@il.pw.edu.pl (M.U.); e.szmigiera@il.pw.edu.pl (E.S.)

* Correspondence: k.protchenko@il.pw.edu.pl

Abstract: The use of sustainable materials is a challenging issue for the construction industry; thus, Fiber Reinforced Polymers (FRP) is of interest to civil and structural engineers for their lightweight and high-strength properties. The paper describes the results of tensile and shear strength testing of Basalt FRP (BFRP) and Hybrid FRP (HFRP) bars. The combination of carbon fibers and basalt fibers leads to a more cost-efficient alternative to Carbon FRP (CFRP) and a more sustainable alternative to BFRP. The bars were subjected to both tensile and shear strength testing in order to investigate their structural behavior and find a correlation between the results. The results of the tests done on BFRP and HFRP bars showed that the mechanical properties of BFRP bars were lower than for HFRP bars. The maximum tensile strength obtained for a BFRP bar with a diameter of 10 mm was equal to approximately 1150 MPa, whereas for HFRP bars with a diameter of 8 mm, it was higher, approximately 1280 MPa. Additionally, better results were obtained for HFRP bars during shear testing; the average maximum shear stress was equal to 214 MPa, which was approximately 22% higher than the average maximum shear stress obtained for BFRP bars. However, HFRP bars exhibited the lowest shear strain of 57% that of BFRP bars. This confirms the effectiveness of using HFRP bars as a replacement for less rigid BFRP bars. It is worth mentioning that after obtaining these results, shear testing can be performed instead of tensile testing for future studies, which is less complicated and takes less time to prepare than tensile testing.

Keywords: Fibers Reinforced Polymer (FRP); Hybrid Fibers Reinforced Polymer (HFRP); Basalt Fibers Reinforced Polymer (BFRP); tensile strength testing; shear strength testing

1. Introduction

Concrete structures built during the previous century will be in need of modernization and renovation, which will lead to an increase in resources and costs associated with these buildings. Hence, it is necessary to develop inexpensive and efficient modernizing techniques in order to prevent structural damage. Throughout the years, researchers have experimented with different materials in various contexts, and Fiber Reinforced Polymers (FRP) have proved to be an applicable material for strengthening and reinforcing concrete structures.

For decades, FRP has been widely used in the construction industry as a substitute for steel due to its lightweight, increased corrosion resistance, and improved strength. FRP can be used in a wide range of applications in construction, such as reinforcement of concrete, formwork, modular structures, bridge decks, and external reinforcement for strengthening structures.

FRP reinforcement exhibits a linear elastic behavior prior to failure but fails in a brittle manner. The weakness of FRP reinforcement in terms of its mechanical behavior is observed in the transverse direction, which can be seen in shear resistance testing, where the strength characteristics of tested samples are significantly reduced. Moreover, the FRP reinforcement lacks plastic behavior. However, FRP reinforcement has several advantages, such as high tensile strength in the longitudinal direction, long durability, high corrosion resistance, and easy application.

Tensile tests have been conducted by several researchers over the past decades on different FRP bars to determine their mechanical and physical properties [1–7]. Conclusions drawn from these earlier studies [4–10] indicate that the mechanical characteristics of FRP bars are mainly dependent on fiber volume and type. Studies have also shown that FRP bars exhibit elastic linear behavior up to failure, with a modulus of elasticity lower than that of steel.

In the studies of Walsh, Chang, and Wallenberger et al. [8–10], extensive research can be found on the most common fibers used to manufacture FRP rebars, such as carbon, aramid, and glass. Hollaway et al. [11] found that fibers exhibited a linear elastic behavior under tensile loading up to failure without showing any yield.

Nanni et al. [12] investigated the tensile properties of hybrid rods for concrete reinforcement. The tested specimens were made of 1.0 m long rods, including the anchorage system. The results of uniaxial tensile testing, conducted in order to determine stress–strain relationships and tensile properties, showed that the stress–strain diagrams displayed a bilinear behavior.

Zhishen Wu et al. found that the tensile modulus of the fibers influenced the fatigue failure mode, where the Glass Fiber Reinforced Polymers (GFRP) and Basalt Fiber Reinforced Polymers (BFRP) composites exhibited transverse cracks that indicated a stress transfer to the adhesive layer with a modulus of about 90 GPa, whereas the Carbon Fiber Reinforced Polymers (CFRP) and Poly(p-phenylene benzobisoxazole) (PBO) composites showed longitudinal cracks along the test coupons with a modulus of approximately 250 GPa [13].

Marta Kosior-Kazberuk performed an investigation on the application of Basalt FRP (BFRP) bars and Hybrid FRP (HFRP) bars as reinforcement for full-scale concrete beams. The results showed that the FRP bars used in the tensile zone worked in a similar way to steel reinforcement. Rapid destruction in the tensile zone with crushing of concrete in the compression zone, deformation of compressed bars, stirrups breaking within cracks, and failure of tensile reinforcement were observed [14].

Shi et al. [15] investigated the tensile behavior of FRP and Hybrid FRP sheets in Freeze–Thaw (FT) cycling environments. Additional degradation of the tensile strength and rupture elongation of FRP sheets were observed as a result of sustained loading during FT cycling, while the elastic modulus of the FRP sheets was not influenced by the sustained loading during FT cycling. Hence, it is necessary to consider service loads during durability tests to reflect the actual conditions faced by FRP composites in harsh civil engineering environments.

Many researchers have studied the shear strength behavior of different FRP bars. The limited application of BFRP in the past decades led to the limited investigation of its shear behavior. However, Benmokrane et al. [16] proved that the type of the resin plays a significant role in the inter-laminar shear strength of BFRP bars, as BFRP bars with epoxy resin showed higher durability than BFRP bars with vinyl ester resin. In addition, the durability of GFRP bars was lower than for BFRP bars with epoxy resin.

Alam et al. [17] studied the shear strength of FRP reinforced members without transverse reinforcement and reported that the normalized shear strength increased linearly with the cube root of the axial stiffness of the reinforcing bars. However, there were similar results for the shear strength of FRP reinforced elements where it was proportional to the axial stiffness of the longitudinal reinforcement obtained by Alam et al. [17], El-Sayed et al. [18,19], and Razaqpur et al. [20]. Hence, the conclusion that the shear strength was proportional to the amount of longitudinal reinforcement was reached by Alkhrdaji et al. [21], and the effect of the reinforcement stiffness and amount of reinforcement on the shear strength of FRP elements, despite no significant influence of the longitudinal reinforcement ratio on the shear strength, was observed by Yost et al. [22].

Khalifa et al. [23], Täljsten [24], and Triantafillou [25] developed a theory that aims at describing FRP stress distribution along a shear crack with closed-form equations, as opposed to the regression-based formula that Triantafillou et al. [26] introduced. FRP contribution to the resisting shear and the FRP resultant across the crack can be computed when this formula is correctly defined.

Based on the literature, axial tensile testing is most commonly used for defining the mechanical properties of composite bars, and some of the researchers have studied the shear properties of the bars through tensile testing. Current research is concentrated on the defining properties of BFRP and HFRP bars through axial-tensile testing and shear testing, as well as finding a correlation between these two testing methods.

The variables in this study were types of FRP bars and the diameter of bars. The outcomes were compared, and a correlation between the results was found. Establishing this correlation allowed the identification of mechanical properties of FRP bars basing on shear testing, which is easier to prepare and takes less time than tensile testing.

Purpose and Novelty of the Work

RC bending elements are the basic structural components that cover the operational load as well as the load from partition walls and transfer them further to vertical elements (such as columns and load-bearing walls). Reinforcement of these elements should be characterized by high strength values and appropriate corrosion resistance. In order to determine these properties, strength tests should be carried out. On the one hand, the tensile strength test for FRP bars is complex, relatively labor-intensive, and expensive. On the other hand, testing the shear strength of FRP bars is simple, low-cost, and fast. By carrying out a quick and simple shear test, it is possible to determine the tensile strength and stiffness (modulus of elasticity) of both homogeneous BFRP bars and HFRP bars on the basis of the method presented in the article. Additionally, the method allows optimizing the tensile strength and stiffness based on the determined correlation between shear strength and tensile strength.

2. The Concept of Hybrid FRP Bars

In spite of the fact that FRP has numerous prevalent material properties, such as high corrosion resistance, high specific stiffness, high specific strength, and durability, the brittle nature and high cost of FRP limits its extensive usage in many industries. To overcome these obstacles, combinations of FRP and ordinary materials are being explored by several researchers [27]. Appropriate properties can be obtained by utilizing a combination of FRP and steel [28,29] or by combining different sorts of FRP materials in one structure [30]. The advantages of hybrid structural systems include cost-effectiveness and the ability to optimize the cross-section based on the material properties of each constituent material [31].

HFRP bars can be an optimal solution since hybrid bars have a better combination of mechanical properties and are less expensive than FRP bars with homogeneous fibers. On the other hand, they are corrosion resistant and have significantly better strength characteristics than conventional steel reinforcements [32]. Hybridization between different constituents focuses on picking out the advantages of each constituent, while the disadvantages can be improved [33].

Due to the high cost of carbon, the ideal solution is the usage of HFRP bars, where the part of low-cost basalt fibers was substituted by high-cost carbon fibers. Hence, the hybridization of carbon/basalt fibers is less expensive than CFRP; concurrently, this combination is characterized by better mechanical properties in comparison with BFRP [34–36].

Carbon fibers are chosen due to their high properties in the longitudinal direction as well as their strong anisotropy. However, basalt fibers were chosen for their environmentally friendly producing process and low-cost. In addition, basalt fibers are significantly less brittle when used in composites [37]. An additional reason for such selection was similar strain parameters for both types of fibers. Mechanical properties of the constituents utilized for preparing HFRP bars are represented in Table 1.

Table 1. Properties of constituents utilized for preparing Hybrid Fiber Reinforced Polymers (HFRP) rebars [34].

Properties	Units	Carbon Fibers LS	Basalt Fibers	Epoxy Resin
Density	g/cm^3	1.90–2.10	2.60–2.80	1.16
Diameter	μm	7.00–11.00	11.20–13.40	–
E_{11}	GPa	232.00	89.00	3.45
E_{22}	GPa	15.00	89.00	3.45
v_{12}	–	0.28	0.26	0.35
v_{23}	–	0.49	0.26	0.35
G_{12}	GPa	24.00	21.70	1.28
G_{23}	GPa	5.03	21.70	1.28
σ_{11}	MPa	2500–3500	1153–2100	55–130

E_{ii} is the modulus of elasticity along axis i, v_{ij} is the Poisson ratio that corresponds to a contraction in direction j when an extension is applied in direction i, G_{ij} is the shear modulus in direction j on the plane whose normal is in direction i, and σ_{ii} is the tensile strength in the direction i [34].

Composite materials properties can be calculated according to the Rule of Mixtures (ROM) (axial loading–Voigt model) based on the literature [34–36]. Based on the Rule of Mixtures (ROM) equation, Young's modulus and other parameters were determined for the Hybrid Carbon/Basalt FRP (HC/BFRP) and for different combinations depending on carbon to basalt fiber volume fractions. Different levels of fibers substitution (carbon-to-basalt C/B) were proposed: 1:1, 1:2, 1:3, 1:4, and 1:9, which is represented in Figure 1 [36].

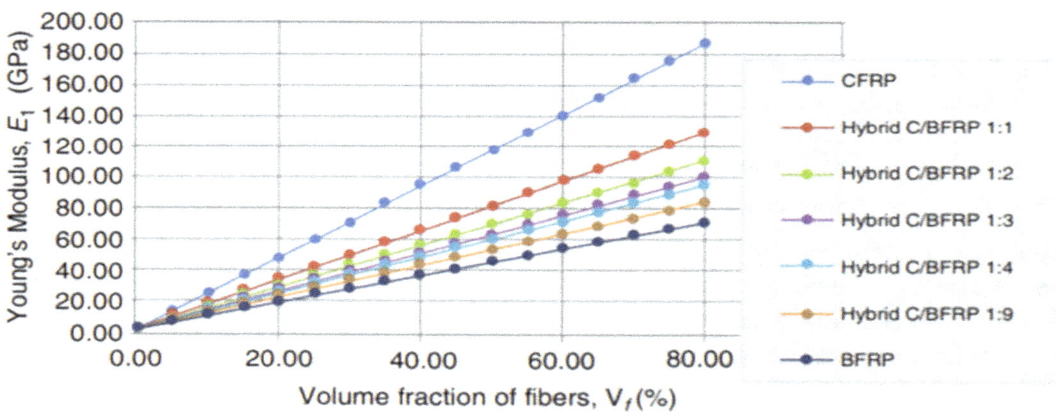

Figure 1. Theoretical relationship of fiber volume fraction and Young's modulus (obtained from Voigt's Model) for Hybrid Carbon/Basalt FRP [36,37].

The Voigt model does not consider the location of the fibers; therefore, numerical analysis was made for two configurations: (i) carbon fibers in the core region and basalt fibers in the near-surface region, (ii) carbon fibers in the near-surface region and basalt fibers in the core region of the bar section. The numerical simulation of the tensile strength test for HFRP bars was performed by Finite Element Methods (FEM). The bars consisted of two parts, the core region (cylindrical form) and the surface region (tube form), which were perfectly interconnected. The constant pressure was applied on both sides along the longitudinal axis. The obtained results from numerical modeling were compared with analytical considerations and were found to be in a good convergence with each other.

Numerical results indicated that the arrangement of the fibers was not of high importance for the final mechanical properties of HFRP bars with different combinations of fibers. More on the analytical and numerical investigation of Hybrid FRP bars can be found in companion papers [36,37]. Two different HFRP bars' configurations were produced by the manufacturing company. For the first configuration, carbon fibers were placed in the core region, while the second configuration was when carbon fibers were in the near-surface region. However, some technological issues were observed while placing carbon near-surface, increasing heterogeneity in fiber distribution, and local scorching of carbon fibers caused by rising temperature. Finally, carbon fibers were placed mostly in the core region of the bar.

HFRP bars utilized for this work were composed of epoxy matrix and basalt fibers, with a volume ratio of 1:4 (i.e., 16% carbon fibers, 64% basalt fibers, and 20% matrix). Due to this substitution, HFRP bars are characterized by a much higher stiffness, which enables more efficient application as reinforcement for concrete elements subjected to bending and compression [37–39]. More about the hybridization of FRP bars used in work can be found in these companion papers [38,39].

In the tensile and shear strength tests conducted, two types of FRP were used: BFRP and HFRP. Twisting FRP braids were done in order to improve adhesion with concrete (the equivalent of ribbing for steel bars) [36].

3. Tensile Strength Test

3.1. Experimental Procedure

The tensile strength testing of different BFRP and HFRP bars was carried out according to the ACI 440.3R and ASTM D7205/D7205M methodologies [40].

In the tensile strength test, anisotropic FRP materials of different diameters were tested according to the methodologies. The two different types of FRP bars had different diameters, specifically Ø6, Ø8, Ø10, Ø12, Ø14, and Ø18 mm. Hence, for each type of BFRP and HFRP bar, five samples of each diameter were tested. In total, tensile tests were carried out on sixty bars.

Basalt fiber strength in the transverse direction is very low compared to its extremely high longitudinal strength. Hence, the use of appropriate anchorage at both ends of the bar subjected to the tensile test on the testing apparatus is a requirement. To meet this requirement, two steel pipes with a length of 400 mm, an external diameter of 40 mm, and a thickness of 5 mm each were designed. The steel caps at the end of the steel pipes were designed with an opening in the center for the bar.

The free space between the bar and the pipe was filled with a special adhesive layer. The material filling the anchoring pipes was determined on the basis of previous tests done at the Warsaw University of Technology [39,40]. Figure 2a shows the BFRP bars prepared for tensile strength testing; Figure 2b shows the HFRP bars in the tensile testing machine.

(a)

(b)

Figure 2. (**a**) Basalt Fiber Reinforced Polymers (BFRP) bars prepared for testing, (**b**) Hybrid Fiber Reinforced Polymers (HFRP) bars in the tensile testing machine.

The tensile strength test was carried out in accordance with [40] standard for pultruded FRP bars. However, the tensile strength of the specimens can be calculated by dividing the measured load by the transverse cross-sectional area of FRP bars of the corresponding type, $A_{i,FRP}$. To experimentally determine the modulus of elasticity of FRP bars, Equation (1) can be used.

$$E_{11,iFRP} = \frac{(P_1 - P_2)}{(\varepsilon_1 - \varepsilon_2) \cdot A_{i,FRP}} \quad (1)$$

where:

- $E_{11,iFRP}$ —modulus of elasticity of corresponding FRP bars along the longitudinal axis,
- P_1 and P_2—applied loads corresponding to 50% and 25% of the ultimate load, respectively,
- ε_1 and ε_2—strains corresponding to 50% and 25% of the ultimate strains, respectively, $A_{i,FRP}$—cross-sectional area for BFRP and HFRP bars, respectively.

3.2. Calculations and Results

The deformation measurement was performed up to 70% of the breaking load of the bars due to the possibility of strain gauge damage.

The average values obtained for BFRP and HFRP bars of tensile strength, tensile strain, and tensile modulus for six different diameters of each type are presented in Tables 2–4, respectively.

Table 2. Tensile strength results of FRP bars.

Bar	Parameter	Ø6	Ø8	Ø10	Ø12	Ø14	Ø18
BFRP	f_t, MPa	1148.81	1103.33	1152.54	1111.00	1101.94	877.98
	SD *, MPa	41.18	22.87	35.27	40.23	28.54	11.17
	COV **, %	3.58	2.07	3.06	3.62	2.59	1.27
HFRP	f_t, MPa	1083.78	1277.92	1138.65	1008.52	1160.06	809.44
	SD *, MPa	59.64	55.41	34.89	36.51	44.75	19.03
	COV **, %	4.72	4.34	3.06	3.62	3.86	2.35

* SD–Standard Deviation; ** COV–Coefficient of Variation.

Table 3. Modulus of elasticity of FRP bars.

Bar	Parameter	Ø6	Ø8	Ø10	Ø12	Ø14	Ø18
BFRP	E_f, GPa	46.47	43.87	44.27	47.63	46.02	42.20
	SD, GPa	2.58	0.86	1.71	0.89	0.23	1.58
	COV, %	5.55	1.95	3.56	1.87	0.49	3.74
HFRP	E_f, GPa	72.71	73.89	64.77	53.35	72.12	52.43
	SD, GPa	1.67	3.07	3.50	3.93	2.20	2.46
	COV, %	2.29	4.15	4.76	5.78	3.05	4.70

Table 4. Strains at FRP bars rupture.

Bar	Parameter	Ø6	Ø8	Ø10	Ø12	Ø14	Ø18
BFRP	ε_t, %	2.48	2.52	2.40	2.33	2.39	2.08
	SD, GPa	0.15	0.05	0.15	0.11	0.06	0.10
	COV, %	5.92	2.09	6.09	4.60	2.71	4.84
HFRP	ε_t, %	1.74	1.73	1.55	1.49	1.61	1.55
	SD, GPa	0.08	0.07	0.10	0.10	0.06	0.04
	COV, %	4.38	4.33	6.65	6.70	3.73	2.59

Figure 3 displays the stress–strain relationships of tension FRP bars tested with mean values for diameters Ø6, Ø8, Ø14, and Ø18 mm.

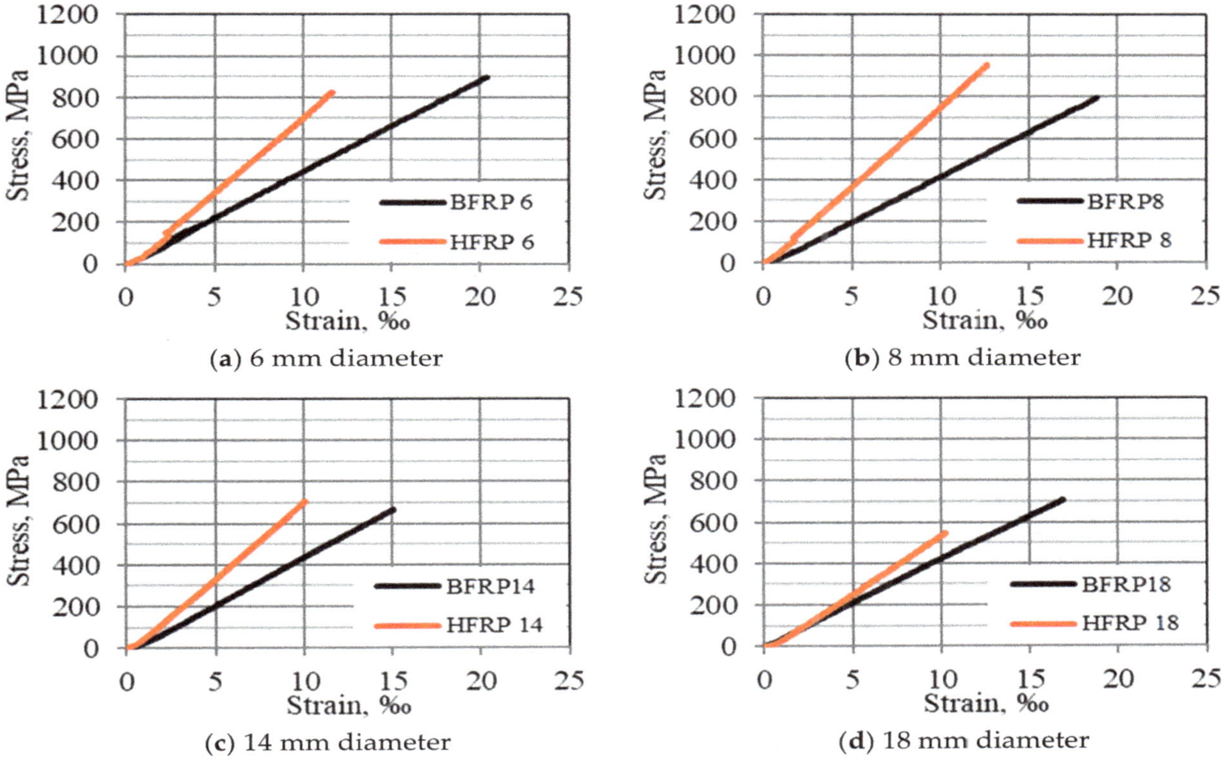

Figure 3. Stress–strain relationships of tension Fiber Reinforced Polymers (FRP) bars with mean values for diameters: (**a**) Ø6, (**b**) Ø8, (**c**) Ø14, (**d**) Ø18 mm.

In the tensile property values obtained for BFRP and HFRP, the deformation values obtained for HFRP were lower than the analogous deformation values for BFRP at rupture.

Additionally, comparing the (Coefficient of Variation, COV) values of the two types of FRPs used in the test, much higher (COV) values were obtained for HFRP bars compared to the tested BFRP bars. However, this phenomenon might occur due to technical issues indicated in the previous chapter, which might occur during the manufacturing of bars composed of several FRP roving. Hence, the (COV) value depends on the diameter of the bars. The average (COV) values for BFRP bars were twice as small as the corresponding COV values for HFRP. The changes in mechanical properties for the different diameters and FRP types are represented in Figure 4.

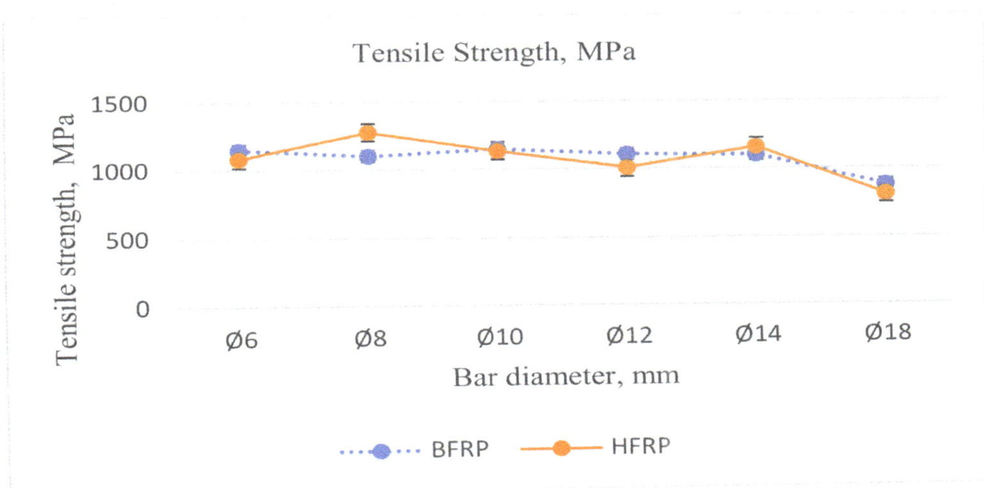

(**a**)

Figure 4. *Cont.*

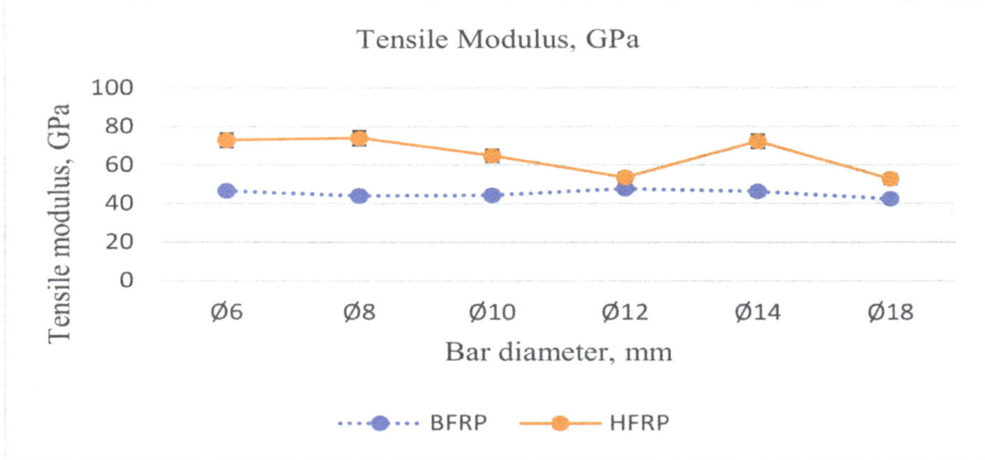

(b)

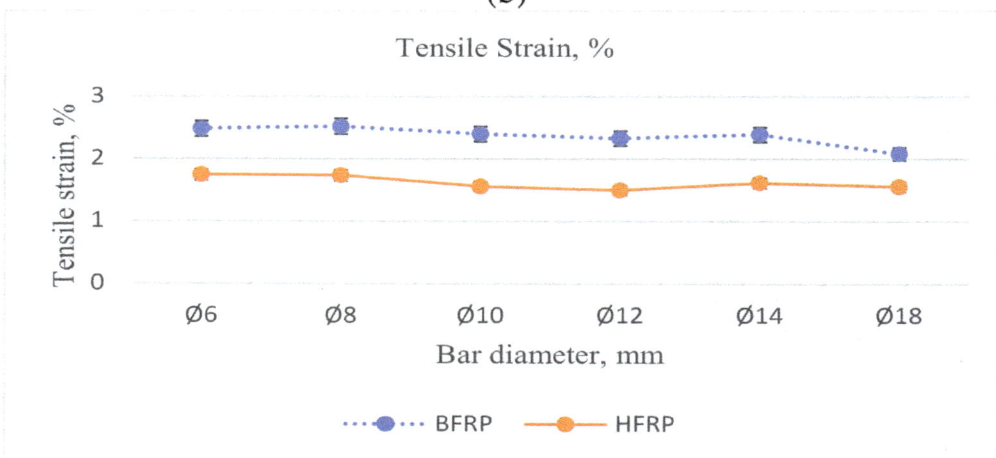

(c)

Figure 4. Mechanical properties of FRP bars obtained from tensile testing: (**a**) Tensile strength, (**b**) Modulus of elasticity, (**c**) Tensile strain.

Comparing the tensile strength of BFRP and HFRP bars with diameters of Ø8 and Ø14 mm, the values for BFRP bars of diameter Ø8 mm were smaller than HFRP bars of the same diameter by 15% and slightly smaller for bars of diameter Ø14 mm by approximately 5%. However, the tensile strength values for BFRP bars with diameters Ø6, Ø10, Ø12, and Ø18 mm were slightly higher than HFRP bars with the same diameters. Values for BFRP bars with diameters Ø6, Ø12, and Ø18 mm were slightly higher than HFRP bars by approximately 5%, and for diameter Ø10 mm by approximately 2%.

The rupture strain for all HFRP bar diameters is about 1/3 lower than the rupture strain of the BFRP bars. This is due to the smaller ultimate tensile strains of the carbon fiber (compared to the strain of basalt fiber) contained in the HFRP bars.

Observations made during the study confirm theoretical forecasts. The ultimate strain of hybrid bars corresponds to the strain of its fibers, which are characterized by the lowest ultimate strain among the fibers contained in the HFRP bar. The values obtained for the tensile modulus of elasticity for HFRP bars for all diameters were greater than the values obtained for the tensile modulus of elasticity for BFRP bars for all diameters.

The results obtained during this test confirm the viability of using HFRP bars as a replacement for less rigid BFRP bars.

4. Shear Strength Testing

4.1. Experimental Procedure

The shear strength test of FRP bars was carried out in accordance with the methodology described in ACI 440.3R-04 and ASTM D7617/D7617M–11 [40].

The shear strength testing was carried out on BFRP and HFRP bars with diameters Ø6, Ø8, Ø10, Ø12, Ø14, and Ø18 mm. For the shear strength test, in order to ensure that the bar sample shears in two planes simultaneously, a steel device was constructed with blades converging along a surface perpendicular to the sample axis. The device consists of one upper blade, two lower blades, and a sample holder. The sample holder has a V-shaped groove for placing FRP samples and a rectangular notch for holding the lower and upper blades in the center of the upper part of the device. The sample holder has 100 mm width, 110 mm height, and 230 mm length. It is attached to the instrument stand with screws that stabilize it in order to eliminate horizontal movement of the bar. Figure 5 represents the components of the shear testing device and the shear failure of the FRP bar. A total of 60 specimens of BFRP and sixty specimens of HFRP bar were tested, where five test specimens were produced for each diameter of every FRP bar type used during the shear strength test.

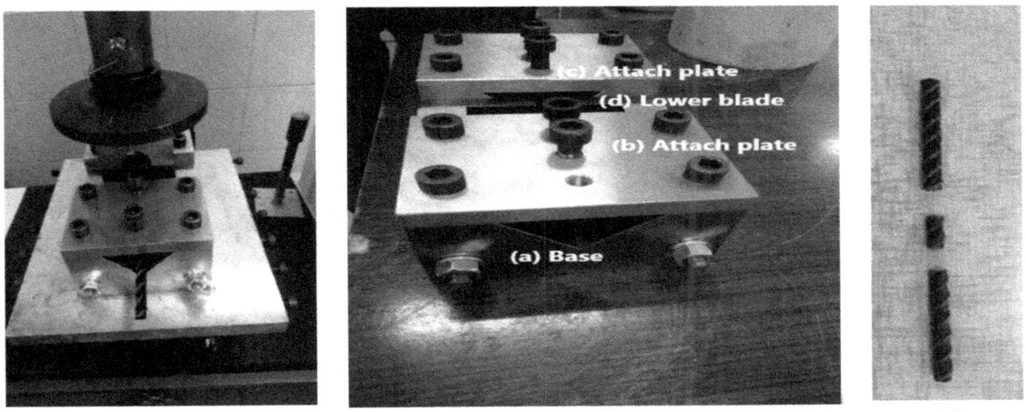

(a) (b) (c)

Figure 5. (**a**) Double shear testing device, Instron 3382, (**b**) Description of Instron 3382 testing machine, (**c**) Shear failure of the FRP bar.

All FRP bars that were tested for shear strength were composed of a spiral braid of basalt fiber. All the FRP bars were tested at speeds up to 50 MPa/min for a range of time between 3.5 min and 4.5 min, with the test ending when a sharp decrease in shear stress was observed.

According to ACI 440.6-08 and ASTM A615/A615M [40] methodologies, shear stress fs was calculated according to Equation (2) using an equivalent cross-section of bars.

$$f_s = \frac{F_u}{2 \cdot A_{i,FRP}} \quad (2)$$

where:

- F_u —the maximum shear force, N.
- $A_{i,FRP}$ —the equivalent cross-section of the tested bar, mm^2.

4.2. Calculations and Results

Failure of the tested bars due to shearing could occur in two ways; two failure modes are shown in the Figure 6. The blue "B" line (straight line), which occurred most often, was typical for simultaneous shearing of a bar in two planes. In contrast, the red "R" (dot line) graph showed the situation when shearing occurred in one and then slightly later in the other cross-section of the tested bar.

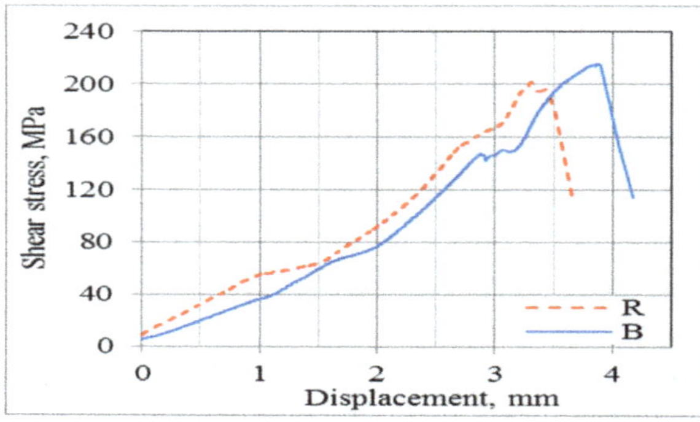

Figure 6. Stress-displacement diagram of transverse shear test for two HFRP bars with a mean of 14 mm.

Figure 7 shows a relationship of shear stress to strain for three types of BFRP and HFRP bars with nominal diameters of 6, 8, 10, 12, 14, 18 mm.

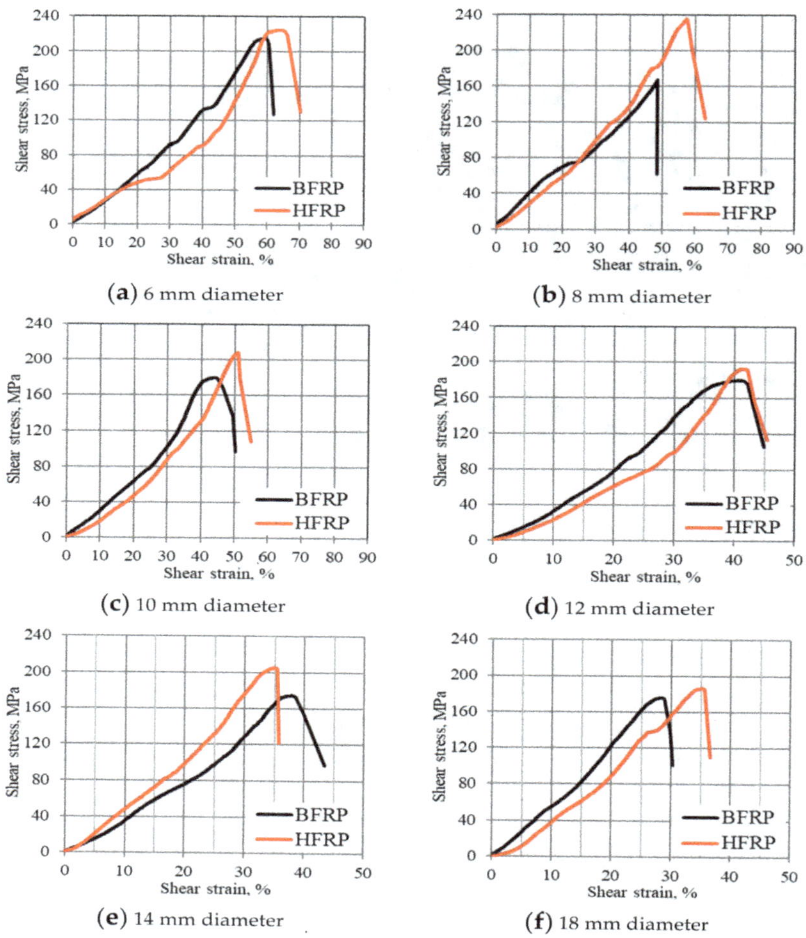

Figure 7. Shear stress–strain relationship of FRP bars tested with means. **(a)** Ø6, **(b)** Ø8, **(c)** Ø10, **(d)** Ø12, **(e)** Ø14, **(f)** Ø18 mm.

It was impossible to clearly determine the values of the shear modulus due to the non-uniform slope of the shear curves, in contrast to the stress–strain relationship. Comparing BFRP bars to HFRP bars, it was noticed that BFRP bars had greater shear deformation and less shear strength compared to HFRP bars. Table 5 shows shear strength values obtained for different FRP bars.

Table 5. Shear strength testing results of FRP bars for different diameters.

Bar	Parameter	Unit	Ø6	Ø8	Ø10	Ø12	Ø14	Ø18
BFRP	f_s	MPa	210.66	205.37	175.33	185.46	172.85	172.25
	SD	MPa	3.84	11.57	6.46	4.70	8.59	6.90
	COV	%	1.82	5.63	3.68	2.54	4.97	4.01
HFRP	f_s	MPa	202.04	229.44	198.31	187.09	196.84	176.35
	SD	MPa	17.51	4.06	6.17	5.99	12.06	9.75
	COV	%	8.67	1.77	3.11	3.20	6.13	5.53

The type of FRP bar affects the COV variation for shear strength. The quite significant spread of FRP bar test results confirms that they are heterogeneous materials consisting of a matrix and fibers. Studies show that the average lateral shear strength of BFRP bars ranges from 170 to 210 MPa. While for HFRP bars, it is up to 20% higher due to the participation of carbon fibers.

The transverse shear strength displays a slight downward trend as the bar diameter increases, most evidently in HFRP bars. However, minor variation is observed in the shear strength data, with the COV greater than in the rebar tension tests. Figure 8 shows the comparison of shear strength of FRP bars for different diameters.

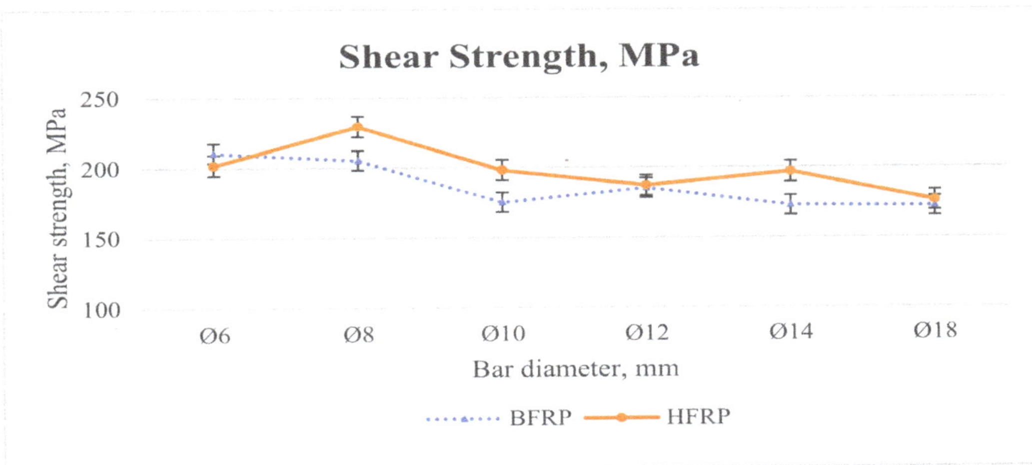

Figure 8. Comparison of average shear strength testing results of different FRP bars for all diameters.

Comparing the average shear strength of BFRP bars to HFRP bars, the values for BFRP bars were lower than the average shear strength values of HFRP bars except for diameter Ø6 mm, where it was slightly higher by approximately 1%. However, regarding the other diameters, the average shear strength of BFRP bars was lower for diameters Ø8, Ø10, and Ø14 mm by approximately 19% and for diameters Ø12 and Ø18 mm by approximately 1%. This might suggest that the shear strength of BFRP bars was lower than for HFRP.

5. Results and Discussions

From the results obtained from the tensile and shear strength testing of BFRP and HFRP with different diameters, a correlation was made between the obtained values in order to find a method for calculating tensile strength from the data obtained from shear strength experimental testing.

In Table 6, the correlation between shear strength and tensile strength of different types of FRP bars with different diameters is shown.

Table 6. Shear strength to tensile strength ratio.

Bar Type	Parameter	Ø6	Ø8	Ø10	Ø12	Ø14	Ø18
BFRP	f_s/f_t	0.183	0.186	0.152	0.167	0.157	0.196
HFRP	f_s/f_t	0.186	0.180	0.174	0.186	0.170	0.218

From Table 6, the average ratio (k_d) according to bar diameter can be calculated from Equation (3):

$$k_d = \frac{\sum \frac{f_s}{f_t}}{n_t} \tag{3}$$

where n_t–number of tested bar types ($n_t = 2$).

Based on Equation (3), the coefficient of accuracy corresponding to a given diameter can be identified. The values are presented in Table 7. Tables 8 and 9 represent comparison between the tensile strength values of experimental and calculated results.

Table 7. Coefficient of accuracy (k_d) of FRP bars according to diameter.

FRP Bars	Ø6	Ø8	Ø10	Ø12	Ø14	Ø18
Coefficient of Accuracy, k_d	0.185	0.183	0.163	0.176	0.163	0.207

Table 8. Tensile strength calculation according to diameter ratios ($f_t = f_s/k_d$).

Bar Diameter	k_d	BFRP	HFRP
mm	–	MPa	MPa
Ø6	0.185	1138.70	1092.11
Ø8	0.183	1122.24	1253.77
Ø12	0.163	1075.64	1216.63
Ø14	0.176	1053.75	1063.01
Ø16	0.163	1060.43	1207.61
Ø18	0.207	832.12	851.93

Table 9. Comparison between the tensile strength values of experimental and calculated results.

Bar type	Parameter	Units	Ø6	Ø8	Ø10	Ø12	Ø14	Ø18
BFRP	f_t(exp)	MPa	1148.81	1103.33	1152.54	1111.00	1101.94	877.98
	f_t(Ø)	MPa	1138.70	1122.24	1075.64	1053.75	1060.43	832.12
HFRP	f_t(exp)	MPa	1083.78	1277.92	1138.65	1008.52	1160.06	809.44
	f_t(Ø)	MPa	1092.11	1253.77	1216.63	1063.01	1207.61	851.93
BFRP Accuracy	f_t(ex)/f_t(Ø)	%	0.88%	−1.71%	6.67%	5.15%	3.77%	5.22%
HFRP Accuracy	f_t(ex)/f_t(Ø)	%	−0.77%	1.89%	−6.85%	−5.40%	−4.10%	−5.25%

Accuracy represented in Table 9 was calculated according to Equation (4):

$$FRP_{accuracy} = \frac{f_t(exp) - f_t(\varnothing)}{f_t(exp)} \cdot 100 \tag{4}$$

where:

- $f_t(exp)$ —Tensile strength obtained experimentally,
- $f_t(\varnothing)$ —Tensile strength obtained theoretically.

Figure 9 presents results obtained from Equation (3) according to the diameter of FRP bars proposed in order to calculate the tensile strength of FRP bars from the shear strength obtained experimentally; the accuracy of the values obtained theoretically using coefficient (k_d) varied between −7% and +7% in comparison with the tensile strength experimental values, using the Equation (5) for BFRP and HFRP:

$$f_{tensile} = \frac{f_{shear}}{k_d} \tag{5}$$

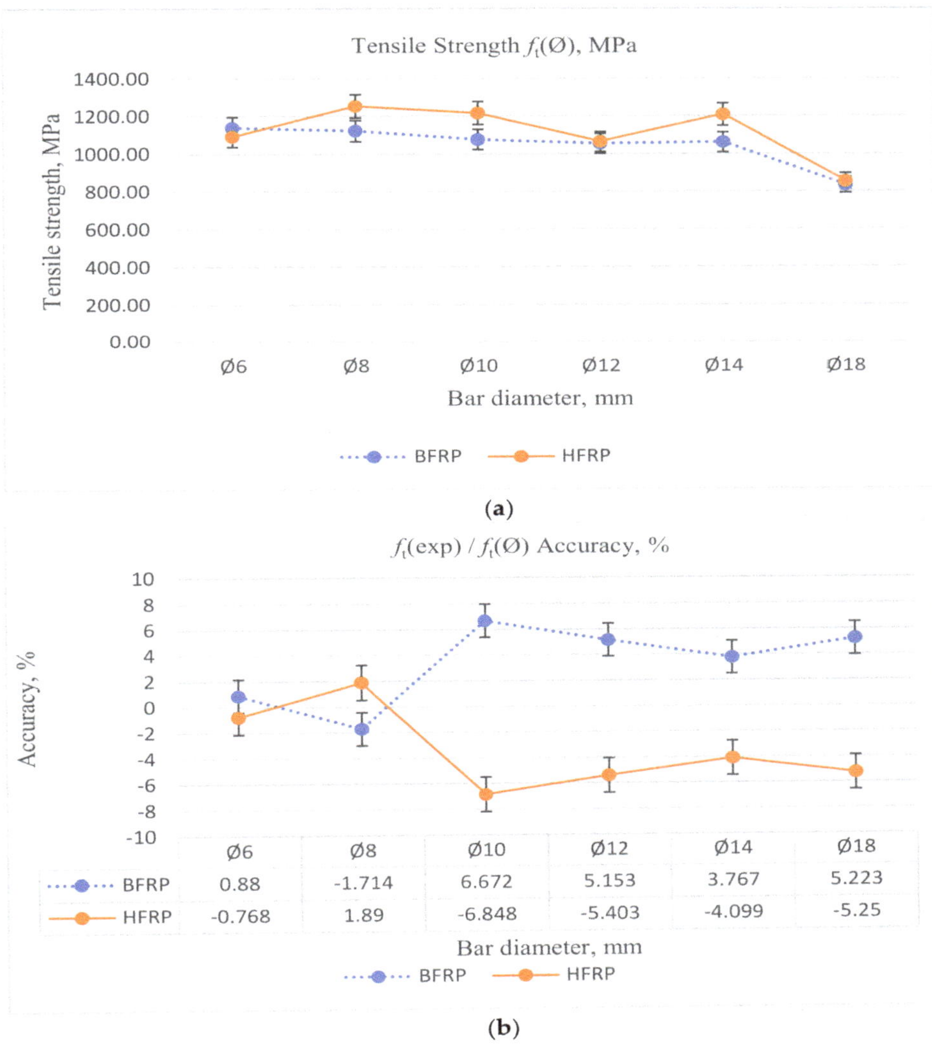

Figure 9. (**a**) Tensile strength obtained theoretically. (**b**) Accuracy according to coefficient (k_d).

Thus, it is possible to obtain the tensile strength of FRP bars by applying shear test strength only using Equation (5), taking into consideration the additional number of samples to be tested in order to eliminate errors and obtaining better accuracies. The coefficient (k_d) can be identified for any specific type of FRP, which will allow us to determine tensile strength based on shear testing.

The modulus of elasticity for the tested bar diameters depends mostly on the volume fraction of carbon and basalt fibers. The differences in volume fraction can be caused due to a various number of rovings used in the pultrusion process, which do not correspond to the bar diameter. In the case of

BFRP bars with diameters of 10, 12, and 14 mm, the volume fraction of basalt fibers was greater than the average value by about 10%, while for the diameters of 18 mm, it was lower by over 10%. On the other hand, for the HFRP bars with diameters of 14 mm, the volume fraction of fibers was higher by more than 10% than the average value, while for the diameters of 18 mm, it was 20% lower.

The tensile strength, in turn, is influenced by the shear lag effect, which causes that as the diameter of the bar increases, there is an increasing disproportion of the stress of the fibers in the cross-section of the bar.

Summarizing, the values of modulus of elasticity and tensile strength, as well as shear strength, depend on a combination of two factors, specifically the volume fraction of fibers and the shear lag effect.

6. Conclusions

The tests conducted in this study provided insight into the mechanical behavior of the types of FRP bars analyzed. Considering the results observed during tests done on BFRP bars for all diameters, it can be seen that the modulus of elasticity was quite low compared to HFRP bars, lower, on average, by approximately 18%. However, the main objective of this research was to check the maximum strength of HFRP and how it is compared to BFRP, as well as to find the stress–strain correlation. The influence of material hybridization was explored in order to examine the possibility of using less complicated shear testing as an alternative to tensile testing.

Based on experimental tests and numerical and analytical considerations for BFRP and HFRP bars, the following conclusive remarks can be drawn:

1. The tensile stress–strain correlation of BFRP bars was tested for given diameters. The average maximum stress obtained equaled to approximately 700 MPa, with an average strain approximately equal to 17%. However, the HFRP bars tested exhibited average maximum stress approximately equal to 725 MPa with the average strain approximately equal to 11%. This confirmed the viability of using HFRP bars as a replacement for less rigid BFRP bars.
2. Based on the tests performed, HFRP bars exhibited better results in terms of their tensile stress–strain relationship. However, comparing the obtained results for HFRP to BFRP bars, the tensile strength and modulus of elasticity of HFRP bars were higher by approximately 68% and 16%, respectively. The 6% lower elongation of HFRP bars could be explained by the superior extensibility properties of carbon fiber.
3. Additionally, according to the shear stress–strain relationship, better results were obtained for HFRP bars with average maximum shear stress equal to 214 MPa, which was approximately 22% higher than the average maximum shear stress obtained for BFRP bars. However, HFRP bars obtained a lower shear strain, 57% compared to BFRP bars.
4. Comparing experimental to analytical/numerical calculations done by the authors, the predicted analytical/numerical results were much better than those obtained experimentally. The difference in mechanical properties of FRP bars could be explained by the exact chemistry of the polymer matrix and the strength of polymer/BFRP or polymer/HFRP interactions.
5. Comparing the tensile stress–strain relationship to shear stress–strain for all bars and diameters, it was observed that BFRP and HFRP bars could withstand more tensile force and had a lower accompanying strain in tensile tests than shear tests.

The preparation for shear testing is less complicated than for tensile testing; therefore, on the basis of the results and observations, shear testing can substitute tensile testing for this class of composite materials.

Author Contributions: Conceptualization, K.P. and E.S.; Data curation, M.U.; Formal analysis, M.U.; Investigation, K.P. and F.Z.; Methodology, M.U. and E.S.; Supervision, M.U.; Validation, E.S.; Writing—original draft, K.P. and F.Z.; Writing—review & editing, E.S. All authors have read and agreed to the published version of the manuscript.

Acknowledgments: This research was funded by the National Centre for Research and Development (NCRD), grant number NCBR: PBS3/A2/20/2015, and the APC was funded by the Warsaw University of Technology. Special thanks to Andrzej Garbacz for useful private communication.

References

1. Pleimann, L. Strength, modulus of elasticity, and bond of deformed FRP rods. In *Advanced Composites Materials in Civil Engineering Structures: Proceedings of the Specialty Conference*; ASCE: New York, NY, USA, 1991; pp. 99–110.
2. Chaallal, O.; Benmokrane, B. Glass-fiber reinforcing rod: Characterization and application to concrete structures and grouted anchors. In *Materials: Performance and Prevention of Deficiencies and Failures: Proceedings of* the Materials Engineering Congress; ASCE: New York, NY, USA, 1992; pp. 606–617.
3. Mukae, K.; Kamagai, S.; Nakai, H.; Asai, H. Characteristics of aramid FRP rods. In *Proceedings of the Symposium on Fiber Reinforced Polymer Reinforcement for Concrete Structures*; American Concrete Institute: Detroit, MI, USA, 1993.
4. Yamasaki, Y.; Masuda, Y.; Tanano, H.; Shimizu, A. Fundamental properties of continuous fiber bars. In *Proceedings of the Symposium on Fiber Reinforced Polymer Reinforcement for Concrete Structures; ACI SP 138*; American Concrete Institute: Detroit, MI, USA, 1993.
5. Nanni, A.; Okamoto, T.; Tanigaki, M.; Osakada, S. Tensile properties of braided FRP rods for concrete reinforcement. *Cem. Concr. Compos.* **1993**, *15*, 121–129.
6. Porter, M.; Barnes, B. Tensile testing of glass fiber composite rod. In *Advanced Composites Materials in Civil Engineering Structures: Proceedings of the Specialty Conference*; ASCE: New York, NY, USA, 1991; pp. 123–131.
7. Uomoto, T.; Hodhod, H. Properties of fiber reinforced plastic rods for prestressing tendons. In *Proceedings of the Symposium on Fiber Reinforced Polymer Reinforcement for Concrete Structures; ACI SP 138*; American Concrete Institute: Detroit, MI, USA, 1993.
8. Walsh, P.J. Carbon Fibres. In *ASM Handbook Volume 21-Composites*; ASM International: Material Park, OH, USA, 2001; pp. 35–40.
9. Chang, K.K. Aramid Fibers. In *ASM Handbook, Volume 21 Composites*; ASM International: Material Park, OH, USA, 2001; pp. 41–45.
10. Wallenberger, F.T.; Watson, J.C.; Hong, L. Glass Fibers. In *ASM Handbook Volume 21 Composites*; ASM International: Material Park, OH, USA, 2001; pp. 27–34.
11. Hollaway, L. *Polymer Composites for Civil and structural Engineering*; Blackie Academic & Professional: Glasgow, UK, 1993.
12. Nanni, A.; Henneke, M.J.; Okamoto, T. Tensile properties of hybrid rods for concrete reinforcement. *Constr. Build. Mater.* **1994**, *8*, 27–34. [CrossRef]
13. Wu, Z.; Wang, X.; Iwashita, K.; Sasaki, T.; Hamaguchi, Y. Tensile fatigue behaviour of FRP and hybrid FRP sheets. *Compos. Part B Eng.* **2010**, *41*, 396–402. [CrossRef]
14. Kosior-Kazberuk, M. Application of basalt-FRP bars for reinforcing geotechnical concrete structures. *MATEC Web Conf.* **2019**, *265*, 05011. [CrossRef]
15. Shi, J.-W.; Zhu, H.; Wu, G.; Wu, Z.-S. Tensile behavior of FRP and hybrid FRP sheets in freeze–thaw cycling environments. *Compos. Part B Eng.* **2014**, *60*, 239–247. [CrossRef]
16. Benmokrane, B.; Elgabbas, F.; Ahmed, E.A.; Cousin, P. Characterization and comparative durability study of glass/vinylester, basalt/vinylester, and basalt/epoxy FRP bars. *J. Compos. Constr.* **2015**, *19*, 04015008. [CrossRef]
17. Alam, M.S.; Hussein, A. Experimental investigation on the effect of longitudinal reinforcement on shear strength of fibre reinforcement on shear strength of fibre reinforced polymer reinforced concrete beams. *Can. J. Civil. Eng.* **2011**, *38*, 243–251.
18. El-Sayed, A.K.; El-Salakawy, E.F.; Benmokrane, B. Shear strength of FRP-reinforced concrete beams without transverse reinforcement. *ACI Struct. J.* **2006**, *103*, 235–243.
19. El-Sayed, A.K.; El-Salakawy, E.F.; Benmokrane, B. Shear capacity of high-strength concrete beams reinforced with FRP bars. *ACI Struct. J.* **2006**, *103*, 383–389.
20. Razaqpur, A.G.; Isgor, B.O.; Greenaway, S.; Selley, A. Concrete contribution to the shear resistance of fiber reinforced polymer reinforced concrete members. *J. Compos. Constr.* **2004**, *8*, 452–460.

21. Alkhrdaji, T.; Wideman, M.; Belarbi, A.; Nanni, A. Shear strength of RC beams and slabs. In *Composites in Construction Figueiras*; Juvandes, L., Faria, R., Eds.; Balkema Publishers: Lisse, The Netherlands, 2001; pp. 409–414.
22. Yost, J.R.; Gross, S.P.; Dinehart, D.W. Shear strength of normal strength concrete beams reinforced with deformed GFRP bars. *J. Compos. Constr.* **2001**, *5*, 263–275. [CrossRef]
23. Khalifa, A.; Gold, W.J.; Nanni, A.; Aziz, A.M.I. Contribution of externally bonded FRP to shear capacity of RC flexural members. *ASCE J. Compos. Constr.* **1998**, *2*, 195–202. [CrossRef]
24. Täljsten, B. Strengthening of concrete structures for shear with bonded CFRP-fabrics. In *Recent Advances in Bridge Engineering, Advanced Rehabilitation, Durable Materials, Nondestructive Evaluation and Management*; Meier, U., Betti, R., Eds.; Columbia Univ. Press: Dübendorf, Switzerland, 1997; pp. 57–64.
25. Triantafillou, T.C. Shear strengthening of reinforced concrete beams using epoxy bonded FRP composites. *ACI Struct. J.* **1998**, *95*, 107–115.
26. Triantafillou, T.C.; Antonopoulos, C.P. Design of concrete flexural members strengthened in shear with FRP. *ASCE J. Compos. Constr.* **2000**, *4*, 198–205. [CrossRef]
27. Alnahhal, W.; Chiewanichakorn, M.; Aref, A.; Alampalli, S. Temporal thermal behavior and damage simulations of FRP deck. *J. Bridge Eng.* **2006**, *11*, 452–465. [CrossRef]
28. Aiello, M.A.; Ombres, L. Structural performances of concrete beams with hybrid (fiber reinforced polymer-steel) reinforcements. *J. Compos. Constr.* **2002**, *6*, 133–140. [CrossRef]
29. Newhook, J.P. Design of under-reinforced concrete T-sections with GFRP reinforcement. In Proceedings of the 3rd International Conference on Advanced Composite Materials in Bridges and Structures, Canadian Society for Civil Engineering, Ottawa, ON, Canada, 15–18 August 2000.
30. Wang, X.; Wu, Z. Integrated high-performance thousand-meter scale cable-stayed bridge with hybrid FRP cables. *Compos. Part B Eng.* **2011**, *41*, 166–175. [CrossRef]
31. Urbański, M. Compressive Strength of Modified FRP Hybrid Bars. *Materials* **2020**, *13*, 1898. [CrossRef]
32. Lau, D. Hybrid fiber-reinforced polymer (FRP) composites for structural applications. In *Developments in Fiber-Reinforced Polymer (FRP) Composites for Civil Engineering*; Woodhead Publishing: Cambridge, UK, 2013; pp. 205–225. [CrossRef]
33. Garbacz, A.; Szmigiera, E.; Protchenko, K.; Urbanski, M. On Mechanical Characteristics of HFRP Bars with Various Types of Hybridization. In *International Congress on Polymers in Concrete (ICPIC 2018)*; Springer: Washington, DC, USA, 2018; pp. 653–658. [CrossRef]
34. Protchenko, K.; Młodzik, K.; Urbański, M.; Szmigiera, E.; Garbacz, A. Numerical estimation of concrete beams reinforced with FRP bars. In *MATEC Web of Conferences*; EDP Sciences: Moscow, Russia, 2016; Volume 86, p. 02011.
35. Protchenko, K.; Szmigiera, E.D. Post-Fire Characteristics of Concrete Beams Reinforced with Hybrid FRP Bars. *Materials* **2020**, *13*, 1248. [CrossRef]
36. Szmigiera, E.D.; Protchenko, K.; Urbański, M.; Garbacz, A. Mechanical Properties of Hybrid FRP Bars and Nano-Hybrid FRP Bars. *Arch. Civ. Eng.* **2019**, *65*, 97–110. [CrossRef]
37. Protchenko, K.; Dobosz, J.; Urbański, M.; Garbacz, A. Wpływ substytucji włókien bazaltowych przez włókna węglowe na właściwości mechaniczne prętów B/CFRP (HFRP). [Influence of substitution of basalt fibres by carbon fibres on mechanical properties of B/CFRP (HFRP)]. *J. Civ. Eng. Environ. Archit.* **2016**, *63*, 149–156.
38. Garbacz, A.; Urbanski, M.; Lapko, A. BFRP bars as an alternative reinforcement of concrete structures–Compatibility and adhesion issues. *Adv. Mater. Res.* **2015**, *1129*, 233–241. [CrossRef]
39. Urbański, M.; Lapko, A.; Garbacz, A. Investigation on Concrete Beams Reinforced with Basalt Rebars as an Effective Alternative of Conventional R/C Structures. *Procedia Eng.* **2013**, *57*, 1183–1191. [CrossRef]
40. American Concrete Institute (ACI). *Guide test methods for fiber reinforced polymers (FRPs) for reinforcing or strengthening concrete structures*; ACI: Farmington Hills, MI, USA, 2006; ISBN 9780870317811.

2

Effects of Cellulose Nanocrystals and Cellulose Nanofibers on the Structure and Properties of Polyhydroxybutyrate Nanocomposites

Bobo Zhang, Chongxing Huang *, Hui Zhao *, Jian Wang, Cheng Yin, Lingyun Zhang and Yuan Zhao

College of Light Industry & Food Engineering, Guangxi University, Nanning 530004, China; Zbobo05@163.com (B.Z.); wjcj125521@163.com (J.W.); yc666yc@163.com (C.Y.); zhang@163.com (L.Z.); zy199113@163.com (Y.Z.)

* Correspondence: Huangcx@gxu.edu.cn (C.H.); zhh@gxu.edu.cn (H.Z.)

Abstract: One of the major obstacles for polyhydroxybutyrate (PHB), a biodegradable and biocompatible polymer, in commercial applications is its poor elongation at break (~3%). In this study, the effects of nanocellulose contents and their types, including cellulose nanocrystals (CNCs) and cellulose nanofibers (CNFs) on the crystallization, thermal, and mechanical properties of PHB composites were systematically compared. We explored the toughening mechanisms of PHB by adding CNCs and cellulose CNFs. The results showed that when the morphology of bagasse nanocellulose was rod-like and its content was 1 wt %, the toughening modification of PHB was the best. Compared with pure PHB, the elongation at break and Young's modulus increased by 91.2% and 18.4%, respectively. Cellulose nanocrystals worked as heterogeneous nucleating agents in PHB and hence reduced its crystallinity and consequently improved the toughness of PHB. This simple approach could potentially be explored as a strategy to extend the possible applications of this biopolymer in packaging fields.

Keywords: polyhydroxybutyrate (PHB); cellulose nanocrystals (CNCs); cellulose nanofibers (CNFs); mechanical properties

1. Introduction

In the past few decades, the negative environmental impacts caused by petroleum-based polymers and the depletion of petroleum resources have been an important issue. Therefore, the research and development of materials based on renewable resources, such as biodegradable polymers, is imperative [1,2].

Polyhydroxybutyrate (PHB)—part of the polyhydroxyalkanoate (PHA) aliphatic polyester class—is a semi-crystalline polymer produced through numerous mechanisms, including the bacterial fermentation of sugars [3]. PHB is a water-insoluble biodegradable polymer with high crystallinity, biocompatibility, biodegradability, permeability, and processability [4]. PHB is potentially useful for biodegradable materials, such as mulch film, absorbable surgical sutures for drug delivery, and tissue engineering due to its biodegradability and biocompatibility [5]. Nevertheless, these potential applications of PHB have been limited due to its severe brittleness, which is caused by its low crystallization nucleation density, high brittleness, and poor toughness [6]. Therefore, the reinforcing of PHB has been frequently carried out, and natural fibers are always involved due to its biocompatibility and biodegradability [7]. In addition, as both nanocellulose and PHB are renewable materials, PHB-based biodegradable nanocomposites have great application prospects in the packaging field.

Nanocellulose is used as a nanotoughening agent to improve the toughness of polymers because of its large specific surface area, high reactivity, good dispersion in the polymer matrix, easy formation of hydrogen bonds between the nanocellulose and the polymer, and ability to reduce the crystallinity of the polymer [8]. Until now, researchers have focused on the preparation and properties of nanocellulose/bio-based composite materials, such as starch [9], polyvinyl alcohol (PVA) [10], waterborne polyurethane (WPU) [11], polycaprolactone (PCL) [12], polylactic acid (PLA) [13], and epoxy resin (EP) [14]. Studies have shown that the mechanical properties of nanocellulose/bio-based composites can be better improved than for other reinforcing materials such as nanoclay or carbon-based materials.

The main methods of preparing nanocellulose are acidolysis and physical methods. Physical methods such as high-pressure homogenization [15], grinding/refining [16], and ultrasonic treatment [17] have been used to extract cellulosic fibers from wood fibers (WF) and plant fibers (PF), microcrystalline cellulose (MCC), cysts, algae, and bacteria. These methods use high shear force to expand and separate cellulose molecules into nanometer-sized microfibrils, which are called cellulose nanofibrils (CNFs) [18]. Acid hydrolysis is mainly used to extract crystalline particles from various kinds of cellulose (PF, WF, MCC, cysts, algae, and bacteria). The nanocellulose prepared by hydrolyzing amorphous areas in cellulose microfibers is called cellulose nanocrystals (CNCs) [19,20]. It has already been utilized for many potential applications such as paper, newspaper, textiles fibers, paper board, construction, etc. [21]. Some researchers have extracted CNCs as nanoreinforcement into bionanocomposite for biomedical and other value-added products in industrial applications [22]. Until now, two kinds of nanocellulose have been effectively used during the processing of reinforced PHA composites. Srithep et al. [23] prepared poly(3-hydroxybutyrate-co-3-hydroxyvalerate) (PHBV)/CNFs composites by liquid ammonia cooling and melt blending or by blending PHBV solution with a suspension of CNFs. Studies have shown that the addition of CNFs increases the Young's modulus of PHBV/CNFs nanocomposites by nearly two times, and the tensile strength is slightly enhanced, but the elongation at break is reduced because of the agglomeration of CNFs in the PHBV matrix and the degradation of some PHBV. Cherpinski et al. [24] prepared a PHB/CNF/PHB three-layer composite film by an electrospinning coating technique and reported that in comparison with pure PHB, the elongation at break of the multilayer film was increased by 135.7%, and the Young's modulus and tensile strength were decreased by 11.5% and 35.6%, respectively. This was due to the good adhesion property of the PHB film to CNF, which resulted in a multilayer film with a higher elongation at break than the pure PHB film. The decrease in tensile strength was probably related to the delamination failure during the fracture process. Seoane et al. [25] studied the effect of CNCs on PHB characteristics and stated that at 6% CNC concentration, the tensile modulus and strength of the nanocomposites were increased about 50% and 35%, respectively.

Nanocellulose with different morphologies has different effects on PHB performance. At present, the research on the improvement of PHB performance with nanocellulose only considers a single morphology (CNCs or CNFs) of nanocellulose and does not systematically compare the effects of the two morphologies on PHB performance. Therefore, in this paper, CNCs and CNFs were used as nanotougheners to blend PHB with nanocellulose. The effects of different contents of CNCs and CNFs on the mechanical properties, thermal properties, crystallization process, crystal structure, and surface morphology of PHB were systematically studied and compared. In addition, the process optimization of PHB toughening by nanocellulose and its mechanism was also evaluated in this paper.

2. Materials and Methods

2.1. Materials

Bleached pulp board was obtained from Guangxi Guitang Co., Ltd. (Nanning, China), sulfuric acid (98%, analytical grade) was obtained from Linyi Mengyang Chemical Co., Ltd. (Linyi, China), acetone was obtained from Yangzhou Tianda Chemical Co., Ltd. (Yangzhou China), Phosphotungstic acid was obtained from Xi'an Tianmao Chemical Co., Ltd. (Xi'an, China), while PHB powder (model: ENMAT Y3000P) and chloroform (analytical grade) were purchased from Ningbo Tianan Biomaterial Co., Ltd. (Ningbo, China), and Shanghai Honghe Industrial Co., Ltd. (Shanghai, China), respectively.

2.2. Preparation of Cellulose Nanocrystals (CNCs)

First, 10 g slurry of was weighed in a beaker to which 200 g (64 wt %) sulfuric acid was added while keeping the beaker in an ice bath. After complete dissolving, the beaker was taken out from the ice bath and hydrolyzed at room temperature for 3 h at a stirring speed of 1000 r/min, followed by dilution with 1000 ml of deionized water to terminate the reaction. The diluted dispersion was centrifuged for 15 min in a high-speed refrigerated centrifuge (Himac CR21N, Tokyo, Japan) until the pH of the dispersion was 7. Then, the sample was subjected to ultrasonic treatment for 5 min to obtain a uniform nanocellulose suspension (The experimental chart of the best choice of acid hydrolysis temperature, acid hydrolysis time, and acid hydrolysis concentration can be found in the Supplementary Material).

2.3. Preparation of Cellulose Nanofibers (CNFs)

A quantity of 200 g of dry pulp was soaked in distilled water until the slurry became soft. After balancing the water for 24 h, water was added to the slurry at a rate of 2% and soaked for 2 h. The soaked slurry was ground 5 to 6 times using an ultrafine pulverizer (MKZA10-15J, Tokyo, Japan). Distilled water was added to continue the swelling sufficiently, and the concentration was controlled at 0.5%. Then, the slurry dispersion was transferred to a high-pressure homogenizer (M-110EH30, Westwood, MA, USA) for 15 cycles to obtain a uniform nanocellulose suspension.

2.4. Preparation of Nanocellulose Chloroform Suspension

The obtained nanocellulose suspension was repeatedly centrifuged at a rotation speed of 12,000 r/min for 45 min at 10 °C. Acetone (3 mL) was added to the centrifuge tube to displace the nanofiber and centrifuged again at a rotation speed of 12000 r/min and at 10 °C for 15 min. To obtain an acetone suspension, 3 mL of chloroform was added to the centrifuge and centrifuged three times at a rotation speed of 12,000 r/min at 10 °C for 15 min to replace the acetone in the nanocellulose. Finally, the chloroform suspension of nanocellulose was obtained by stirring with a shear dispersing emulsifier for 5 min.

2.5. Preparation of Nanocellulose/PHB Composite Film

The PHB (2 g) was dissolved in chloroform (50 mL) at 90 °C and mechanically stirred at a speed of 500 rpm under reflux for 3 h. After completely dissolving, the nanocellulose suspension was added, stirred at a temperature of 60 °C, a rotation speed of 1000 r/min for 30 min, and then coated on a 25 cm × 25 cm glass plate by automatic film application machine (ZEHNTNER, ZAA2300, Basel, Switzerland). After the sample was volatilized by the solvent at room temperature, the residual solvent was further removed at 40 °C for 3 days. The composition of each component is presented in Table 1.

Table 1. Mass ratio of nanocellulose/polyhydroxybutyrate (PHB) composites.

Sample Serial Number	PHB (wt %)	Percentage of Absolute Dry Mass and PHB Mass of Nanocellulose (wt %)
PHB	100	0
PHBCNC1	100	1
PHBCNC3	100	3
PHBCNC5	100	5
PHBCNF1	100	1
PHBCNF3	100	3
PHBCNF5	100	5

2.6. Characterization

2.6.1. Scanning Electron Microscope (SEM) Analysis

The surface morphology of the film was observed by SEM (Philips Phenom™, Amsterdam, Netherlands) at a voltage of 10 kV.

2.6.2. Transmission Electron Microscopy (TEM) Analysis

The samples were observed by TEM (Gao Tech HT7700, Tokyo, Japan). The acceleration voltage was 80 kV. The samples were mixed with 0.1% CNCs or CNFs suspension. After magnetic stirring for 4 hours, ultrasound was used for 30 min. A drop of 0.1% nanocellulose suspension was dripped onto the copper net. In order to increase the contrast, the samples were dyed with 1% phosphotungstic acid aqueous solution for 5 min [26].

2.6.3. Dynamic Light Scattering (DLS) Analysis

Dynamic light scattering (DLS) measurements were carried out on a Malvern Nanozetasizer (Zetasizer Nano ZSP, Malvern, United Kingdom). Samples were diluted in deionized water to 10^{-2} wt %. Five measurements were conducted from each sample, and their average was presented. All data were taken at 25 °C. For each sample, the final values represent an average of at least 5 measurements.

2.6.4. X-Ray Diffraction (XRD) Test

The crystalline structure of the PHB and nanocellulose was analyzed using X-ray diffractometer (Rigaku MiniFlex600, Tokyo, Japan). All the experiments were conducted at a voltage of 40 kV and a current of 30 mA using nickel-filtered Cu–Kα radiation between 5 and 40°, at a rate of 5 °C/min.

2.6.5. Thermal Characterization

The calorimetry analysis (DSC) was carried out by using a Nerzsch DSC instrument (Nerzsch, TA, Germany). A 10 mg sample was heated from room temperature to 200 °C at 10 °C/min under a nitrogen atmosphere for 3 min to eliminate the heat history, then cooled to −50 °C at 10 °C/min, kept for 3 min, and then again heated to 200 °C at 10 °C/min. Crystallinity of the composite film was calculated according to [27] using the following equation,

$$X_c = \frac{\Delta H_m}{\Delta H_m^0 \Delta w_{PHB}} \times 100\% \quad (1)$$

where ΔH_m is the measured value for PHB melting enthalpy, ΔH_m^0 is the melting heat associated with pure crystalline PHB (146 J/mol) [28], and w_{PHB} is the weight fraction of PHB in the composite.

The thermal stability of the films was analyzed using a synchronous thermogravimetric analyzer (TGA, JingKe ZRY-2P, Shanghai, China). The sample was placed in a crucible, and thermal analysis was performed under a nitrogen atmosphere in a temperature range of 20–600 °C at a heating rate of 10 °C/min.

2.6.6. Mechanical Performance Test

According to ASTM D882-12, the film (50 ± 8 μm) was cut into a rectangle of 15 mm × 100 mm. The tensile strength, elongation at break, and Young's modulus were measured using an electronic universal material testing machine (Instron 3367, Boston, MA, USA) equipped with a load cell of 1 kN. The initial clamping distance and testing speed were kept constant as 50 mm and 100 mm/min, respectively. Five parallel samples were used and the results were averaged.

2.6.7. Barrier Performance Test

The oxygen transmission rates (OTRs) of the films were determined by the pressure method reported by Gali et al. [29], using an oxygen transmission instrument (Brugger GDP-C, Schramberg, Germany) with a low chamber vacuum and vacuum time of 5 min. Three parallel samples were used with a test area of 78.4 cm^2 for each sample. The results of OTRs are presented as $cm^3/(m^2 \cdot 24\ h)$.

The water vapor transmission rates of films were measured according to ASTM F1249 using a water vapor transmission instrument (MOCON PERMATRAN-W Model 3/61, Minneapolis, MN, USA) at a test temperature of 25 °C and a relative humidity of 75 %. Three parallel samples were used with a test area of 1 cm^2 for each sample, and the results are presented as $g/(m^2 \cdot 24\ h)$.

2.6.8. Optical Properties by UV–Vis Spectroscopy

Optical properties were measured at three different points from each sample using a UV–Vis spectrometer (Jena SPECORD plus 50, Jena, Germany) in the range of 400–700 nm. An average of the measurements was presented.

2.6.9. Atomic Force Microscopy (AFM) Test

The suspension of CNCs or CNFs was diluted with distilled water to 0.008 wt % and stirred for 4 hours. The suspension of 25 μL of nanocellulose was absorbed onto the silicon wafer with a pipette gun and dried naturally at room temperature. The silicon wafer was adhered to the iron wafer with double-sided adhesive. The AFM (Gao Tech 5100N, Tokyo, Japan) test was performed in tapping mode, and the cantilever elastic constant was 26 N/m.

3. Results and Discussion

3.1. Morphology of CNCs and CNFs

Figure 1 shows images of the two nanocelluloses. CNCs exhibited a "narrow, two-pointed" rod-like structure, while CNFs exhibited a grid-like structure, which were intertwined with each other. In addition, the diameters of both CNCs and CNFs were below 100 nm, which is in line with the existing size requirements for nanocellulose. The results obtained in our study were similar to those reported by predecessors [30].

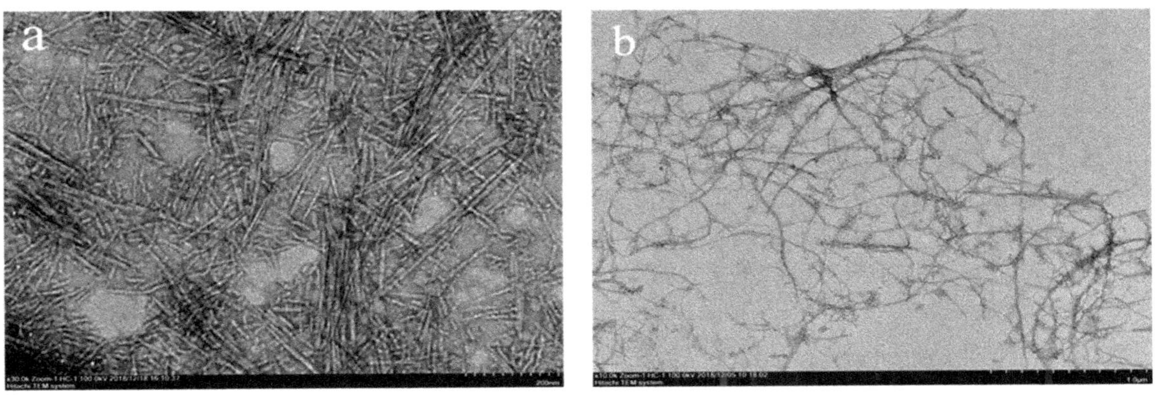

Figure 1. TEM images of the two nanocelluloses: (**a**) cellulose nanocrystals (CNCs); (**b**) cellulose nanofibers (CNFs).

Figure 2 shows the dispersion of bagasse fiber and two kinds of 1 wt % nanocellulose. The dispersion of bagasse fiber showed a clear phase separation, while the suspension of nanocellulose remained stable and did not show any phase separation or precipitation. This indicated a complete dispersion of nanocellulose and resulted in a colloidal structure. According to Dong et al. [31], sulfuric acid can induce the grafting of negatively charged sulfuric acid groups on the surface of nanocrystals during acid hydrolysis, which helps to stabilize colloidal suspensions through repulsive inter-particle forces. The CNFs prepared by the mechanical method form a colloid due to the strong shear, impact, and cavitation of bagasse bleached pulp under high pressure. Cavitation includes the formation, expansion, and implosion of microscopic bubbles when molecules in the liquid absorb ultrasonic energy. In the cavitation bubble and the surrounding area, there will be a violent shock wave, which can be used to separate the fiber from the cellulose fiber, so that the dispersion of bagasse fiber can be refined and the stable colloidal suspension can be obtained [17,32].

Figure 2. Appearance of bagasse fibers and two nanocelluloses: (**a**) Dispersion of bagasse fibers; (**b**) Suspension of CNCs; (**c**) Suspension of CNFs.

The DLS result showed a single peak with a hydrodynamic diameter of ~60 nm, indicating the average size of the CNCs (Figure 3a). The CNFs suspension showed a nearly uniform particle size distribution with an average hydrodynamic diameter of ~106 nm (Figure 3b). Using the particle size analysis Nano Measurer drawing software, 50 nanocellulose from three different TEM images of CNCs and CNFs were selected, and the diameter and length of the nanocellulose were calculated. Figure 3c shows the diameter distribution of CNCs; their size distribution was from 2 to 18 nm, with the average diameter of 8.8 nm. Figure 3d shows the length distribution of CNCs, which ranged from 80 to 240 nm with an average length of 131.6 nm. Figure 3e shows the diameter distribution of CNFs, which were distributed between 6 and 26 nm with an average diameter of 12.6 nm. From the TEM diagram of CNFs, it can be seen that CNFs show a network structure at the microlevel, and the length range was identified. It can also be seen that the size of cellulose nanoparticles prepared by acid hydrolysis was clearly smaller than that of nanoparticles prepared by grinding and homogenization. The average size of the nanocellulose examined by TEM was about six or nine times smaller than that measured by DLS. Similar results were observed while extracting and characterizing nanocellulose from yerba mate sticks and wood pulp [33,34]. This discrepancy could be attributed to the dehydration of the nanocellulose during the sample preparation for TEM [35].

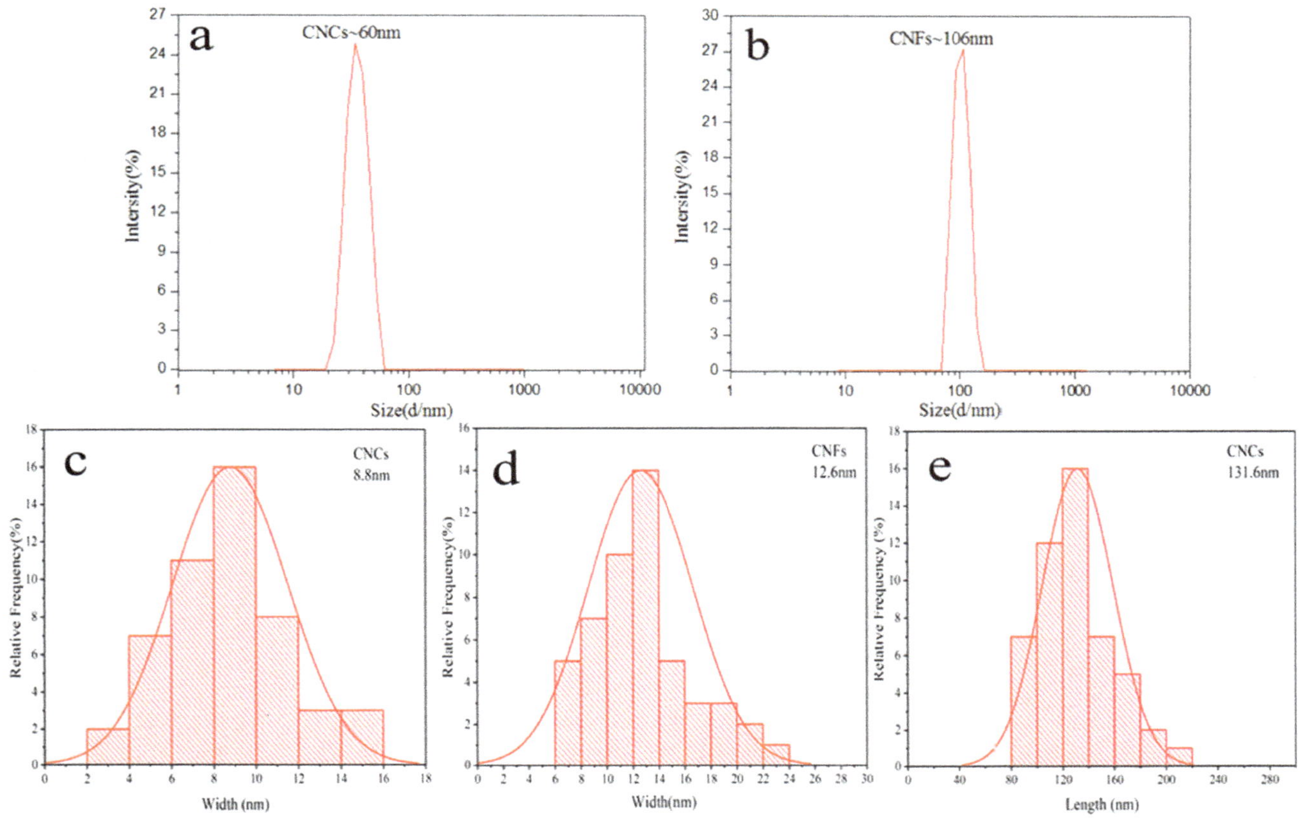

Figure 3. (**a**) Dynamic light scattering (DLS) measurement of the cellulose nanocrystals (CNCs) suspension, (**b**) DLS measurement of the cellulose nanofibers (CNFs) suspension, (**c**) TEM measurement of the CNCs—width, (**d**) TEM measurement of the CNCs—length, (**e**) TEM measurement of the CNFs—width.

AFM can evaluate the dispersion performance of a nanomaterial and detect the surface morphology below the nanometer scale. Figure 4a–d shows examples of tapping mode AFM images from CNCs and CNFs particles deposited onto freshly cleaved mica surfaces. Despite a sonication of the suspensions before spreading on mica, some aggregates were observed on the surface. Interestingly, AFM results supported the TEM images of CNCs and CNFs, which showed that the CNCs have a rod-like structure while CNFs have a grid-like structure. Furthermore, the diameter of the two nanofibers prepared were below 100 nm.

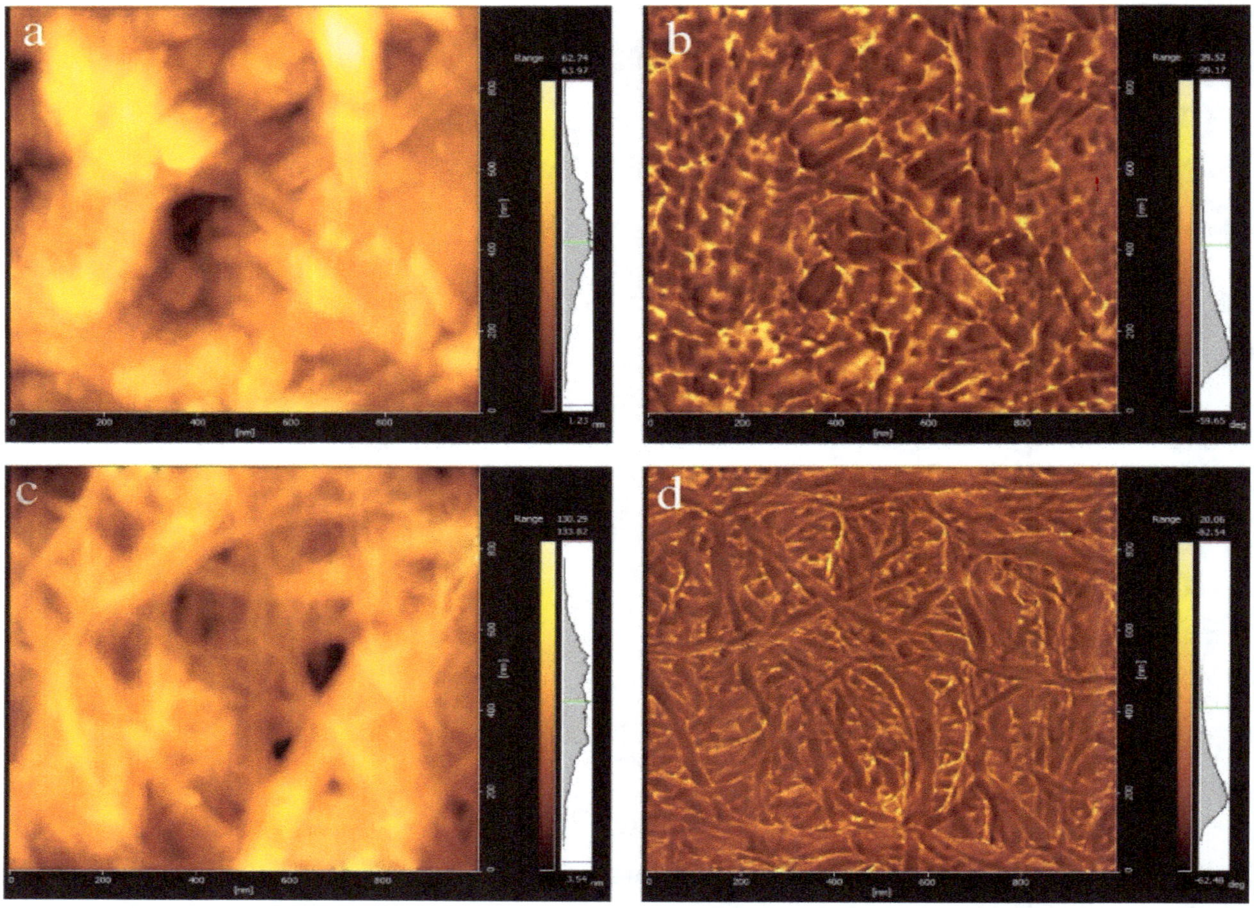

Figure 4. Atomic force microscopy (AFM) diagrams of the two nanocelluloses: (**a**) topography of CNCs, (**b**) phase diagram of CNCs, (**c**) topography of CNFs, and (**d**) phase diagram of CNFs.

3.2. Crystal Structure Analysis of CNCs and CNFs

The X-ray diffraction patterns of bagasse cellulose (BC), CNCs, and CNFs are shown in Figure 5. When the non-cellulose was removed and the amorphous regions were dissolved, the fibers showed an increased orientation along a specific axis. Typical cellulose I crystal diffraction peaks were 16.4°, 22.5°, and 34.5° at diffraction angles of 2θ [36]. Three diffraction peaks of the CNCs obtained by the acid treatment and the CNFs obtained by the mechanical method were at the same position, which corresponded to the characteristic absorption peaks of the crystal plane of the cellulose I-type crystal structure. This suggested that the acid and mechanical treatments of cellulose did not change the crystal form of cellulose. The diffraction peaks of the two nanocellulose at 22.5° were sharper, and the relative peak intensities were significantly ($p \leq 0.05$) enhanced, which proves that the crystallinity has been increased. The possible reason for this increase in the crystallinity of cellulose by acid treatment could be the hydrogen ions that can enter into the amorphous area of cellulose and destroy the amorphous area [37]. Mechanical grinding and homogenization increased the crystallinity of cellulose due to the non-knotting portion of bagasse bleaching pulp being reduced correspondingly after the high-intensity shearing treatment in the preparation of nanocellulose and the more regular arrangement of fiber in the crystallization area [38,39].

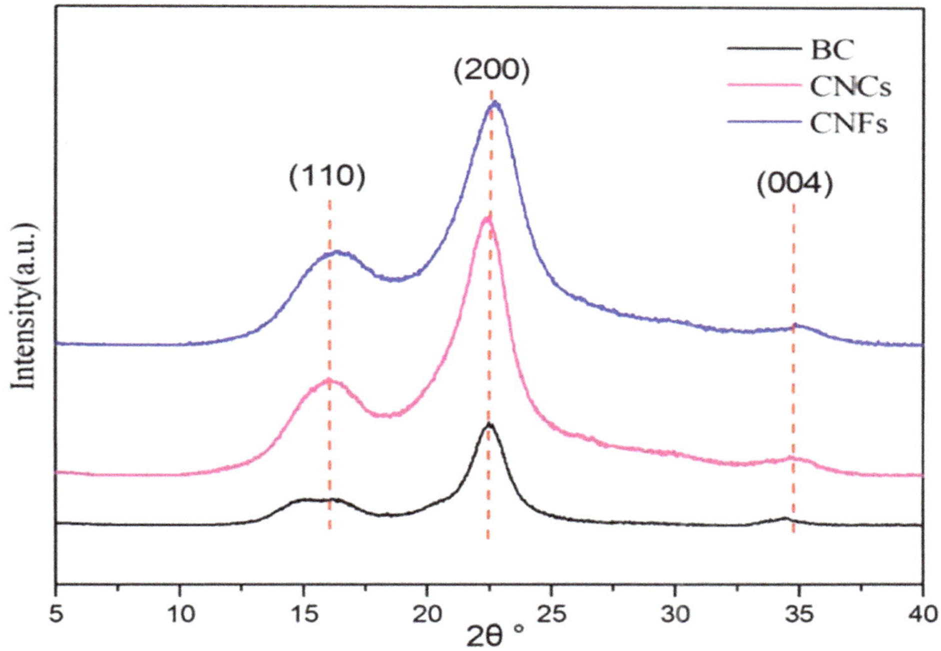

Figure 5. X-ray diffraction pattern of bagasse cellulose (BC), CNCs, and CNFs.

3.3. Fracture Morphology Analysis of Nanocellulose/PHB Composite Film

Figure 6 shows the fracture morphology of PHB and PHB/nanocellulose composite films. The fracture of pure PHB was relatively smooth and had no concavity, but it did have convexity, which belongs to brittle fracture. When 1 wt % CNCs was added, the fracture surface of the composite become rough, and many evenly distributed holes appeared (circles in Figure 6b), which might be left by the CNCs when they were pulled out [40]. Compared with pure PHB, the nanocellulose prepared by the acid method was better wrapped by PHB. In addition, there was no obvious fiber nudity, and the agglomeration was less, which indicated that the interaction force between hydrophilic CNCs and hydrophobic PHB was stronger and the compatibility was better. This orientation of CNCs in the PHB matrix could be attributed to the adsorption of the PHB chain on the CNCs hydroxyl group at lower load [41]. Most of the nanocellulose was embedded on the polymer surface to provide more adhesive contact areas [42], which can enhance the mechanical properties of the PHB matrix and correspond to the test results of mechanical properties. When the content of CNCs was more than 1 wt %, the defects on the fracture surface expanded, especially the volume and number of holes, and the distribution of these defects became uneven (circles in Figure 6c). In contrast, the composite film shows brittle fracture. When the content of CNCs was increased to 5 wt %, the CNCs showed a serious agglomeration phenomenon, and the pore size and fracture surface size of CNCs increased sharply due to the strong intramolecular hydrogen bonds of the nanocellulose (circles in Figure 6d).

Figure 7 is a cross-sectional view of the different concentrations of cellulose nanocrystals added to the PHB/CNFs composite film. With the increase of CNFs content, the voids on the fracture surface changed from small to large, and the distribution becomes inhomogeneous. These holes were due to the pulling out of CNFs at the time of PHB/CNFs composite breakage in liquid nitrogen (Figure 7b–d arrow). At the same time, some blocks were observed, which were caused by agglomerated CNFs and single CNFs. CNFs' agglomeration can lead to the weakening of the interface bonds between CNFs and PHB, and the composite film breaks near the agglomeration of CNFs during drawing, so it does not have its due toughness effect (see Figure 11).

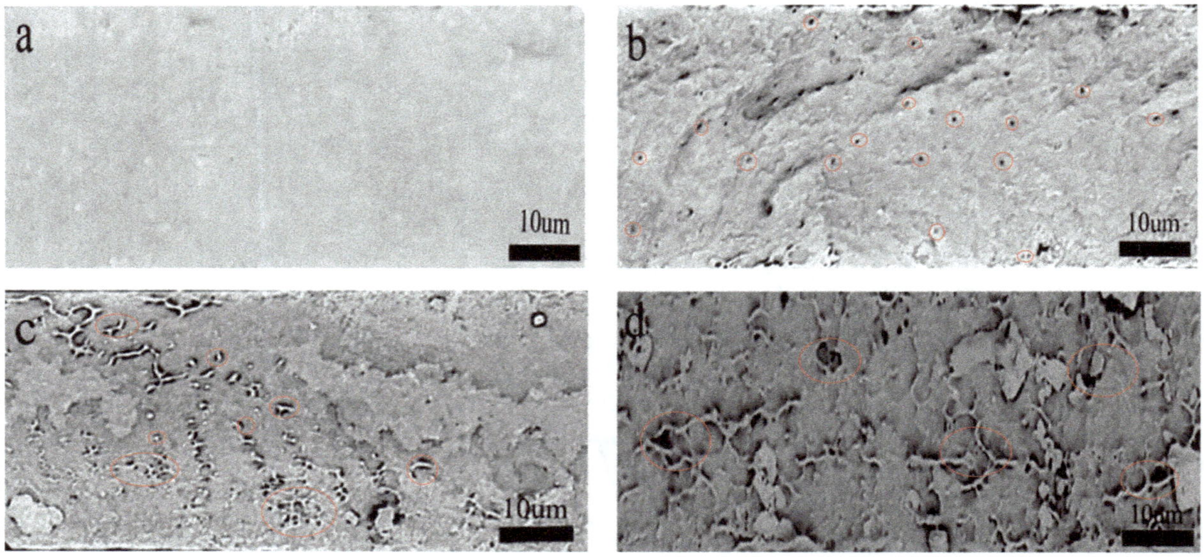

Figure 6. Effect of the different concentrations of cellulose nanocrystals on the impact cross-section of PHB/CNCs (SEM image): (**a**) pure PHB, (**b**) 1 wt %, (**c**) 3 wt %, (**d**) 5 wt %.

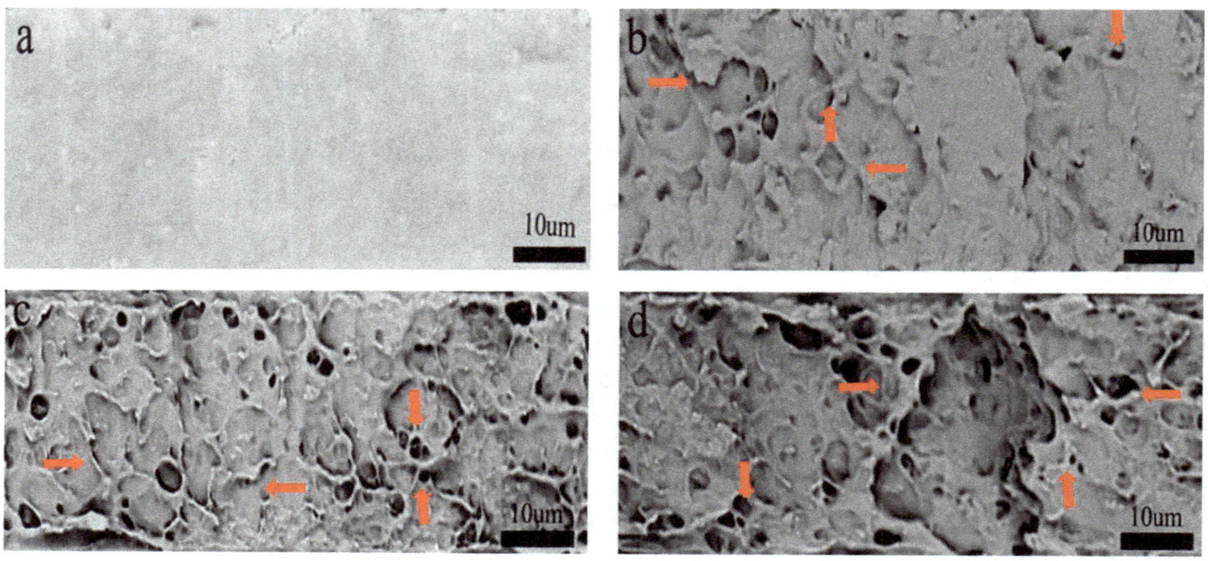

Figure 7. Effect of the different concentrations of cellulose nanofibrils on the impact cross-section of PHB/CNFs (SEM image): (**a**) pure PHB, (**b**) 1 wt %, (**c**) 3 wt %, (**d**) 5 wt %.

3.4. Crystal Structure Analysis of Nanocellulose/PHB Composite Film

The XRD curves of pure PHB thin films, PHB/CNCs, and PHB/CNFs nanocomposite films with different nanocellulose contents are shown in Figure 8. The characteristic lattice planes of PHB were clearly observed in the diffraction pattern, and the 2θ were 13.8°, 16.5°, 20.3°, 22.3°, 25.1°, and 30.6°, respectively [43]. Among them, two strong scattering intensity peaks were detected near $2\theta = 13.8°$ and 16.5°, which were designated as orthorhombic cells (020) and (110), respectively [44].

The weak reflection near 21.50° and 22.50° corresponded to (101) and (111). The stronger reflection near 25.1° and 30.6° corresponded to (121) and (200) [45]. It can be seen from the diagram that the main diffraction peaks of PHB/CNCs and PHB/CNFs composites were similar to those of pure PHB thin films, indicating that the addition of CNCs and CNFs has no effect on the crystal structure of PHB. Nanocellulose can be embedded in PHB crystals by hydrogen bonds, which leads to the shift of the XRD peak to a lower angle at $2\theta = 13.5°$. The crystallization degree of the composites was calculated by Jade software, as shown in Table 2. When the content of CNCs was 1 wt %, the crystallinity of the composite was lower than that of PHB, which indicates that CNCs can be better embedded into the PHB matrix at the lower content. This might be due to the intertwining of PHB macromolecular chains and CNFs. The rigid structure of nanocellulose restricts the movement of molecular chains in the crystallization process of PHB, resulting in the difficulty of crystallization and the decrease in crystallinity. When the content of CNCs was increased than 1 wt %, the main diffraction peak strengths of the nanocomposite were higher than those of pure PHB. This indicated that the crystallinity of the nanocomposites becomes larger at higher concentrations of CNCs. Furthermore, the 2θ of crystal planes (020) and (110) gradually becomes smaller, which was because CNCs and CNFs act as nucleating agents in the composite films. Increasing the number of nucleating agents and crystals in the system can increase the crystallinity. Our results were in agreement with the work of Ten et al. [46].

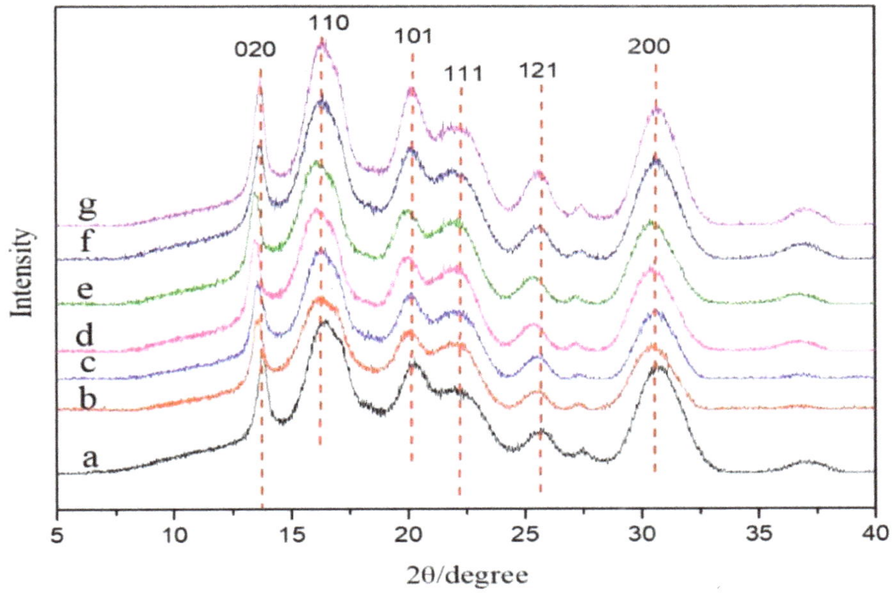

Figure 8. X-ray diffraction (XRD) patterns of PHB/CNCs and PHB/CNFs composite films: (**a**) PHB; (**b**) PHB/CNC1; (**c**) PHB/CNC3; (**d**) PHB/CNF5; (**e**) PHB/CNF1; (**f**) PHB/CNF3; (**g**) PHB/CNF5.

Table 2. Crystallinity of nanocellulose/PHB composite film.

Sample	Crystallinity (%)
PHB	56.61
PHB/CNC1	54.6
PHB/CNC3	56.85
PHB/CNC5	59.37
PHB/CNF1	57.9
PHB/CNF3	59.5
PHB/CNF5	60.1

3.5. Melting and Non-Isothermal Crystallization Behavior of Nanocellulose/PHB Composite Film

Crystallinity is an important parameter that primarily affects the physical properties of biodegradable polymers. The non-isothermal crystallization behavior of the polymer is presented in in Figures 9 and 10. Figure 9 shows the second DSC heat curve for pure PHB, PHB/CNCs, and PHB/CNFs composites. The thermal performance characteristic parameters including the melting temperature (T_m), cold crystallization temperature (T_c), melting enthalpy (ΔH_m), and crystallinity of the material (X_c) are summarized in Table 3.

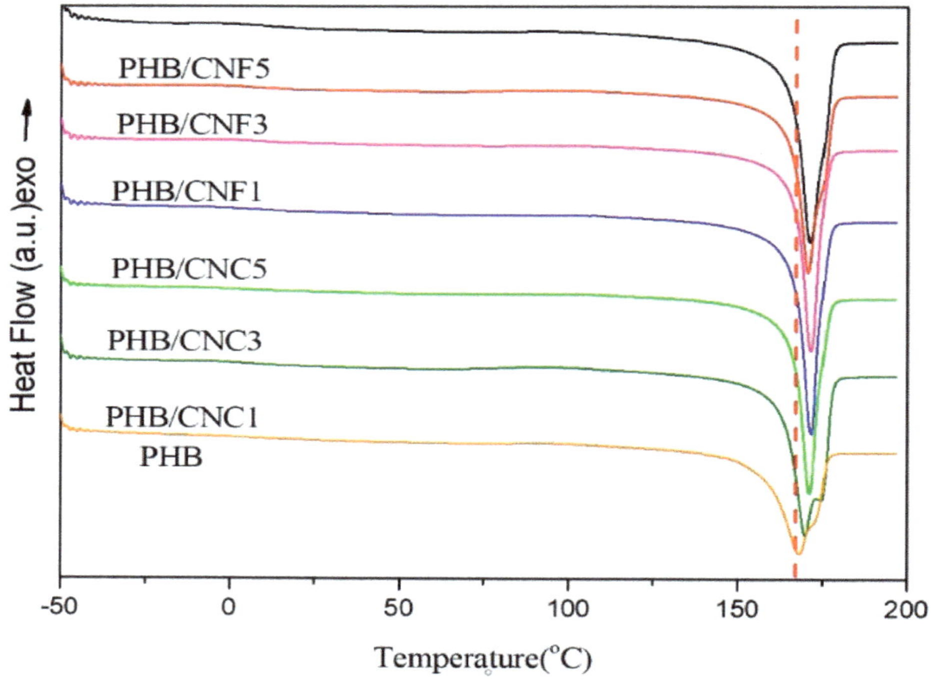

Figure 9. DSC second heating curve of pure PHB, PHB/CNCs, and PHB/CNFs composite films.

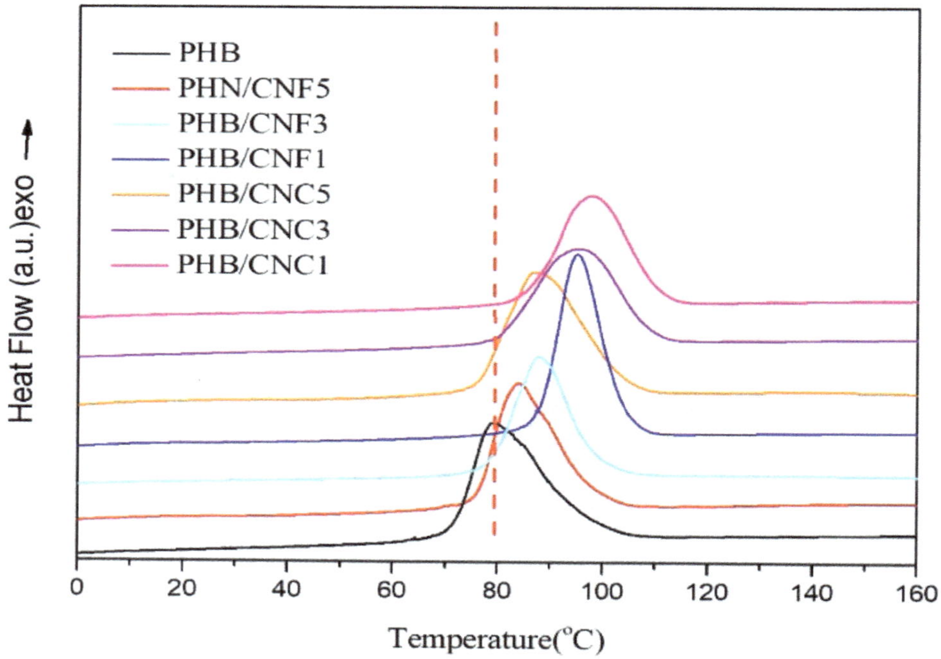

Figure 10. DSC first cooling curve for pure PHB, PHB/CNCs, and PHB/CNFs composite films.

From the secondary heat curve of DSC in Figure 9, it can be seen that with the addition of CNCs to PHB, the Tm increased from 168.2 °C to 171.0 °C, and with the increase of CNFs content from 1 wt % to 5 wt %, the T_m of PHB/CNCs composites increased from 168.2 °C to 172.5 °C. Interestingly, when the CNFs and CNCs were used at the same concentrations, the T_m of PHB/CNFs was higher than that of PHB/CNCs. This indicates that the nucleation of nanoscale CNFs was significantly stronger which further improved the integrity of PHB crystallization than that of CNCs. The toughness of PHB reduced by the improvement of the PHB crystal structure, which corresponded to the test results of mechanical properties.

Table 3. Melting parameters of PHB, PHB/CNCs, and PHB/CNFs composite films.

Sample	Second Melting			First Melting
	T_m/°C	ΔH_m/(J/g)	X_c/%	T_c /°C
PHB	168.2	94.6	64.8	79.2
PHB/CNC1	169.5	88.3	61.1	97.5
PHB/CNC3	170.9	90.8	64.2	96.1
PHB/CNC5	171.0	93.1	67.1	86.8
PHB/CNF1	170.8	91.2	63.1	94.8
PHB/CNF3	171.3	94.8	66.9	87.5
PHB/CNF5	172.5	96,8	69.8	83.5

Figure 10 shows the DSC first cooling curve of pure PHB, PHB/CNCs, and PHB/CNFs composite films. It can be seen from Figure 10 and Table 3 that when 1 wt % CNCs was added to PHB, the T_c of the composite film increased from 79.2 °C to 97.5 °C, which further indicated that the crystallization of PHB becomes easier. This increase in temperature was due to the addition of CNCs to PHB, which played the role of a heterogeneous nucleating agent, reduced the energy barrier of PHB nucleation, significantly increased the nucleation rate, and promoted melting crystallization. However, with the continued increase in CNCs content, the T_c of PHB/CNCs decreased from 97.5 °C to 86.8 °C; this decrease of 10.7 °C indicated that the nucleation ability of CNCs decreased, which was due to agglomeration of nanocellulose in the matrix. When CNFs were added into the PHB substrate, the T_c of the composite film increased from 79.2 °C to 94.8 °C and then decreased to 83.5 °C. When the contents of CNCs and CNFs in the PHB substrate were equal, the T_c of the CNCs/PHB composite film was higher than that of the CNFs/PHB composite film, which indicates that CNCs are more beneficial to the crystallization of PHB than CNFs.

3.6. *Effect of Nanocellulose on the Mechanical Properties of PHB*

A uniaxial tensile test was conducted to determine the effect of cellulose nanocellulose on the mechanical properties of PHB, as shown in Figure 11. Figure 11a–c show the tensile strength, elongation at break, and Young's modulus of the composite films with different nanocellulose concentrations. The results showed that a low proportion of CNCs has a significant effect on the mechanical properties of PHB. At 1 wt % of CNCs, all the mechanical properties of the composite films were the best. The elongation at break increased from 3.40% to 6.5%, and Young's modulus increased from 1.4 GPa to 1.7 GPa, which represents an increase of 91.2% and 18.4%, respectively, compared with pure PHB. On the other hand, the tensile strength decreased from 31.2 MPa to 27.1 MPa, which represents a decrease of 3.5%. The Young's modulus of CNCs/PHB composite films increased from 1.7 GPa to 2.0 GPa, which was 17.5% higher than that of pure PHB. However, the tensile strength and elongation at break of the composite film decreased to 22.8 MPa and 2.6%, respectively. This is because of the CNCs contents, which can make the molecular arrangement more compact, reduce the free volume, and increase the elongation at break. As shown in Figure 6, the CNCs at low concentration uniformly dispersed into PHB, and the surface defects were relatively low. From the perspective of crystallization behavior, CNCs played the role of a heterogeneous nucleating agent in PHB, accelerated the crystallization process, reduced the crystallization degree of PHB, and basically solved the problem of poor toughness of PHB (as shown in Table 3). When the concentration of CNCs increased from 1 wt %, the tensile strength and elongation at break of the composite films decreased, but the Young's modulus continuously increased. This was mainly due to the fact that CNCs were derived from natural biomass with an organic rigid material having a stiffness of 140–220 GPa [47,48]. The dispersion of CNCs can reach about 1 wt % by the solvent exchange method; however, the agglomeration of CNCs can be observed under a higher load of CNCs, and their effect on the mechanical properties of PHB decreases gradually. At this time, the interfacial binding force between hydrophilic CNCs and the hydrophobic polymer decreases, and the intermolecular hydrogen bond binding force increases [49].

Other researchers [50] stated that under higher CNCs loading, CNCs begin to aggregate in the form of clusters on the surface of the polymer, and the side effects on mechanical properties are more significant.

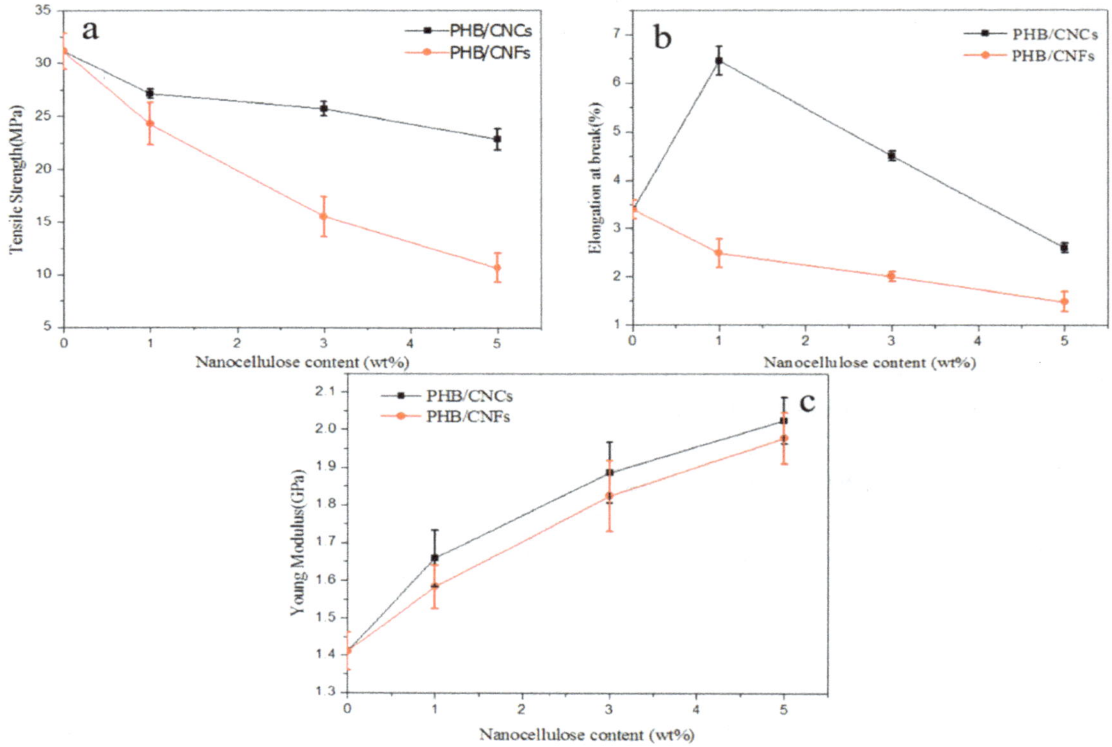

Figure 11. Effect of nanocellulose content on the mechanical properties of composites: (**a**) tensile strength, (**b**) elongation at break, and (**c**) Young's modulus.

When the concentrations of CNFs increased from 1 wt % to 5 wt %, the tensile strength and elongation at break of PHB continuously decreased; the tensile strength decreased from 31.2 MPa to 10.6 MPa, and the elongation at break decreased from 3.4% to 1.4%, which was 66.0% and 58.8% lower than that of pure PHB, respectively. However, Young's modulus increased from 1.7 GPa to 1.97 GPa, which was 15.9% higher than that of pure PHB. This was due to the agglomeration of CNFs in the PHB matrix, which was consistent with the SEM diagram of CNFs/PHB (Figure 6).

3.7. Barrier Properties of Nanocellulose/PHB Composite Film

The shelf life, safety, and health of packaged food are strongly affected by the humidity in the package, because humidity is one of the important factors that provide a suitable environment for the growth of unnecessary microorganisms. The water vapor and oxygen permeability of composite films can be reduced by incorporating nanoparticles into the polymer matrix. Therefore, the water vapor and oxygen transmission rates of pure PHB, PHB/CNCs, and PHB/CNFs nanocomposite films were investigated. The results are shown in Figures 12 and 13.

The water vapor and oxygen transmission rates of PHB were significantly reduced with the addition of the CNCs nanoparticles. The water vapor and oxygen transmittance of pure PHB membrane were 298.7 and 301.2 $cm^3/(m^2{\cdot}24\ h)$, respectively. With the addition of 1% CNCs, the water vapor and oxygen transmission rates greatly decreased to 157.9 and 140.4 $cm^3/(m^2{\cdot}24\ h)$, representing a decrease of 52.9% and 46.6%, respectively. When the CNCs concentration increased to 3% CNCs, the water vapor and oxygen transmittance rates of the composite membrane decreased to 85.2 $g/(m^2{\cdot}24\ h)$ and 87.1 $g/(m^2{\cdot}24\ h)$, respectively. This indicates that the resistance of the composite film to oxygen molecular penetration has increased by the nanometer size effect of CNCs. The rheological study of Bharadwaj et al. [51] shows that the instantaneous penetration network of CNCs formed in the polymer matrix by hydrogen bonds increases the resistance of CNCs' penetration to gas permeability. The increased permeability resistance provided by the CNCs' percolation network in the polymer matrix might be the cause of this situation [52]. Furthermore, the embedding of CNCs into the PHB matrix might also be the reason for the decrease in oxygen transmittance rate. When the content of CNCs was increased to 5%, the water vapor transmittance and oxygen transmittance increased to 120.4 g/(m^2·24 h) and 101.5 cm^3/(m^2·24 h), respectively. This was due to CNCs' intercalation effect and the osmotic network that can be ignored at a higher CNCs content, which leads to a higher oxygen transmittance rate. This might also be due to weak dispersion and agglomeration of CNCs nanoparticles.

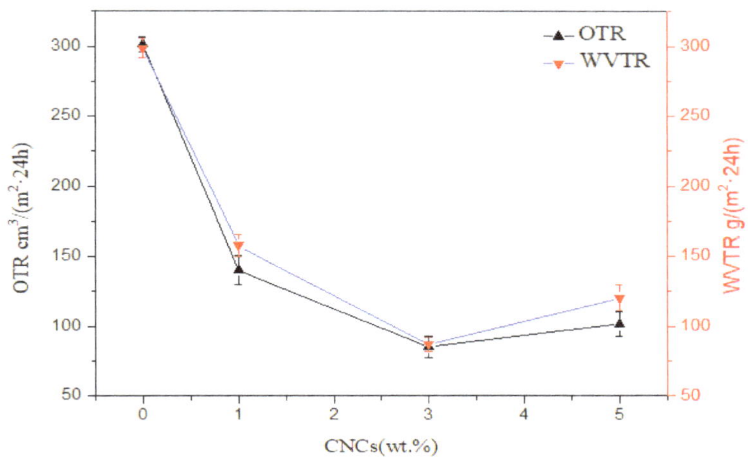

Figure 12. Water vapor and oxygen transmission rates against CNCs content for neat PHB and PHB/CNCs nanocomposite film.

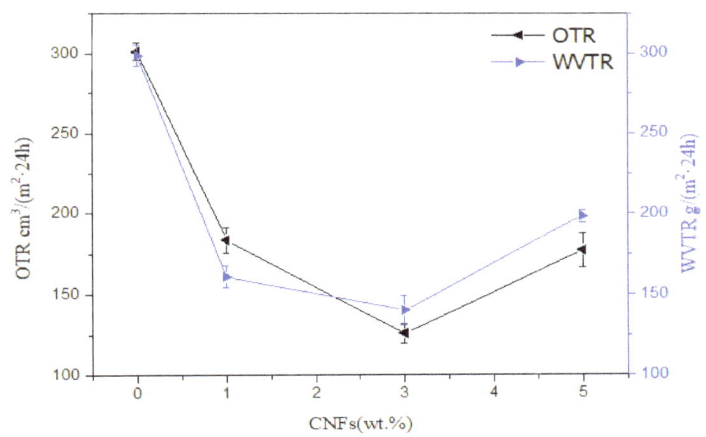

Figure 13. Water vapor and oxygen transmission rate against CNFs content for neat PHB and PHB/CNFs nanocomposite film.

Water vapor and oxygen transmittance rates of PHB also significantly ($p \leq 0.05$) decreased with the addition of CNFs nanoparticles. The water vapor transmittance of the composite film decreased from 298.7 to 140.4 g/(m^2·24 h) and oxygen transmittance decreased from 301.2 to 125.8 g/(m^2·24 h), representing a decrease of 47.0% and 41.8%, respectively. This effect was similar as found in the addition of CNCs. Due to the hydrogen bonding between CNFs and PHB in the composite film, a physical crosslinked network was formed in the system, where the gas dissolution process becomes more difficult. At this time, according to the tortuous path theory of composites, water molecules have to go through more tortuous paths in nanocomposites; therefore, the addition of CNFs improves the water vapor barrier performance of composite membranes. At 5% CNFs content, the water vapor and oxygen transmittance rates of the composite membrane increased to 198.6 and 177.2 cm^3/(m^2 24 h), respectively, and this increase was greater than that of CNCs. On the one hand, due to the existence of gaps in the cross-structure of CNFs, under the action of pressure, gas molecules can pass through this gap; hence, the agglomeration of CNFs with the same addition amount was more obvious.

3.8. Transparency of Nanocellulose/PHB Composite Films

Transmittance is one of the indexes to characterize the transparency of materials, and it is also one of the means to judge the compatibility of composite films. The transparency of the nanocomposite film is determined by UV–visible spectroscopy, and the spectra are shown in Figure 14. The specific values of the transmittance of nanocellulose/PHB composite films at the main wavelengths of visible light, at different nanocellulose mass fractions, are summarized in Table 4.

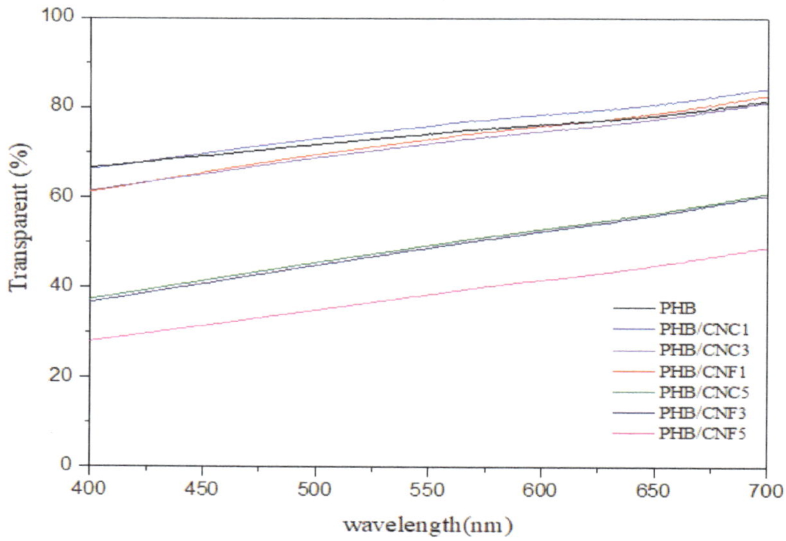

Figure 14. Transmittance curves of nanocellulose/PHB nanocomposite films with UV–visible light at different nanocellulose mass fractions.

Table 4. Transmittance of PHB, PHB/CNCs, and PHB/CNFs composite films.

Sample	Transmittance (%)			
	400 nm	500 nm	600 nm	700 nm
PHB	66.69	71.84	76.25	82.56
PHB/CNC1	66.3	73.04	78.43	84.14
PHB/CNC3	61.46	69.47	75.83	81.42
PHB/CNC5	37.36	45.52	52.96	60.85
PHB/CNF1	61.2	68.76	74.71	81.06
PHB/CNF3	36.69	44.85	52.45	60.44
PHB/CNF5	28.08	34.93	41.6	48.88

With the increase in wavelength, the transmittance of the film is lower than that of pure PHB films except for 1% PHB/CNCs samples (Figure 14). With the addition of 1% nanocellulose, the transmittance of the composite film slightly increased. This might be uniformly dispersion of low nanocellulose in the PHB matrix. Another possible explanation could be the size of nanocellulose, which is smaller than the wavelength of visible light, and therefore the transmission of visible light did not affect it.

As presented in Table 4, the transmittance of the composite film decreased with the increase in nanocellulose concentration. This was probably due to the poor compatibility between hydrophilic nanocellulose and hydrophobic PHB, as well as the fact that the aggregation of nanocellulose in the PHB matrix was not conducive to light transmission; hence, the transmittance of the film began to decline.

3.9. Thermal Stability of Nanocellulose/PHB Composite Films

Figure 15 shows the thermogravimetric (TG) and differential thermogravimetric (DTG) curves of PHB/CNCs composite films. The main mass loss range of the PHB/CNCs composite film was 250–300 °C, and the whole process shows a "Z" font. The initial decomposition temperature (T_0) of pure PHB was 254 °C, and that of the composite films was almost the same (Table 5). With the increase of CNCs concentration, the peak of the fastest decomposition temperature (T_{max}) of the PHB/CNCs composite film was shifted to the right, from 286.5 °C to 291.7 °C. This shows that the addition of a small amount of CNCs does not significantly improve the thermal performance of PHB. This might be due to the strong interaction between CNCs and PHB, which changed the thermal stability of PHB. This might also be due to the formation of hydrogen bonds between the carbonyl group of PHB and the hydroxyl group of CNCs, which slowed down the random breaking chain of PHB, thus improving the thermal stability of PHB [42].

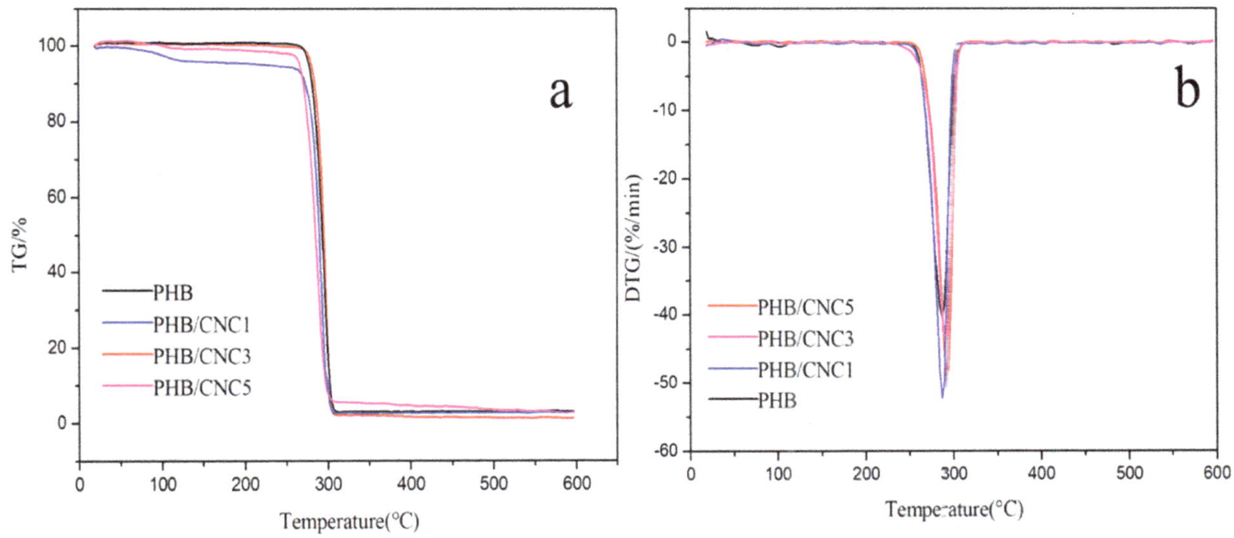

Figure 15. (**a**) Thermogravimetric (TG) and (**b**) differential thermogravimetric (DTG) curves of PHB/CNCs composite films.

Table 5. PHB, PHB/CNCs and PHB/CNFs thermogravimetric data.

Sample	T_0/°C	T_{max}/°C	600 °C Residual Ash (%)
PHB	254.0	286.5	1.12
PHB/CNC1	254.1	288.4	0.97
PHB/CNC3	250.7	290.2	1.01
PHB/CNC5	252.7	291.7	1.76
PHB/CNF1	254.4	291.4	2.42
PHB/CNF3	255.2	294.6	2.86
PHB/CNF5	255.6	298.2	1.17

Figure 16 shows the TG and DTG curves of PHB/CNFs composite films. As can be seen from Figure 16a, the main thermal weight loss region and the whole process of PHB/CNFs were consistent with those of PHB/CNCs composite films. Figure 16b shows the DTG curves of PHB/CNFs composite thin films. PHB/CNFs composite films also showed single-peak weightlessness, which was similar to the process of PHB/CNCs composite films.

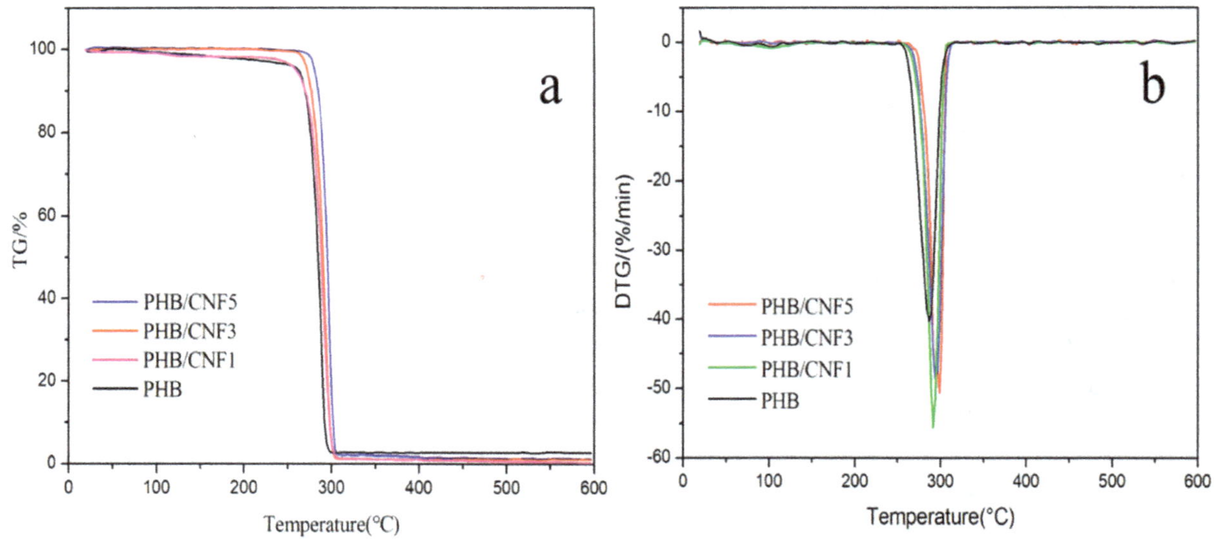

Figure 16. (**a**) TG and (**b**) DTG curves of PHB/CNCs composite films.

The initial decomposition temperature (T_0) of the composite films was not affected by CNFs (Table 5). With the increase in CNFs content, the peak of the highest T_{max} of PHB/CNF composite films shifted to the right from 286.5 to 294.2 °C. Compared with the thermal stability of the two kinds of nanocomposite films, it was found that CNFs were more favorable than CNCs in improving the thermal stability of PHB. This could be because the reticular cross-structure formed by CNFs at the microlevel can inhibit the degradation of PHB [53].

4. Conclusions

In this paper, the effects of the morphology and content of nanocellulose on the properties of PHB were systematically compared. It was concluded that the morphology of bagasse nanocellulose was rod-shaped, and the toughening effect of bagasse nanocellulose on PHB was the highest at 1 wt %. Compared with pure PHB, the elongation at break and Young's modulus of CNCs/PHB composite film increased by 91.2% and 18.4%, respectively, while the tensile strength decreased by 3.5%. With the increase in CNCs content, the elongation at break and T_c of CNCs/PHB composite films first increased and then decreased, while the crystallinity and gas permeability first decreased and then increased. With the addition of CNCs, the tensile strength and light transmittance decreased, and the Young's modulus, T_m, and T_{max} increased. Similarly, with the increase in CNFs content, the T_c of CNFs/PHB composite films first increased and then decreased while the barrier properties initially decreased and then increased. The elongation at break, tensile strength, and light transmittance were reduced, and the Young's modulus, crystallinity, T_m, and T_{max} increased.

At 1 wt % CNCs content, the toughness of PHB increased; as in the crystallization behavior, CNCs act as heterogeneous nucleating agents in PHB, thereby reducing the crystallinity of PHB. In the aspect of interfacial bonding, CNCs were bound to PHB by hydrogen bonding. In addition, at 1 wt % CNCs content, most of the nanocellulose was embedded on the surface of the polymer, which provided more bonding contact areas, and hence the interface compatibility between CNCs and PHB was better. The bio-based nanocomposites prepared in this study are expected to replace petroleum-based non-degradable materials in the packaging field. In addition, the application of agricultural and forestry by-products has been broadened.

Supplementary Materials:.
Figure S1. Effects of sulfuric acid concentration, acid hydrolysis time, and acid hydrolysis temperature on the yield of CNCs. Figure S2. Machine stick on automatic film application machine. Figure S3. (a) PHB/CNCs film, (b) Actual picture of PHB / CNCs film fracture during mechanical test.

Author Contributions: B.Z. was in charge of all the experiments and wrote the paper; C.H. conceived the project and edited the work; J.W. helped with scientific writing; C.Y. was involved with the writing and helped with data analysis; H.Z. helped with reviewing and editing; L.Z. and Y.Z. were responsible for the appointment and accounting of experimental instruments. All the authors contributed to the scientific discussion.

Acknowledgments: The authors are grateful for the financial support from the National Natural Science Foundation of China (grant number 2156070033), Guangxi Natural Science Foundation Program (2019GXNSFBA185006) and Guangxi Young Teachers Promotion Program (2019KY0045). The authors also gratefully acknowledge the Devices Sharing Platform Testing Center of Guangxi University. B.Z Thanks Jiaxue You from Shaanxi Normal University, China, for his help of revising the paper.

References

1. Utekar, P.; Gabale, H.; Khandelwal, A.; Mhaske, S.T. Hot-Melt Adhesives from Renewable Resources: A Critical Review. *Rev. Adhes. Adhes.* **2016**, *4*, 104–118. [CrossRef]
2. Miao, C.; Hamad, W.Y. Cellulose reinforced polymer composites and nanocomposites: A critical review. *Cellulose* **2013**, *20*, 2221–2262. [CrossRef]
3. Lopera-Valle, A.; Caputo, J.V.; Leao, R.; Sauvageau, D.; Luz, S.M.; Elias, A. Influence of Epoxidized Canola Oil (eCO) and Cellulose Nanocrystals (CNCs) on the Mechanical and Thermal Properties of Polyhydroxybutyrate (PHB)-Poly(lactic acid) (PLA) Blends. *Polymers* **2019**, *11*, 933. [CrossRef]
4. Carvalho, K.C.C.D.; Montoro, S.R.; Cioffi, M.O.H.; Voorwald, H.J.C. Chapter 13—Polyhydroxyalkanoates and Their Nanobiocomposites With Cellulose Nanocrystals. In *Design & Applications of Nanostructured Polymer Blends & Nanocomposite Systems*; William Andrew Publishing: Boston, MA, USA, 2016; pp. 261–285.
5. Chiulan, I.; Mihaela, P.D.; Nicoleta, F.A.; Teodorescu, M.; Nicolae, C.A.; Căşărică, A.; Tofan, V.; Sălăgeanu, A. Biocompatible polyhydroxyalkanoates/bacterial cellulose composites: Preparation, characterization and in vitro evaluation. *J. Biomed. Mater. Res. Part A* **2016**, *104*, 2576–2584. [CrossRef]
6. Teresita Seoane, I.; Beatriz Manfredi, L.; Paola Cyras, V.; Torre, L.; Fortunati, E.; Puglia, D. Effect of Cellulose Nanocrystals and Bacterial Cellulose on Disintegrability in Composting Conditions of Plasticized PHB Nanocomposites. *Polymers* **2017**, *9*, 561.
7. Chen, J.; Xu, C.; Wu, D.; Pan, K.; Qian, A.; Sha, Y.; Wang, L.; Tong, W. Insights into the nucleation role of cellulose crystals during crystallization of poly(β -hydroxybutyrate). *Carbohydr. Polym.* **2015**, *134*, 508–515. [CrossRef]
8. Du, J.; Zhao, G.; Pan, M.; Zhuang, L.; Li, D.; Rui, Z. Crystallization and mechanical properties of reinforced PHBV composites using melt compounding: Effect of CNCs and CNFs. *Carbohydr. Polym.* **2017**, *168*, 255.
9. Prakobna, K.; Galland, S.; Berglund, L.A. High-Performance and Moisture-Stable Cellulose-Starch Nanocomposites Based on Bioinspired Core-Shell Nanofibers. *Biomacromolecules* **2015**, *16*, 904–912. [CrossRef]
10. Cheng, Q.; Wang, S.; Rials, T.G. Poly(vinyl alcohol) nanocomposites reinforced with cellulose fibrils isolated by high intensity ultrasonication. *Compos. Part A Appl. Sci. Manuf.* **2009**, *40*, 218–224. [CrossRef]
11. Santamaria-Echart, A.; Ugarte, L.; Arbelaiz, A.; Barreiro, F.; Angeles Corcuera, M.; Eceiza, A. Modulating the microstructure of waterborne polyurethanes for preparation of environmentally friendly nanocomposites by incorporating cellulose nanocrystals. *Cellulose* **2017**, *24*, 823–834. [CrossRef]
12. Lu, Z.; Liu, M.; Gao, Q.; Yang, D.; Zhang, Z.; Xiong, X.; Yuan, J.; Xiang, Y.L. Design of Heterogeneous Nuclei Composed of Uniaxial Cellulose Nanocrystal Assemblies for Epitaxial Growth of Poly(ε-caprolactone). *Macromolecules* **2017**, *50*, 3355–3364. [CrossRef]
13. Mariano, M.; Pilate, F.; de Oliveira, F.B.; Khelifa, F.; Dubois, P.; Raquez, J.-M.; Dufresne, A. Preparation of Cellulose Nanocrystal-Reinforced Poly(lactic acid) Nanocomposites through Noncovalent Modification with PLLA-Based Surfactants. *Acs Omega* **2017**, *2*, 2678–2688. [CrossRef]
14. Saba, N.; Mohammad, F.; Pervaiz, M.; Jawaid, M.; Alothman, O.Y.; Sain, M. Mechanical, morphological and structural properties of cellulose nanofibers reinforced epoxy composites. *Int. J. Biol. Macromol.* **2017**, *97*, 190–200. [CrossRef]
15. Zimmermann, T.; Bordeanu, N.; Strub, E. Properties of nanofibrillated cellulose from different raw *materials and its reinforcement potential. Carbohydr. Polym.* **2010**, *79*, 1086–1093. [CrossRef]
16. Iwamoto, S.; Nakagaito, A.N.; Yano, H. Nano-fibrillation of pulp fibers for the processing of transparent nanocomposites. *Appl. Phys. A* **2007**, *89*, 461–466. [CrossRef]
17. Wang, S.; Cheng, Q. A novel process to isolate fibrils from cellulose fibers by high-intensity ultrasonication, Part 1: Process optimization. *J. Appl. Polym. Sci.* **2009**, *113*, 1270–1275. [CrossRef]

18. Nechyporchuk, O.; Belgacem, M.N.; Bras, J. Production of cellulose nanofibrils: A review of recent advances. *Ind. Crop. Prod.* **2016**, *93*, 2–25. [CrossRef]
19. Bai, W.; Holbery, J.; Li, K. A technique for production of nanocrystalline cellulose with a narrow size distribution. *Cellulose* **2009**, *16*, 455–465. [CrossRef]
20. Samira, E.H.; Yoshiharu, N.; Jean-Luc, P.; Laurent, H.; Frédéric, D.; Cyrille, R. The shape and size distribution of crystalline nanoparticles prepared by acid hydrolysis of native cellulose. *Biomacromolecules* **2008**, *9*, 57–65.
21. Kumar, A.; Negi, Y.S.; Choudhary, V.; Bhardwaj, N. Sugarcane Bagasse: A Promising Source for the Production of Nano-Cellulose. *J. Polym. Compos.* **2014**, *2*, 23–27.
22. Kumar, A.; Negi, Y.S.; Choudhary, V.; Bhardwaj, N.K. Characterization of Cellulose Nanocrystals Produced by Acid-Hydrolysis from Sugarcane Bagasse as Agro-Waste. *J. Mater. Phys. Chem.* **2014**, *2*, 1–8.
23. Srithep, Y.; Ellingham, T.; Peng, J.; Sabo, R.; Clemons, C.; Turng, L.S.; Pilla, S. Melt compounding of poly (3-hydroxybutyrate-co-3-hydroxyvalerate)/nanofibrillated cellulose nanocomposites. *Polym. Degrad. Stab.* **2013**, *98*, 1439–1449. [CrossRef]
24. Cherpinski, A.; Torres-Giner, S.; Vartiainen, J.; Peresin, M.S.; Lahtinen, P.; Lagaron, J.M. Improving the water resistance of nanocellulose-based films with polyhydroxyalkanoates processed by the electrospinning coating technique. *Cellulose* **2018**, *25*, 1291–1307. [CrossRef]
25. Seoane, I.T.; Fortunati, E.; Puglia, D.; Cyras, V.P.; Manfredi, L.B. Development and characterization of bionanocomposites based on poly(3-hydroxybutyrate) and cellulose nanocrystals for packaging applications. *Polym. Int.* **2016**, *65*, 1046–1053. [CrossRef]
26. Youssef, H.; Lucia, L.A.; Rojas, O.J. Cellulose nanocrystals: Chemistry, self-assembly, and applications. *Chem. Rev.* **2010**, *110*, 3479–3500.
27. Seoane, I.T.; Cerrutti, P.; Vazquez, A.; Manfredi, L.B.; Cyras, V.P. Polyhydroxybutyrate-Based Nanocomposites with Cellulose Nanocrystals and Bacterial Cellulose. *J. Polym. Environ.* **2017**, *25*, 586–598. [CrossRef]
28. Barham, P.J.; Keller, A.; Otun, E.L.; Holmes, P.A. Crystallization and morphology of a bacterial thermoplastic: Poly-3-hydroxybutyrate. *J. Mater. Sci.* **1984**, *19*, 2781–2794. [CrossRef]
29. Kuroda, M.; Ohta, T.; Uchiyama, I.; Baba, T.; Yuzawa, H.; Kobayashi, I.; Cui, L.; Oguchi, A.; Aoki, K.; Nagai, Y. Whole genome sequencing of meticillin-resistant Staphylococcus aureus. *Lancet* **2001**, *357*, 1225–1240. [CrossRef]
30. De Oliveira, F.B.; Bras, J.; Borges Pimenta, M.T.; da Silva Curvelo, A.A.; Belgacem, M.N. Production of cellulose nanocrystals from sugarcane bagasse fibers and pith. *Ind. Crop. Prod.* **2016**, *93*, 48–57. [CrossRef]
31. Dong, X.M.; Revol, J.F.; Gray, D.G. Effect of microcrystallite preparation conditions on the formation of colloid crystals of cellulose. *Cellulose* **1998**, *5*, 19–32. [CrossRef]
32. Chen, W.; Yu, H.; Liu, Y.; Chen, P.; Zhang, M.; Hai, Y. Individualization of cellulose nanofibers from wood using high-intensity ultrasonication combined with chemical pretreatments. *Carbohydr. Polym.* **2011**, *83*, 1804–1811. [CrossRef]
33. Gicquel, E.; Bras, J.; Rey, C.; Putaux, J.-L.; Pignon, F.; Jean, B.; Martin, C. Impact of sonication on the rheological and colloidal properties of highly concentrated cellulose nanocrystal suspensions. *Cellulose* **2019**, *26*, 7619–7634. [CrossRef]
34. Dahlem, M.A.; Borsoi, C.; Hansen, B.; Catto, A.L. Evaluation of different methods for extraction of nanocellulose from yerba mate residues. *Carbohydr. Polym.* **2019**, *218*, 78–86. [CrossRef] [PubMed]
35. Goswami, N.; Lin, F.; Liu, Y.; Leong, D.T.; Xie, J. Highly Luminescent Thiolated Gold Nanoclusters Impregnated in Nanogel. *Chem. Mater.* **2016**, *28*, 4009–4016. [CrossRef]
36. Li, Q.; Renneckar, S. Supramolecular Structure Characterization of Molecularly Thin Cellulose I Nanoparticles. *Biomacromolecules* **2011**, *12*, 650–659. [CrossRef]
37. Bodin, A.; Backdahl, H.; Risberg, B.; Gatenholm, P. Nano cellulose as a scaffold for tissue engineered blood vessels. *Tissue Eng.* **2007**, *13*, 1–30.
38. Abraham, E.; Deepa, B.; Pothen, L.A.; Cintil, J.; Thomas, S.; John, M.J.; Anandjiwala, R.; Narine, S.S. Environmental friendly method for the extraction of coir fibre and isolation of nanofibre. *Carbohydr. Polym.* **2013**, *92*, 1477–1483. [CrossRef]
39. Chirayil, C.J.; Joy, J.; Mathew, L.; Mozetic, M.; Koetz, J.; Thomas, S. Isolation and characterization of cellulose nanofibrils from Helicteres isora plant. *Ind. Crop. Prod.* **2014**, *59*, 27–34. [CrossRef]

40. Liu, Q.-S.; Zhu, M.-F.; Wu, W.-H.; Qin, Z.-Y. Reducing the formation of six-membered ring ester during thermal degradation of biodegradable PHBV to enhance its thermal stability. *Polym. Degrad. Stab.* **2009**, *94*, 18–24. [CrossRef]
41. Lin, N.; Dufresne, A. Physical and/or Chemical Compatibilization of Extruded Cellulose Nanocrystal Reinforced Polystyrene Nanocomposites. *Macromolecules* **2013**, *46*, 5570–5583. [CrossRef]
42. Dhar, P.; Bhardwaj, U.; Kumar, A.; Katiyar, V. Poly (3-hydroxybutyrate)/cellulose nanocrystal films for food packaging applications: Barrier and migration studies. *Polym. Eng. Sci.* **2015**, *55*, 2388–2395. [CrossRef]
43. Lai, M.D.; Li, J.; Yang, J.; Liu, J.J.; Tong, X.; Cheng, H.M. The morphology and thermal properties of multi-walled carbon nanotube and poly(hydroxybutyrate-co-hydroxyvalerate) composite. *Polym. Int.* **2004**, *53*, 1479–1484. [CrossRef]
44. Sun, X.; Guo, L.; Sato, H.; Ozaki, Y.; Yan, S.; Takahashi, I. A study on the crystallization behavior of poly(β-hydroxybutyrate) thin films on Si wafers. *Polymers* **2011**, *52*, 3865–3870. [CrossRef]
45. Mottin, A.C.; Ayres, E.; Oréfice, R.L.; Câmara, J.J.D. What Changes in Poly(3-Hydroxybutyrate) (PHB) When Processed as Electrospun Nanofibers or Thermo-Compression Molded Film? *Mater. Res.* **2016**, *19*, 57–66. [CrossRef]
46. Ten, E.; Jiang, L.; Wolcott, M.P. Preparation and properties of aligned poly(3-hydroxybutyrate- co -3-hydroxyvalerate)/cellulose nanowhiskers composites. *Carbohydr. Polym.* **2013**, *92*, 206–213. [CrossRef]
47. Janne, L.; Junji, S.; Malin, G.; Linda, F.; Markus, L.; Teeri, T.T. The binding specificity and affinity determinants of family 1 and family 3 cellulose binding modules. *Proc. Natl. Acad. Sci. USA* **2003**, *100*, 484–489.
48. Fortunati, E.; Peltzer, M.; Armentano, I.; Torre, L.; Jiménez, A.; Kenny, J.M. Effects of modified cellulose nanocrystals on the barrier and migration properties of PLA nano-biocomposites. *Carbohydr. Polym.* **2012**, *90*, 948–956. [CrossRef]
49. Berthet, M.A.; Angellier-Coussy, H.; Chea, V.; Guillard, V.; Gastaldi, E.; Gontard, N. Sustainable food packaging: Valorising wheat straw fibres for tuning PHBV-based composites properties. *Compos. Part A Appl. Sci. Manuf.* **2015**, *72*, 139–147. [CrossRef]
50. Dhar, P.; Bhardwaj, U.; Kumar, A.; Katiyar, V. Investigations on Rheological and Mechanical Behavior of Poly(3-Hydroxybutyrate)/Cellulose Nanocrystal Based Nanobiocomposites. *Polym. Compos.* **2017**, *38*, E392–E401. [CrossRef]
51. Bhardwaj, U.; Dhar, P.; Kumar, A.; Katiyar, V. Polyhydroxyalkanoates (PHA)-Cellulose Based Nanobiocomposites for Food Packaging Applications. In *Food Additives and Packaging*; American Chemical Society: New York, NY, USA, 2014; pp. 275–314.
52. Fraga, A.; Ruseckaite, R.A.; Jiménez, A. Thermal degradation and pyrolysis of mixtures based on poly(3-hydroxybutyrate-8%-3-hydroxyvalerate) and cellulose derivatives. *Polym. Test.* **2005**, *24*, 526–534. [CrossRef]
53. Mokhena, T.C.; Sefadi, J.S.; Sadiku, E.R.; John, M.J.; Mochane, M.J.; Mtibe, A. Thermoplastic Processing of PLA/Cellulose Nanomaterials Composites. *Polymers* **2018**, *10*, 1363. [CrossRef] [PubMed]

3

Structure/Function Analysis of Cotton-Based Peptide-Cellulose Conjugates: Spatiotemporal/Kinetic Assessment of Protease Aerogels Compared to Nanocrystalline and Paper Cellulose

J. Vincent Edwards [1,*], Krystal Fontenot [1], Falk Liebner [2], Nicole Doyle nee Pircher [2], Alfred D. French [1] and Brian D. Condon [1]

1 Southern Regional Research Center, USDA, New Orleans, LA 70124, USA; krystal.fontenot@ars.usda.gov (K.F.); Al.French@ars.usda.gov (A.D.F.); brian.condon@ars.usda.gov (B.D.C.)
2 University of Natural Resources and Life Sciences Vienna, Konrad-Lorenz-Straße 24, A-3430 Tulln an der Donau, Austria; falk.liebner@boku.ac.at (F.L.); nicole.pircher@boku.ac.at (N.D.n.P.)
* Correspondence: vince.edwards@ars.usda.gov

Abstract: Nanocellulose has high specific surface area, hydration properties, and ease of derivatization to prepare protease sensors. A Human Neutrophil Elastase sensor designed with a nanocellulose aerogel transducer surface derived from cotton is compared with cotton filter paper, and nanocrystalline cellulose versions of the sensor. X-ray crystallography was employed along with Michaelis–Menten enzyme kinetics, and circular dichroism to contrast the structure/function relations of the peptide-cellulose conjugate conformation to enzyme/substrate binding and turnover rates. The nanocellulosic aerogel was found to have a cellulose II structure. The spatiotemporal relation of crystallite surface to peptide-cellulose conformation is discussed in light of observed enzyme kinetics. A higher substrate binding affinity (K_m) of elastase was observed with the nanocellulose aerogel and nanocrystalline peptide-cellulose conjugates than with the solution-based elastase substrate. An increased K_m observed for the nanocellulosic aerogel sensor yields a higher enzyme efficiency (k_{cat}/K_m), attributable to binding of the serine protease to the negatively charged cellulose surface. The effect of crystallite size and β-turn peptide conformation are related to the peptide-cellulose kinetics. Models demonstrating the orientation of cellulose to peptide O6-hydroxymethyl rotamers of the conjugates at the surface of the cellulose crystal suggest the relative accessibility of the peptide-cellulose conjugates for enzyme active site binding.

Keywords: nanocellulose; protease sensor; human neutrophil elastase; peptide-cellulose conformation; aerogel

---

1. Introduction

Nanocellulose-Based Protease Sensors

The sophistication and sensitivity of point of care diagnostic detectors underlies much of the progress in the rapidly growing field of biosensors as is found with point of care protease sensors [1]. The sensor's transducer serves to ground and facilitate transformation of the sensor molecule (in this study, a peptide fluorophore) into a signature signal, which in this study is a fluorescence response indicating proteolytic activity. However, the composition of the transducer surface material and its interaction at the biomolecular interface (protease binding on the transducer surface) is of equal value. Another important consideration is the detectors interface with the biological environment, which can complicate and dampen the sensitivity of detection. This is especially the case when considering the design of in situ protease sensors interfaced with dressings to detect harmful levels of proteolytic enzymes.

Proteolytic enzymes including, matrix metalloproteases (MMPs) and serine proteases such as the neutrophil serine protease (human neutrophil elastase, HNE), are responsible for proteolytic degradation of growth factors [2] and the extracellular matrix (ECM) proteins in chronic wounds. In general, acute wounds have normal levels of MMPs and HNE [3–5], which facilitates the clearance of cellular debris. However, chronic wounds have elevated levels of MMPs (0.1–0.2 U/mL) and HNE (0.02–0.1 U/mL) depending on the type, e.g., diabetic, venous, pressure, and arterial ulcers [5]. Thus, elevated concentrations of MMPs and HNE delay the healing process, and are also considered biomarkers for chronic wound treatment evaluation [6–8].

Biosensor materials should be compatible with the complexity of the bio-system they are applied to. This is especially relevant to the chronic wound environment, which varies in exudative fluid levels, bacterial burden, and biochemical diversity. Biocompatibility is imparted to resist nonselective surface adsorption by lipids, proteins, polysaccharides, cellular debris, and breakdown of the transducer surface. Due to cellulosic's hydrophilic properties, protein binding on the transducer surface for sensor detection is relatively attenuated when compared with more hydrophobic materials.

We have recently examined the relative contributions of cellulosic and nanocellulosic sensor transducer surfaces to detection sensitivity of the neutrophil protease, HNE, demonstrating that different forms of cellulose and nanocellulose materials have varying surface area, porosity, and biocompatible properties affording sensor/transducer surface functionality [9–14].

Notably it is worth considering that nanocellulose is a highly crystalline biopolymer with a hydrophilic and high specific surface area that possesses reactive hydroxyls, which can be derivatized to covalently append a wide range of biologically active molecules. Here, the focus is on comparison of the relative structure/function contributions of three different cotton transducer surfaces (nanocellulose crystals (CNC), nanocellulosic aerogels (NA), and cellulose filter paper (CFP) (Figure 1), each having unique properties contributing to protease sensor sensitivity, yet varying features of sensor and dressing applicability. The sensor molecule is a tripeptide substrate of elastase. Accordingly, we examine here a nanocellulosic aerogel material in comparison with two other cellulosic and nanocellulosic designs for protease-lowering activity, and sensor/transducer surface performance with the aim of understanding how the molecular and physical properties of the materials may provide benefit for a dressing/sensor platform useful to simultaneously test and treat targeted elevated protease levels in chronic wounds.

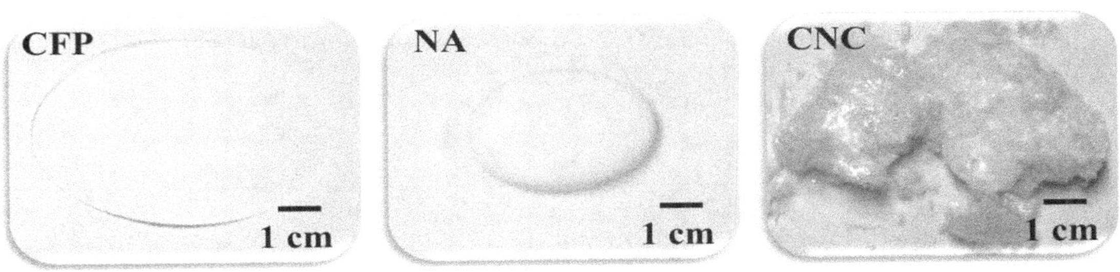

Figure 1. Images of the cellulosic (CFP) and nanocellulosic (NA and CNC) transducers.

In this study, we evaluate the relative structure/function contributions of the protease substrate, which is immobilized to a nanocellulosic aerogel. The structure function approach analyzes the enzyme kinetic turnover rate and binding affinity of the enzyme to structural findings of the crystal structure that influences the immobilized peptide in a β turn conformation on the cellulose crystallite surface in its interaction with the protease. Accordingly, the structure/function relations of two nanocellulosic and a cellulosic sensor analogs are contrasted, and also demonstrate interesting differences in protease enzymology.

2. Results

2.1. Structure/Function and Physical Property Considerations

A depiction of the physical structure of the materials of this study is shown in Figure 1. The intended application of the nanocellulose aerogel sensor is as an interface with chronic wound dressings to detect protease levels in situ. The functional properties of a protease sensor are determined by the composition of the transducer surface, the bioactivity of the enzyme substrate attached, and how the substrate is immobilized. A cellulose-based transducer surface, to which the peptide protease substrate is covalently attached, may be characterized for its crystalline structure and physical property composition [13,15]. In this study, conjugation of the tripeptide elastase substrate (Suc-Ala-Pro-Ala-AMC) to the cellulosic and nanocellulosic transducers via the amide bond formation between the carboxyl terminus of the tripeptide and the reactive hydroxyls on the transducer surfaces afforded the peptide-cellulose conjugates, pCFP, pNA and pCNC, which has previously been outlined [9,12].

2.2. Structure/Function Relations of the Protease Activity at a Molecular Level

2.2.1. Cellulose Crystal Structure

X-ray crystallography of the different transducer surfaces was undertaken and used to compare the relative contributions of cellulose structure to the peptide-cellulose conjugates and structural relationships of the cellulosic transducer surfaces to enzyme kinetics.

Figure 2 shows patterns from the X-ray diffraction (XRD) analyses. In the case of the CFP and the CNC, which are cotton based materials, the patterns were typical cellulose I_β patterns. The pattern from the aerogel transducer, seemingly a mostly amorphous material, is also shown in Figure 2A. That pattern was subjected to Rietveld analyses to do what was reasonable to confirm the initial diagnosis of amorphous material. The results of the Rietveld analysis are presented in Figure 2B.

Table 1 shows the larger values of Segal CrI (Crystallinity Index) and crystallite size for the CFP, compared to CNC. Apparently, the acid hydrolysis of the cotton reduced somewhat the size of the crystallites.

Table 1. The crystallinity index values, crystallite size values, and crystallite model for the cellulosic filter paper (CFP), cotton nanocrystalline cellulose (CNC), and nanocellulosic aerogel (NA) transducer.

Name	Segal CrI (%)	Crystallite Size (Å)	Crystallite Model	Layers Molecules	Surface Charge (mV)
CFP	87	69.3	18	144	−10.2
NC	79	54.3	14	98	−68
NA I_β	-	26	6	24	−19
NA II	-	65	16	122	−19
NA amorphous	-	19	5	24	−19

A more careful visual analysis of the NA pattern indicates a mixture of cellulose I, II and amorphous structures. [16]. The indications are the small peak at about 12° 2θ (indicating II) as well as the convex intensity in the area of 15°–17° 2θ which indicated some contributions from cellulose I. The presence of some crystalline material was supported by the poor fit of a calculated pattern from very small cellulose II crystals. Earlier, Nam et al. [17] found that such patterns with a very broad peak closely resembled exhaustively ball-milled cotton. Contributions of calculated patterns from a small cellulose I crystal improved the fit, as did one from a relatively large cellulose II crystal. Ultimately, a close fit was obtained between the calculated and experimental patterns, but it was necessary to include a substantial contribution from preferred orientation to obtain the fit depicted in

Figure 2B. Although the analysis indicated 44% of the small cellulose I crystallites, 48% of amorphous as modeled by small cellulose II crystallites, and 8% of large crystallites of II, these values should be taken as indications of the presence of three phases, not precise values. In addition, it is notable that crystallinity index values, i.e., Segal CrI (shown in Table 1) for CFP and CNC, are consistent with previously determined values for filter paper and cotton nanocrystalline cellulose. Such analyses are difficult; to get a visually satisfactory fit, 17 variables were used in the refinement with the most severe being the March–Dollase compensation for preferred orientation applied to the cellulose I component. This distorted the cellulose I pattern in Figure 2A. Because of the limited, somewhat noisy data, the obtained fit is not likely to be unique; other models could also give plausible analyses.

It is noted that these values for the composition were calculated from the exported data from the MAUD (materials analysis using diffraction) Rietveld software, not the values reported during the analysis. However, indications available are that there is considerably more surface area available for the NA samples than for the CFP and CNC materials because of the mostly smaller crystallites of NA.

The presence of the cellulose II crystallites is in good agreement with a previous study comparing the nanomorphology of various types of cellulose II aerogels obtained by dissolution-coagulation of cotton linters using different cellulose solvent systems. For calcium thiocyanate octahydrate/lithium chloride, i.e., the solvent used to prepare the three-dimensional NA scaffolds in this study, small angle X-ray spectroscopy had revealed considerably high cellulose II crystallinity (ca. 50%) and a crystal diameter of 46 ± 1 Å [18]. Notably, the crystallite size impacts the sensitivity of the sensor by increasing the specific surface area of the sensor which is correlated with sensor sensitivity [12].

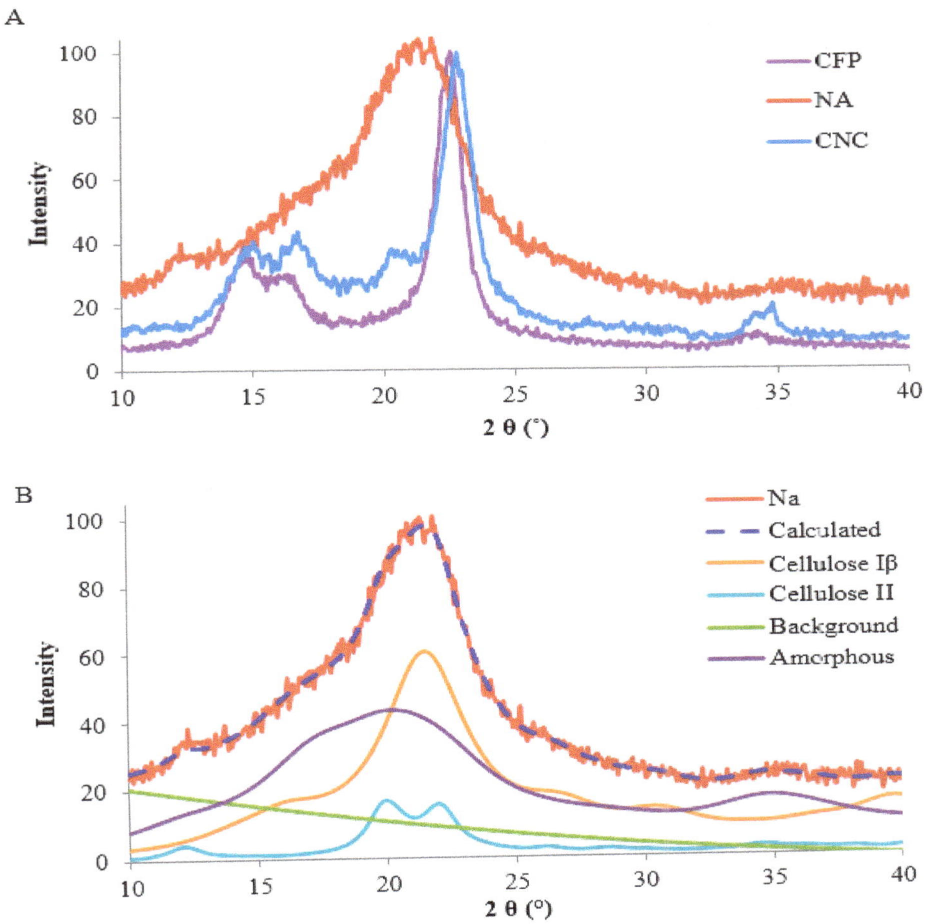

Figure 2. X-ray observed diffraction patterns (**A**) of the CFP, NA, and NC transducers; and (**B**) the calculated pattern of the NA as analyzed using the Rietveld method.

The crystallite models (shown in Figure 3) that are based on the respective cellulose crystallite sizes reveal the relative number of cellulose layers and chains which are CFP (18 layers and 144 chains), NA (12 layers and 78 chains), and CNC (14 layers and 98 chains). The number of chains (36, 28, and 24) on the 1–10 and 110 surfaces for the CFP, CNC, and NA, respectively, indicate half of the primary alcohol groups on those surface chains are in the interior structure of the cellulose or nanocellulose materials [11]. As discussed above, this structural feature effects the available primary hydroxyls for covalent attachment of the sensor molecule.

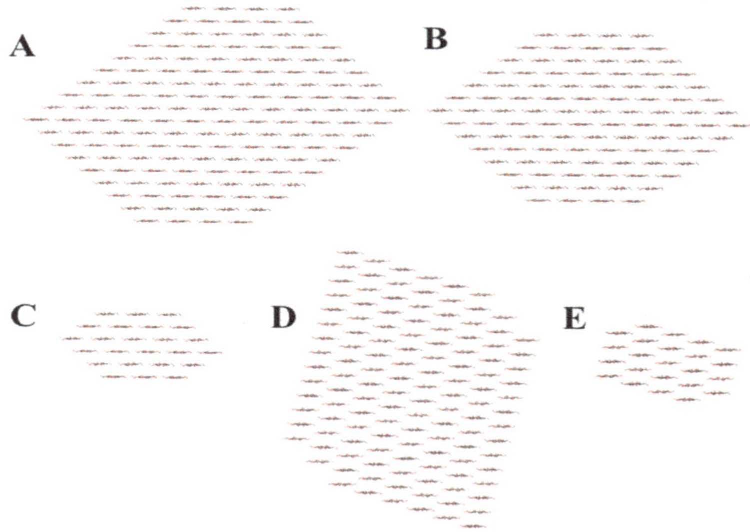

Figure 3. Models of the cotton cellulose and nanocellulose transducer materials based on the X-ray diffraction patterns. (**A**) CFP; (**B**) CNC; and (**C**–**E**) cellulose I, cellulose II, and amorphous contributions to the diffraction pattern for NA.

2.2.2. Peptide Conformation Consideration Based on Circular Dichroism

Here, we provide experimental evidence for secondary structure in the peptide portion of the conjugate. As shown in Figure 4, circular dichroism spectra of the peptide portion of the conjugate, i.e., Suc-Ala-Pro-Ala-AMC, are consistent with a β-turn prediction. Reflected in the CD are spectral minima at 200 (lit 208) and 214 (lit 222) nm, which are typical for alpha helices, indicate a propensity for type I β-turn formation [19], and with strong evidence for the type I β-turn formation as previously characterized for Pro-Ala [20].

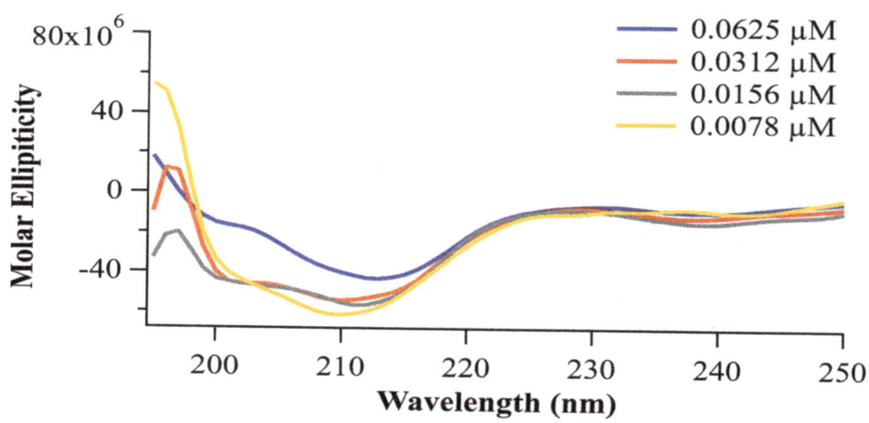

Figure 4. Circular Dichroism of the tripeptide (Suc-APA-AMC) at various concentrations ranging from 0.0625 to 0.0078 μM.

2.2.3. Bioactivity and Kinetic Evaluation of Peptide-Cellulose Conjugates with Elastase

Here, the sensor's ability to detect protease activity and the sensor functionality itself is determined through protease assays typically used to assess the enzyme/substrate recognition and relevant substrate binding and enzyme kinetics. In this study, it is applied to immobilized enzyme substrates i.e., the peptide-cellulose conjugates. The elastase activities of the tripeptide-conjugated sensor analogs, were assessed with succinamidyl-Ala-Pro-Ala-amidylcoumadin peptide-cellulose ester immobilized on the transducer surfaces through a synthesis that has previously been reported [11,13]. The bioactive response was assessed through fluorescence resulting from the protease catalyzed release of the COOH-terminal amino methyl coumarin fluorophore. The results of the bioactivity are shown in Figure 5 by way of reaction progress curves of each sensor's reaction with elastase at a protease concentration of 0.5 U/mL HNE. The relative response curves and detection sensitivity are apparent from the reaction progress and sensitivity curves.

The relative kinetic rates of the pCFP, pNA, and pCNC biosensors with HNE was assessed using traditional Michaelis–Menten kinetic analysis. Table 2 lists the kinetic values for the HNE peptide-cellulose substrates as attached in pCFP, pNA, and pCNC. The kinetic values refer to both the binding affinity of the conjugated enzyme substrate and the turnover rate or rate of product formation; a measurement of the rate of formation of hydrolyzed fluorophore (AMC) from the COOH-terminus of the bound peptide.

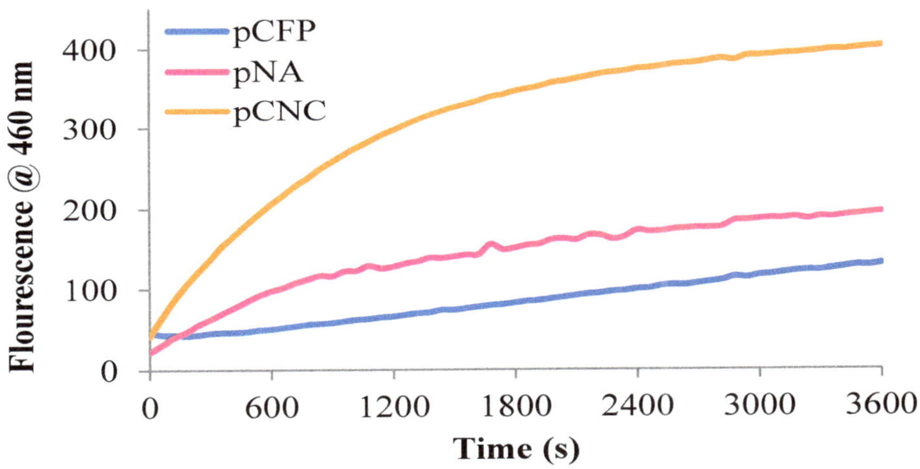

Figure 5. The response curves of 2 mg of pCFP (blue), pNA (pink), and pCNC (green) biosensor upon detection of 7-amino-4-methylcoumarin released with HNE at 0.5 U/mL substrate hydrolysis at 37 °C.

Table 2. Kinetic parameters for human neutrophil elastase (HNE) catalyzed hydrolysis of the tripeptide in solution and on the sensor.

Name	k_{cat} (s^{-1})	K_m (μM)	k_{cat}/K_m ($M^{-1}\cdot s^{-1}$)	V_{max} (s^{-1})	Corr. Coeff. (Correlation Coefficient)
Suc-APA-AMC	2.56	781.4	3272.33	2.17	0.9515
pCFP	0.1201	2.150	5,5860.5	0.1021	0.8645
pNA	1.67	202	8267.33	1.42	0.8935
pCNC	0.7732	23.07	3,3515.39	0.6572	0.9956

3. Discussion

3.1. Structural Features of Peptide-Cellulose Conjugate

Figure 6 shows a model of the crystallite-based structure of the peptide cellulose conjugate representative of the ones reported here. To contrast the properties of the nanocellulosic aerogel-based sensor, we include two other cellulose-based sensors that have similarities and differences. Each of the three forms of cellulose and nanocellulose materials evaluated serves as a transducer surface of the protease sensor. Thus, assessment of the relative transducer surface properties that enable sensor protease detection sensitivity was undertaken in the context of contrasting the nanocellulose aerogel-based sensor with peptide-cellulose conjugate analogs of two other distinctly different cellulose transducer surfaces, i.e., porous cotton-based cellulose filter paper substrate (CFP) and cotton nanocellulose crystals (CNC).

As discussed above, during the generation of the NA, a portion of the cellulose is converted to cellulose II [21]. In agreement with earlier reports of several authors [22], we found that different from all other common cellulose solvent systems, cellulose bodies coagulated from respective solutions in calcium thiocyanate octahydrate suffered from distinctly less shrinkage when subjected to solvent exchange and/or $scCO_2$ drying. This can be explained by a remaining fraction of non-dissolved cellulose I acting as templating and reinforcing scaffold for coagulated cellulose II in the sense of an all-cellulose composite aerogel [23]. It is also worth noting that cellulose II hydrate has been shown to have a larger unit cell than the dehydrated form [24], which is relevant to a further observation here on the influence of nanocellulose aerogel swelling on detection sensitivity and dressing properties.

The crystallite models (Figure 3) which are based on the respective cellulose crystallite sizes reveal the relative number of cellulose layers and chains which are CFP (18 layers and 144 chains), NA (12 layers and 78 chains), and CNC (14 layers and 98 chains). The number of chains (36, 28, and 24) on the 1–10 and 110 surfaces for the CFP, CNC, and NA, respectively indicate half of the primary alcohol groups on those surface chains are in the interior structure of the cellulose or nanocellulose materials [11]. As discussed above, this effects the available primary hydroxyls for covalent attachment of the sensor molecule.

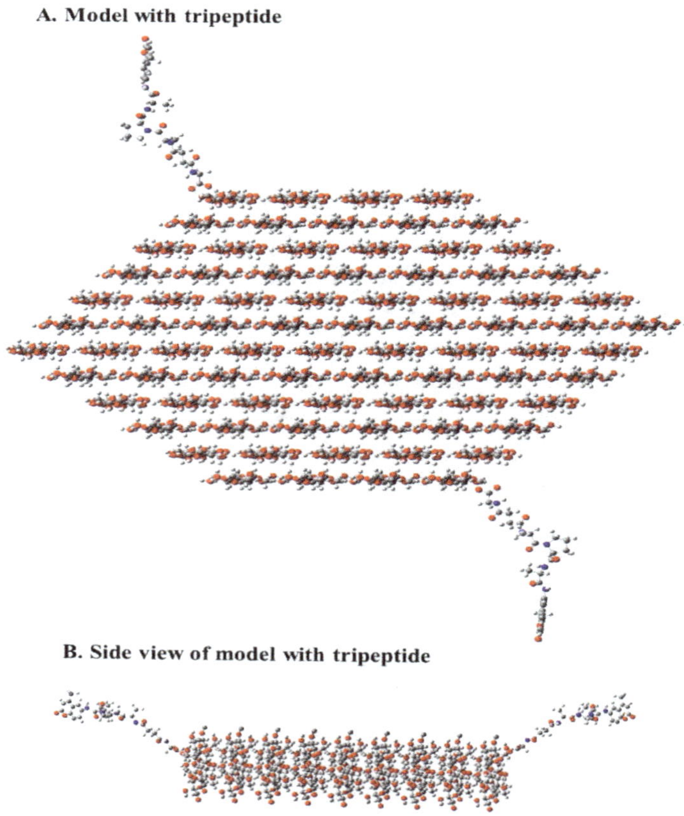

Figure 6. Models of pNA on the X-ray diffraction patterns and Rietveld calculated pattern. The pNA peptide model (**A**) Peptide-cellulose conjugate of aerogel sensor modeled in a single crystal structure with a width of 58.5 Å (78 chains and 12 layers). The side view (**B**) is an image of the same model as in A, but along the longer perimeter of the model.

Previously, we have discussed the role of the elastase substrate's peptide secondary structural conformation relative to the peptide-nanocellulosic conjugate's interaction with the protease from a computational perspective [11] and the potential interaction of the peptide with the crystallite surface. In this regard, the amino acid sequence of this study, Ala-Pro-Ala, has been shown to favor a β-turn conformation [25], which is the smallest element of secondary conformation possible for a protein or peptide. It is also worth noting that β-turns tend to form in more hydrophobic and proteinaceous environments. Thus, the formation of a β-turn in the conjugates of this study may be further enabled by the anionic transducer surface, and upon binding with the protease, which also has a high affinity for the transducer surface, and is likely to populate the transducer surface by way of its positive charge.

3.2. Sensor Bioactivity and Kinetics

The bioactivity of the sensor is dependent on the protease's affinity and interaction with the immobilized substrate and its conjugate surface, which in this study is principally crystalline cellulose. However, we have previously characterized the property of HNE binding to its immobilized substrate from a perspective of adsorption isotherms of elastase binding to the immobilized peptide substrate [26] on crosslinked gels. An assessment of elastase-substrate kinetics and Langmuir adsorption isotherms demonstrated that the protein adsorption at the solid–liquid interface of substrate-bound ethoxyacrylate gels was relatively low i.e., the ratio of maximal adsorbed enzyme to bound ligand was <1%. Thus, in this study the transducer surfaces are negatively charged, and adsorb the HNE marker protein (a positively charged protein) on their crystallite surfaces by way of ionic binding. Accordingly, the negative charge of the transducer surfaces of this study are listed in Table 1, and vary from −10 to −68 mv in order of increasing specific surface area. Moreover, the enhanced HNE binding that these types of transducer surfaces afford is reflected in the Michaelis-Menten kinetics for the immobilized elastase substrate.

An understanding of the relative peptide conformation and crystal structure of the sensor's transducer surface as presented above gives a framework in which understanding enzyme binding and the turnover rates of the cellulose peptide conjugate is possible. Moreover, a perspective of the kinetic turnover and binding affinity between the enzyme and the peptide-cellulose analogs, which possess the enzyme substrate, lends itself to an understanding of the mechanism of action of the sensor with the enzyme (HNE).

The 1.5-fold higher k_{cat} values observed for the peptide substrate in solution compared with the pNA conjugate reflects a generally slower rate of product formation when the enzyme substrates are attached to the nanocellulosic aerogel. This is understandable in light of the two-phase reaction (due to the insolubility of cellulose and nanocellulose in the assay solution) that is occurring between elastase and the peptide bound to the transducer surface, and is also observed for pCFP and pCNC. The lower V_{max} value, which is the maximum reaction rate mediated by the enzyme, observed for the pNA conjugate is consistent with the enzyme turnover rate decreasing in the conjugates.

On the other hand, the enzyme-substrate affinity or ability of the substrate to bind to the enzyme active site, as reflected in the K_m values, was approximately 4-fold greater with the pNA than in solution. Thus, the higher HNE affinity for the tripeptide substrate pNA also gives rise to a higher enzyme efficiency as seen by the higher k_{cat}/K_m assigned to pNA; it is 2.5-fold higher than the k_{cat}/K_m for the analogous enzyme substrate assessed in solution. These increases in binding can be understood in light of the crystallite size, specific surface area, and negative charge, as discussed above.

As shown previously, the macropores of the paper sensor allow passage of the serine protease to the tripeptide substrate attached within the interior structure whereas mesopores as found in the nanocellulose aerogel would not allow passage of the protease or facilitate reaction with the peptide ligand which has a polar surface area of 2 nm^2 (note the peptide substrate may be covalently bonded within the nanocellulose aerogel interior due to its relatively small size) [12]. Thus, access of HNE to the interior of the nanocellulose aerogel may be restricted especially since the pore size of pNA is 30 times less than the polar surface area of elastase (309 nm^2) based on considerations of the material's structure under non-aqueous condition. This may explain the relatively low binding affinity as measured by K_m of the nanocellulose aerogel compared with the paper and nanocrystal. Nonetheless, the turnover rate on the nanocellulose aerogel exceeded that of the paper and nanocrystal. This suggests that increasing the pore size of the aerogel to a volume sufficient for passage of the HNE may result in an increase in the overall enzyme efficiency.

3.3. Spatiotemporal Considerations of the Peptide-Cellulose Conjugate/Elastase Interaction

Spatiotemporal relationships of the peptide-cellulose conjugate to the protease are influenced by charge, protein and peptide-cellulose conjugate structure, and how these components of sensor structure and function come together in time and space. Thus, based on the kinetic properties of the sensors as discussed above it can be inferred that there are strong molecular interactions between elastase and the peptide cellulose conjugates. Specific molecular influences on sensor function are determined by the β turn peptide conformation, the frequency and orientation of peptide-cellulose conjugates on the crystallite surfaces, and the magnitude of the negative charge on the transducer surface.

The negatively charged cellulose surface would favor binding to elastase at a region of the protein where a positively charged cluster occurs and is predominated by arginine residues of HNE. Part of this strongly cationic cluster borders on the S1 active site of the serine protease [27]. Thus, it is feasible that the active site of elastase is brought into close proximity of the cellulose surface through charge transfer interactions between the protein and crystallite surface.

Since the β-turn peptide conformation and its specific resident peptide sequence is required for optimal binding in the oxyanion hole of the active site of the serine protease, stabilization of the conjugate by a negatively charged more hydrophobic surface may increase the enzyme active site binding and in part play a role for the improved K_m values observed for the nanocellulose aerogel as well as paper and nanocrystal forms when compared with the K_m determined in the solution phase.

On the other hand, it is important to recognize that the two different structures of cellulose I and cellulose II also play a role in the sensor activity by virtue of the orientation they confer to the conjugate peptide at the O6 rotamer of cellulose's hydroxymethyl, which is where the peptide is linked to the cellulose. The O6 hydroxymethyl rotamer influences the direction and orientation of the peptide with respect to the cellulose crystallite floor as depicted in Figure 7. For example the conformation of hydroxymethyls in cellulose II has been shown to be predominantly gauche/trans (gt O6) (rotamer oriented from the O5 oxygen through the C4 carbon of anyhydroglucose residues), and conversely the tg O6 rotamer predominates in cellulose I [28]. However, it is worth noting that far less work has been done on cellulose II nanocrystal shapes when they are involved in aerogels compared with cellulose I, and as discussed earlier the coagulation phase of the nanocellulose aerogel preparation may present a unique juxtaposition of minor cellulose polymorphs. Thus, it is likely that the peptide attaches to a gg O6 in both cellulose I and II since the O6 rotamer of the conjugate is not tethered by a hydrogen bond as are the hydroxymethyls in the cellulose crystalline lattice. This is consistent with our examination of a model of a cellulose II crystal where it appears that the top and bottom surfaces have the gt O6 atoms directed to the interior of the crystallite, and may not be favored for the peptide to attach to due to steric hindrance. Accordingly, the gg O6 hydroxymethyl rotamer would orient the peptide outward to the plane of the cellulose crystallite. Thus, we present a minimized model of the peptide cellulose analog in Figure 7 at a hydroxymethyl gg O6 rotamer orientation.

To further examine the relationship of the peptides orientation to the crystal surface the different possible combinations of the peptide-cellulose relation to the crystal surface are shown in Figure 8 and the relative distances of the scissile bond of the sensor peptide (the bond that HNE hydrolyzes upon binding and recognition of the enzyme active site) from the cellulose surface plane are listed in Table 3. The 1–10 and 110 surface planes form two sides of the cellulose crystal in both cellulose I and II. As shown the gg O6 rotamer of the 110 surface plane is considerably closer to the surface than found in the 1–10 surface plane. The orientation of hydroxymethyls in the 110 and 1–10 surface plane is predominantly equatorial [29]. In the case of the gt and tg O6 rotamers the proximity of the peptide to the surface planes is reversed to that observed for the gg O6 rotamer. Nonetheless the gg O6 rotamer on the 110 surface plane of the aerogel may account for the lower binding affinity due to increased steric hindrance to binding of the gg O6 110 surface plane rotamer. However, based on the distances measured for all of the combinations modeled on both surface planes it appears that there is a fairly even variation in the distribution of peptide orientations with respect to the crystal surface, and neither surface the 110 or 1–10 surface would predominately influence enzyme turnover from the standpoint of peptide orientation. On the other hand, the 110 surface plane is more hydrophobic due to exposure of fewer hydroxyls on its surface. Thus, the contribution of this feature to enzyme binding and turnover is not clear other than to point out that elastase binding tends to favor more hydrophobic substrates, and it is likely that the relative polarity and charge of the surfaces and the proximity of the peptide together have a pronounced influence on the peptide-cellulose affinity and turnover of the enzyme.

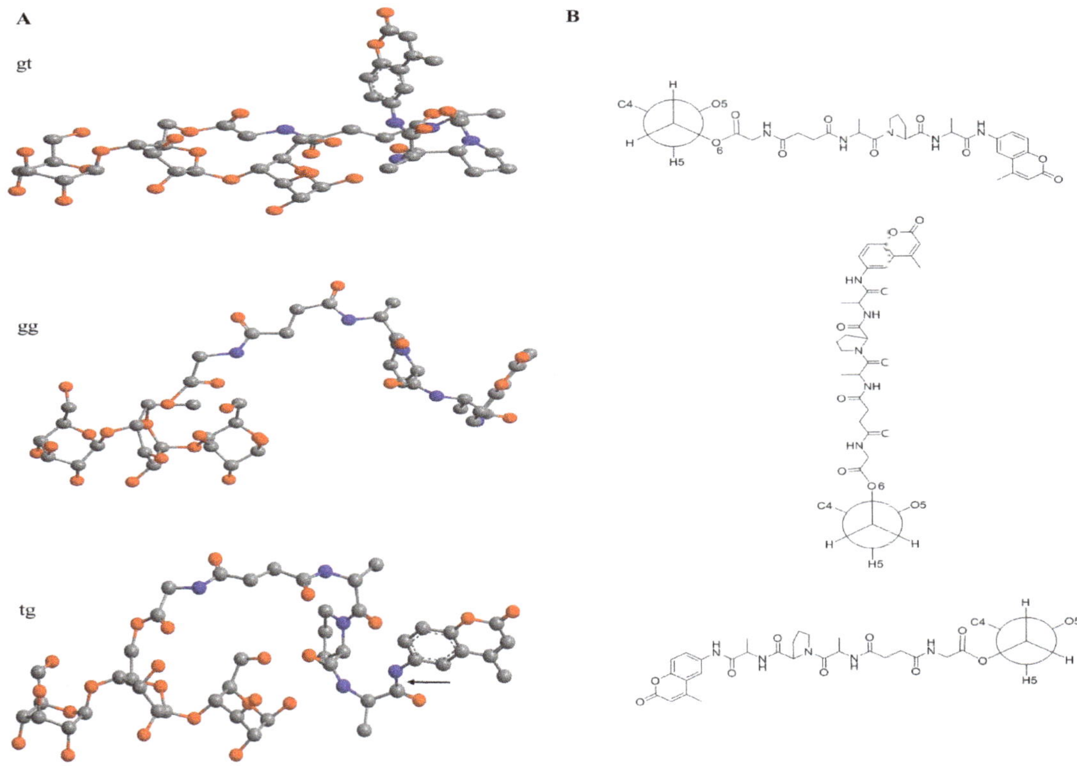

Figure 7. The conformational structures of the tripeptide substrate anchored onto a cellotriose unit at the hydroxymethyl gt O6 rotamer, gg O6 rotamer, and tg O6 rotamer orientation of the cellulose or nanocellulose sensors in: (**A**) minimized model; or (**B**) Newman projections. The black arrow shown for the tg O6 rotamer shows where the HNE protease cleaves the fluorophore. Note: gray is carbon, blue is nitrogen, red is oxygen.

Table 3. Distance of AMC (7-amino-4-methylcoumarin) tripeptide substrate from the five-chain 110 or 1–10 surface of the cellulose floor.

Name	Distance from AMC to Cellulose Floor (A)
110 gg tripeptide	10.7
110 gt tripeptide	22.8
110 tg tripeptide	18.4
1 m 10 gg tripeptide	23.2
1 m 10 gt tripeptide	12.2
1 m10 tg tripeptide	9.9

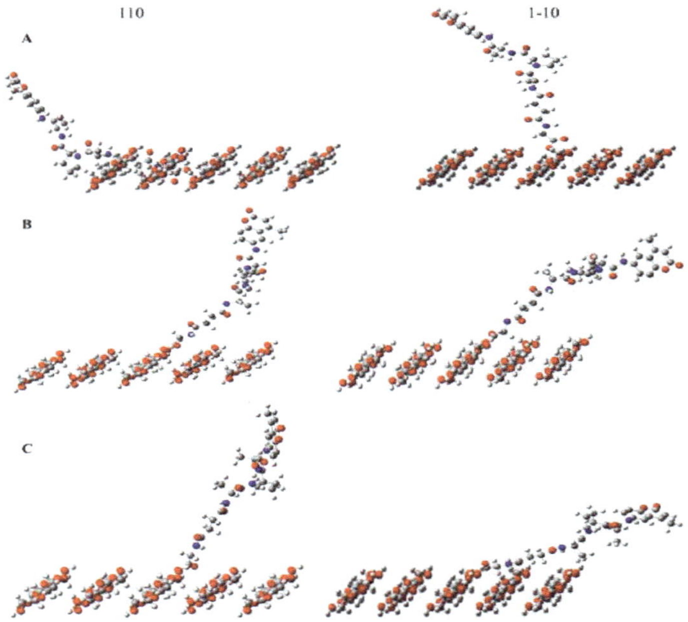

Figure 8. The tripeptide substrate anchored on the five-chain 110 or 1–10 surface of cellulose in the following rotamer positions: (**A**) gauche-gauche; (**B**) gauche-trans; and (**C**) trans-gauche.

4. Materials and Methods

4.1. General

Potassium chloride (KCl), sodium chloride (NaCl), hydrochloric acid (HCl), monosodium phosphate (NaH_2PO_4), trifluoroethanol (TFE), and sodium hydroxide (NaOH) were purchased from VWR (Radnor, PA, USA). Promogran Prisma^{TM} (Systagenix, San Antiono, TX, USA) was purchased from the distributor HighHealthTide. Human neutrophil elastase (HNE) was purchased as a salt-free lyophilized solid from Athens Research and Technology (Athens, GA, USA) without further purification. The peptide n-succinyl-Ala-Pro-Ala-7-amino-4-methylcoumarin (Suc-Ala-Pro-Ala-AMC) was purchased from BACHEM (Torrance, CA, USA) without further purification.

4.2. Synthesis of Nanocellulosic Aerogel and Generation of Biosensors

The nanocellulosic aerogel was prepared and characterized as previously reported by Edwards et al. [12]. The esterification, and peptide immobilization of the cellulosic print cloth (CPC), cotton nanocellulosic crystals (CNC), and nanocellulosic aerogel (NA) were previously reported by Edwards et al. [9,12]. In summary, glycine-esterified cellulose and nanocellulose substrates were prepared by reacting the glycine α–amino functionality with the COOH-terminal group of Suc-Ala-Pro-Ala-AMC by way of a carbodiimide-mediated reaction. Following a simple work-up of rinsing and drying small representative samples of the active peptide component biosensors were cleaved from the matrices by adding a mixture of TFA/water/triisopropylsilane (95/2.5/2.5) to the cellulose-peptide conjugates for three hours, whereupon it was diluted with water (1:10) and submitted for ESI-LC/MS. The intact sequence of the peptide component of the cellulose-peptide conjugate (COOH-Glycine-NH2-Alanine-Proline-Ala-7-amino-4-methylcoumarin) was confirmed for its molecular weight.

4.3. Response and Sensitivity Assay

The fluorescence response and sensitivity assay of the cellulosic and nanocellulosic materials employed was previously outlined by Fontenot et al. [10]. Briefly, duplicates of ~2 mg of each biosensor were placed into a 96 well plate and 100 μL of a phosphate buffer solution (PBS) was added. A standard curve of the tripeptide substrate (1 to 0.0156 μmol/mL) was added to the 96 well plate with a total volume of 100 μL. One well containing only PBS was also included in the standard curve. To start the reaction for the response assay, 50 μL human neutrophil elastase (HNE) was added to the wells containing the biosensors and the standard curve for a total volume of 150 μL. For the sensitivity assay, 50 μL of HNE ranging from 2 to 0.0156 U/mL was added to the standard curve and to the biosensors to provide a total volume of 150 μL, from which the lowest sensitivity detection limit was determined. Fluorescent measurements commenced at 37 °C for 1 hour and measured at 1-min intervals while shaking for 3 s prior to measurement using a Biotech Synergy HT with a tungsten halogen lamp and photomultiplier detection (Winooski, VT, USA). The fluorescence measurements were acquired at 360 nm excitation and 460 nm emission. Fluorescent raw data were processed in Microsoft Excel 2013. The wound like fluid herein comprises of HNE in PBS.

4.4. Surface Charge

The investigation of the surface charge (zeta potential, ζ) values of the cellulose filter paper (CFP) was performed using an Anton Paar Surpass (Ashland, VA, USA). The electrophoretic mobility of the samples was measured using a cylindrical cell. All measurements were performed with a 1 mM KCl solution in deionized water at 21–23 °C. A 0.1M NaOH solution was used to adjust the KCl solution to pH 10.3 for FP. The following parameters were used for titration on the Anton Paar Surpass: aqueous solution of 0.100 M HCl, desired pH increment difference 0.200, volume increments 0.020 mL, pH minimum 2.5, and the pH maximum were slightly higher than the pH of the adjusted KCl solution. The flow rate ranged between 50 and 150 mL/min with a maximum pressure of 300 mbar (for the pressure and rinse program).

4.5. Kinetics Assay

The protocol of the fluorescence response assay was used for the fluorogenic kinetics assay; however, the biosensor concentration varied in duplicate concentrations between 0.5 and 4 mg (dissolved in 100 μL PBS). To start the reaction, 50 μL of 0.5 U/mL of human neutrophil elastase was added to both the free peptide substrate solution used to prepare the standard curve and for the biosensors, providing a total volume of 150 μL.

Fluorescent measurements were conducted as described for the response and sensitivity assay, however, at 20 s intervals for a period of 10 min. The data were imported into GraphPad Prism software (6, GraphPad La Jolla, CA, USA) and analyzed as follows: create XY columns for sample data with enzyme kinetics, non-linear regression for curve fit, enzyme kinetics-substrate vs. velocity, and Michaelis–Menten to determine K_m and V_{max} values.

4.6. Circular Dichroism

Circular dichroism (CD) studies were performed using a Jasco J-815 spectrometer with Spectra Manager 2 Software suite (Jasco, Easton, MD, USA). The CD measurements were carried out using 1 mm path length quartz cell. Peptide solutions were prepared at various concentrations in pH 7.4 PBS (0.5 M NaCl and 0.1 M NaH_2PO_4) containing 10% triflouroethanol (TFE). All spectra correspond to an average of three measurements, which were corrected using the baseline obtained for peptide free solutions. All data were converted to molar ellipticity and smoothed using a binomial function.

4.7. X-ray Diffraction (XRD)

All samples analyzed in the laboratory of Prof. John Wiley at the University of New Orleans Chemistry Department. The CFP, CNC, and NA samples were pressed into pellets and scanned in the θ–2θ reflection mode with a Philips X'pert powder diffractometer using Cu K radiation and a graphite monochromator. The data were obtained in a 2θ scale from 5° and 40° with a step size of 0.05 at 2.5 s per step with an overall run time of 30 min. No background corrections were made. The crystallinity indices (CrI) of the matrices were determined by subtracting the minimum intensity near 18° 2-θ (I_{AM}) from the maximum intensity (at 22.625°–725° 2-θ, I_{200}) and dividing the difference by the maximum intensity (I_{200}) (Equation (1)) [30]. The crystallite size was calculated from the XRD patterns using the Scherrer formula [31] (Equation (2)). The terms of Equation (2) are shape factor (K) of 1, copper K radiation average wavelength (λ) is 1.5418 Å, the full width at half maximum of the (200) peak (β) in radians, and the peak position divided by 2 of the (200) peak (θ) [32]. The crystallite models were drawn using Mercury software 3.5.1 (The Cambridge Crystallographic Data Centre Cambridge, UK), dividing the crystallite sizes from the Scherer equation calculations by the d_{200} interplanar spacing to get the number of chain layers. The tripeptide cellulose analogs were optimized using Chem3DPro 13.0 MM2 minimization energy calculations.

$$CrI = \frac{I200 - IAM}{I200} \times 100 \tag{1}$$

$$t = \frac{K * \lambda}{\beta * COS(\theta)} \tag{2}$$

4.8. Rietveld Refinement Method and Crystallite Models from the NA Diffraction Pattern

The Rietveld method was only used to refine the NA diffraction patterns in order to determine a crystallite size. The standard model diffraction patterns for cellulose I and cellulose II (cif) were based on work by Nishiyama et al. and more recently by French et al. [28,33]. Materials Analysis Using diffraction (Maud) 2.55 software [34] was used to apply the Rietveld method in order to determine the calculated diffraction patterns based on cellulose I, cellulose II, and amorphous patterns in order to resolve the broad overlapping cellulose I and cellulose II peaks and determine the crystallite size. Several parameter were modified in the Maud software during the refining process for the NA: the scale factor, quadratic background, and unit cell length a (layer I: the phase atom); cellulose I β: cell length a, scale factor, crystal size, and cell angle gamma (March-Dollase model: background and crystal size); and cellulose II: cell length a, scale factor, crystal size, and cell angle gamma (March-Dollase model: utilizing background and crystal size).

4.9. Molecular Modeling Studies

Computational studies minimized the energy of the tripeptide substrate anchored on the five-chain 110 or 1–10 surfaces of cellulose in the following rotamer positions: (A) gauche-gauche; (B) gauche-trans; and (C) trans-gauche [11]. The peptide substrate models were built using GaussView 5.0.9 and optimized using a semiempirical Hamiltonian method (PM3) [35] contained in Gaussian 09 revision A.02 molecular orbital software (Gaussian inc., Wallingford, CT, USA).

5. Conclusions

We evaluated cotton-based cellulosic filter paper, nanocellulosic crystals, and nanocellulosic aerogel transducers conjugated to a tripeptide substrate using: (i) X-ray diffraction to determine their diffraction pattern, cellulose orientation, crystallite size, and crystallite model; (ii) Circular dichroism for assessment of the tripeptide substrate conformation, which revealed a β-turn confirmation; (iii) the bioactivity properties of the sensors via fluorescence response and kinetic properties that indicated the sensors provide a faster response to human neutrophil elastase compared to the free tripeptide substrate; and (iv) computational assessment of the cellulose tripeptide substrate analog in different orientations. Therefore, these materials, in particular, the nanocellulosic aerogel, have promising sensitivity, selectivity, and a faster kinetics response for detecting human neutrophil elastase, and are a promising sensor interface for a multilayered chronic wound dressing motif. Furthermore, the nanocellulosic sensors are adaptable for the detection of other biomarkers of clinical interest as well, via substitution of the biomolecule.

Acknowledgments: This project was supported by the U.S. Department of Agriculture. The authors wish to thank Lawson Gary, Chief Operating Officer, T. J. Beall, Co. Inc., for the donation of True Cotton™ samples. Furthermore, the support of project CAPBONE by the Austrian Science Fund is thankfully acknowledged.

Author Contributions: J. Vincent Edwards conceived of the project, guided and mentored the experimental design and is the principle author of the manuscript. Krystal Fontenot designed the experiments, synthesized all peptide modified aerogel derivatives, performed all of the experiments, analyzed the data, conducted literature searches, designed the graphics and helped with a significant part of the writing of the manuscript. Nicole Doyle nee Pircher and Falk Liebner prepared the unmodified aerogels and contributed to editing the manuscript. Alfred D. French analyzed and interpreted X-ray crystallography data, and assisted with the molecular modeling. Brian D. Condon provided funding and management that contributed reagents/materials, analysis tools and personnel resources to the project.

Abbreviations

AMC	7-amino-4-methylcoumarin
HNE	human neutrophil elastase
CFP	(cellulosic filter paper)
NA	nanocellulosic aerogel
CNC	cotton nanocrystalline cellulose
pCFP	peptide-cellulose conjugate on filter paper
pNA	peptide-cellulose conjugate on NA
pCNC	peptide-cellulose conjugate on NC
k_{cat}	enzyme turnover rate
K_m	enzyme binding affinity constant
g	gauche conformation
t	*trans* conformation
MAUD	materials analysis using diffraction

References

1. Dargaville, T.R.; Farrugia, B.L.; Broadbent, J.A.; Pace, S.; Upton, Z.; Voelcker, N.H. Sensors and imaging for wound healing: A review. *Biosens. Bioelectron.* **2013**, *41*, 30–42. [CrossRef] [PubMed]
2. International Consensus. *The Role of Proteases in Wound Diagnostics*; An Expert Working Group Review; Wounds International: London, UK, 2011; pp. 1–13.
3. Yager, D.R.; Chen, S.M.; Ward, S.I.; Olutoye, O.O.; Diegelmann, R.F.; Kelman Cohen, I. Ability of chronic wound fluids to degrade peptide growth factors is associated with increased levels of elastase activity and diminished levels of proteinase inhibitors. *Wound Repair Regen.* **1997**, *5*, 23–32. [CrossRef] [PubMed]
4. Serena, T.E. Development of a novel technique to collect proteases from chronic wounds. *Adv. Wound Care* **2014**, *3*, 729–732. [CrossRef] [PubMed]
5. Serena, T.E.; Cullen, B.M.; Bayliff, S.W.; Gibson, M.C.; Carter, M.J.; Chen, L.; Yaakov, R.A.; Samies, J.; Sabo, M.; DeMarco, D.; et al. Defining a new diagnostic assessment parameter for wound care: Elevated protease activity, an indicator of nonhealing, for targeted protease-modulating treatment. *Wound Repair Regen.* **2016**, *24*, 589–595. [CrossRef] [PubMed]
6. Vandenbroucke, R.E.; Libert, C. Is there new hope for therapeutic matrix metalloproteinase inhibition? *Nat. Rev. Drug Discov.* **2014**, *13*, 904–927. [CrossRef] [PubMed]
7. Synder, R.J.; Driver, V.; Fife, C.E.; Lantis, J.; Peirce, B.; Serena, T.; Weir, D. Using a diagnostic tool to identify elevated protease activity levels in chronic and stalled wounds: A consensus panel discussion. *Ostomy Wound Manag.* **2011**, *57*, 36–46.
8. Moore, K.; Huddleston, E.; Stacey, M.C.; Harding, K.G. Venous leg ulcers—The search for a prognostic indicator. *Int. Wound J.* **2007**, *4*, 163–172. [CrossRef] [PubMed]
9. Edwards, J.V.; Fontenot, K.R.; Haldane, D.; Prevost, N.T.; Condon, B.D. Human neutrophil elastase peptide sensors conjugated to cellulosic and nanocellulosic materials: Part I, synthesis and characterization of fluorescent analogs. *Cellulose* **2016**, *23*, 1283–1295. [CrossRef]
10. Fontenot, K.R.; Edwards, J.V.; Haldane, D.; Graves, E.; Citron, M.S.; Prevost, N.T.; French, A.D.; Condon, B.D. Human neutrophil elastase detection with fluorescent peptide sensors conjugated to cellulosic and nanocellulosic materials: Part II, structure/function analysis. *Cellulose* **2016**, *23*, 1297–1309. [CrossRef]

11. Edwards, J.V.; Prevost, N.T.; French, A.D.; Concha, M.; Condon, B.D. Kinetic and structural analysis of fluorescent peptides on cotton cellulose nanocrystals as elastase sensors. *Carbohydr. Polym.* **2015**, *116*, 278–285. [CrossRef] [PubMed]
12. Edwards, J.V.; Fontenot, K.R.; Prevost, N.T.; Pircher, N.; Liebner, F.; Condon, B.D. Preparation, characterization and activity of a peptide-cellulosic aerogel protease sensor from cotton. *Sensors* **2016**, *16*, 1789. [CrossRef] [PubMed]
13. Edwards, J.V.; Bopp, A.; Graves, E.; Condon, B. *In Vitro Hemostatic, Hydrogen Peroxide Production and Elastase Sequestration Properties of Nonwoven Ultra Clean Greige Cotton Dressing*; Wound Healing Society: Orlando, FL, USA; Wound Repair and Regeneration: Orlando, FL, USA, 2013; p. A21.
14. Edwards, J.V.; Fontenot, K.R.; Prevost, N.T.; Haldane, D.; Nicole, P.; Liebner, F.; French, A.D.; Condon, B.D. Protease biosensors based on peptide-nanocellulose conjugates: From molecular design to dressing interface. *Int. J. Med. Nano Res.* **2016**. [CrossRef]
15. Edwards, J.V.; Prevost, N.; French, A.D.; Concha, M.; DeLucca, A.; Wu, Q. Nanocellulose-based biosensors: Design, preparation, and activity of peptide-linked cotton cellulose nanocrystals having fluorimetric and colorimetric elastase detection sensitivity. *Engineering* **2013**, *5*, 20–28. [CrossRef]
16. Reiner, R.S.; Rudie, A.W. Process scale-up of cellulose nanocrystal production to 25 kg per batch at the forest products laboratory. In *Production and Applications of Cellulose Nanomaterials*; Postek, M.T., Moon, R.J., Rudie, A.W., Bilodeau, M.A., Eds.; TAPPI Press Inc.: Peachtree Corners, GA, USA, 2013; pp. 21–24.
17. Nam, S.; French, A.D.; Condon, B.D.; Concha, M. Segal crystallinity index revisited by the simulation of X-ray diffraction patterns of cotton cellulose Iβ and cellulose II. *Carbohydr. Polym.* **2016**, *135*, 1–9. [CrossRef] [PubMed]
18. Pircher, N.; Carbajal, L.; Schimper, C.; Bacher, M.; Rennhofer, H.; Nedelec, J.-M.; Lichtenegger, H.C.; Rosenau, T.; Liebner, F. Impact of selected solvent systems on the pore and solid structure of cellulose aerogels. *Cellulose* **2016**, *23*, 1949–1966. [CrossRef] [PubMed]
19. Manning, M.C.; Illangasekare, M.; Woody, R.W. Circular dichroism studies of distorted alpha-helices, twisted β-sheets, and β turns. *Biophys. Chem.* **1988**, *31*, 77–86. [CrossRef]
20. Gierasch, L.M.; Deber, C.M.; Madison, V.; Niu, C.-H.; Blout, E.R. Conformations of (X-L-Pro-Y)2 cyclic hexapeptides. Preferred β-turn conformers and implications for β Turns in proteins. *Biochemistry* **1981**, *20*, 4730–4738. [CrossRef] [PubMed]
21. Liebner, F.; Pircher, N.; Schimper, C.; Haimer, E.; Rosenau, T. Aerogels: Cellulose-based. In *Encyclopedia of Biomedical Polymers and Polymeric Biomaterials*; CRC Press: Boca Raton, FL, USA, 2015; pp. 37–75.
22. Hoepfner, S.; Ratke, L.; Milow, B. Synthesis and characterisation of nanofibrillar cellulose aerogels. *Cellulose* **2008**, *15*, 121–129. [CrossRef]
23. Gindl, W.; Keckes, J. All-cellulose nanocomposite. *Polymer* **2005**, *46*, 10221–10225. [CrossRef]
24. Kobayashi, K.; Kimura, S.; Togawa, E.; Wada, M. Crystal transition from cellulose II hydrate to cellulose II. *Carbohydr. Polym.* **2011**, *86*, 975–981. [CrossRef]
25. Liang, G.B.; Rito, C.J.; Gellman, S.H. Thermodynamic analysis of β-turn formation in Pro-Ala, Pro-Gly, and Pro-Val model peptides in methylene chloride. *J. Am. Chem. Soc.* **1992**, *114*, 4440–4442. [CrossRef]
26. Edwards, J.V.; Caston-Pierre, S.; Bopp, A.F.; Goynes, W. Detection of human neutrophil elastase with peptide-bound cross-linked ethoxylate acrylate resin analogs. *J. Pept. Res.* **2005**, *66*, 160–168. [CrossRef] [PubMed]
27. Navia, M.A.; McKeever, B.M.; Springer, J.P.; Lin, T.Y.; Williams, H.R.; Fluder, E.M.; Dorn, C.P.; Hoogsteen, K. Structure of human neutrophil elastase in complex with a peptide chloromethyl ketone inhibitor at 1.84-a resolution. *Proc. Natl. Acad. Sci. USA* **1989**, *86*, 7–11. [CrossRef] [PubMed]
28. Nishiyama, Y.; Langan, P.; Chanzy, H. Crystal structure and hydrogen-bonding system in cellulose Iβ from synchrotron X-ray and neutron fiber diffraction. *J. Am. Chem. Soc.* **2002**, *124*, 9074–9082. [CrossRef] [PubMed]
29. Yamane, C.; Aoyagi, T.; Ago, M.; Sato, K.; Okajima, K.; Takahashi, T. Two Different Surface Properties of Regenerated Cellulose due to Structural Anisotropy. *Polym. J.* **2006**, *38*, 819–826. [CrossRef]
30. Segal, L.; Creely, J.J.; Martin, A.E.; Conrad, C.M. An empirical method for estimating the degree of crystallinity of native cellulose using the X-ray diffractometer. *Text. Res. J.* **1959**, *29*, 786–794. [CrossRef]
31. Scherrer, P. Bestimmung der Inneren Struktur und der Größe von Kolloidteilchen Mittels Röntgenstrahlen. In *Kolloidchemie ein Lehrbuch*; Springer: Berlin/Heidelberg, Germany, 1912; pp. 387–409.
32. French, A.; Santiago Cintrón, M. Cellulose polymorphy, crystallite size, and the segal crystallinity index. *Cellulose* **2013**, *20*, 583–588. [CrossRef]

33. French, A.D. Idealized powder diffraction patterns for cellulose polymorphs. *Cellulose* **2014**, *21*, 885–896. [CrossRef]
34. Lutterotti, L. Total pattern fitting for the combined size-strain-stress-texture determination in thin film diffraction. *Nucl. Instrum. Methods Phys. Res. Sect. B Beam Interact. Mater. Atoms* **2010**, *268*, 334–340. [CrossRef]
35. Stewart, J.J.P. Mopac: A semiempirical molecular orbital program. *J. Comput. Aided Mol. Des.* **1990**, *4*, 1–103. [CrossRef] [PubMed]

4

Polymeric Micelles based on Modified Glycol Chitosan for Paclitaxel Delivery: Preparation, Characterization and Evaluation

Na Liang [1], Shaoping Sun [2,*], Xianfeng Gong [2], Qiang Li [2], Pengfei Yan [2,3,*] and Fude Cui [4]

1. Key Laboratory of Photochemical Biomaterials and Energy Storage Materials, Heilongjiang Province, College of Chemistry & Chemical Engineering, Harbin Normal University, Harbin 150025, China; liangna528@163.com
2. Key Laboratory of Chemical Engineering Process & Technology for High-Efficiency Conversion, College of Heilongjiang Province; School of Chemistry and Material Science, Heilongjiang University, Harbin 150080, China; gongxianfeng@sina.com (X.G.); liqianghlj100@163.com (Q.L.)
3. Key Laboratory of Functional Inorganic Material Chemistry, Heilongjiang University, Harbin 150080, China
4. School of Pharmacy, Shenyang Pharmaceutical University, Shenyang 110016, China; syphucuifude@163.com

* Correspondence: sunshaoping111@163.com (S.S.); yanpf@vip.sina.com (P.Y.)

Abstract: Amphiphilic polymer of α-tocopherol succinate modified glycol chitosan (TS-GC) was successfully constructed by conjugating α-tocopherol succinate to the skeleton of glycol chitosan and characterized by Fourier-transform infrared (FT-IR) and proton nuclear magnetic resonance (^{1}H-NMR). In aqueous milieu, the conjugates self-assembled to micelles with the critical aggregation concentration of 7.2×10^{-3} mg/mL. Transmission electron microscope (TEM) observation and dynamic light scattering (DLS) measurements were carried out to determine the physicochemical properties of the micelles. The results revealed that paclitaxel (PTX)-loaded TS-GC micelles were spherical in shape. Moreover, the PTX-loaded micelles showed increased particle sizes (35 nm vs. 142 nm) and a little reduced zeta potential (+19 mV vs. +16 mV) compared with blank micelles. The X-ray diffraction (XRD) spectra demonstrated that PTX existed inside the micelles in amorphous or molecular state. In vitro and in vivo tests showed that the PTX-loaded TS-GC micelles had advantages over the Cremophor EL-based formulation in terms of low toxicity level and increased dose, which suggested the potential of the polymer as carriers for PTX to improve their delivery properties.

Keywords: glycol chitosan; α-tocopherol succinate; amphiphilic polymer; micelles; paclitaxel

---

1. Introduction

Paclitaxel (PTX), as a powerful anti-tumor drug, has been extensively used in the clinical treatment of several solid tumors, such as refractory ovarian cancer, metastasis breast cancer, non-small cell lung cancer, Acquired Immune Deficiency Syndrome-related Kaposi's sarcoma and other cancers [1,2]. Due to its poor water solubility of approximately <2 μg/mL, PTX is currently solubilized in a 50:50 mixture of Cremophor EL (PEG-35 caster oil) and dehydrated ethanol as Taxol® (Bristol-Myers Squibb, New York City, NY, USA). However, several studies reported that Cremophor EL induced serious side effects such as hypersensitivity, nephrotoxicity, neurotoxicity, and the extraction of plasticizer from the infusion tubes [3]. In light of these drawbacks, a number of alternative preparations were investigated, including liposomes, nanocrystals, micelles, cyclodextrin complexes and PTX conjugates [4–8].

Among these formulations, polymeric micelles have been proven as promising drug delivery systems for PTX administration [9,10], because of their attractive characteristics, such as biocompatibility, high drug-loading content, small size (<200 nm) and propensity to evade scavenging by the mononuclear phagocyte system (MPS) [11,12]. Moreover, the nanoscale dimensions of polymeric micelles permit their selectively accumulation in tumor tissues due to the enhanced permeability and retention (EPR) effect, which is termed "passive targeting" [13,14]. The formation of polymeric micelles is generally considered as the self-assembly of polymeric amphiphiles. In aqueous medium, polymeric amphiphiles form the micelles consisting of the inner hydrophobic core and the outer hydrophilic shell. The hydrophobic core provides a storeroom for loading hydrophobic drugs, and the hydrophilic shell allows retaining the stability of micelles in an aqueous environment [15].

Due to its favorable properties, such as biodegradability, biocompatibility, nontoxicity and bioadhesivity, chitosan has been widely studied as a pharmaceutical carrier for drug delivery [16,17]. However, chitosan is insoluble at pH values above its pKa (6.4) in water, and this obviously limits its biomedical applications.

In recent years, glycol chitosan (GC), which possesses good solubility over a broad range of pH, has been studied to construct drug delivery systems for PTX, such as nanocrystals, nanoparticles, hydrogels and microspheres [18–21]. The hydrophobically modified GC, which could be used as micellar carriers, has been extensively studied [22,23].

α-Tocopherol is a good solvent for many hydrophobic drugs because of its excellent lipophilic nature [24]. Once it is grafted on to the backbone of GC, it may serve as the hydrophobic segment and therefore provide sufficient capacity for poorly soluble drugs.

Inspired by the above investigations, in this study, an amphiphilic polymer α-tocopherol succinate modified glycol chitosan (TS-GC) was designed for PTX delivery. The TS-GC was prepared through amide formation. The preparation, characterization and self-assembling ability of TS-GC were studied. Furthermore, the physicochemical properties, hemolysis, in vitro cytotoxicity and in vivo antitumor activity of PTX-loaded micelles were evaluated deeply.

2. Results and Discussion

2.1. Synthesis and Characterization of TS-GC

In this study, the polymer TS-GC was synthesized via the coupling reaction between carboxyl group of TS and amine group of GC in the presence of water-soluble 1-Ethyl-3-(3-dimethylaminopropyl) carbodiimide hydrochloride (EDC). Firstly, an active ester intermediate was formed between the carboxyl group of TS and EDC. Then the amino bond can easily formed by reaction of primary amino of GC with the intermediate. NHS was used to stabilize the intermediate to achieve higher yield [25]. The scheme of the reaction between GC and TS was shown in Figure 1.

^{1}H-NMR spectra confirmed the grafting of TS chain onto GC as illustrated in Figure 2. In the spectrum of TS-GC, the proton peaks of TS, including methyl (0.77–0.88 ppm) and methylene (0.99–1.54 ppm) that belonged to the protons of the long-chain alkyl group of TS were observed. Moreover, the new-emerged signals at 2.45–2.63 ppm were attributed to the methene hydrogen ($-COCH_2CH_2-$) of the succinyl group of TS [26]. These spectra proved the formation of TS-GC.

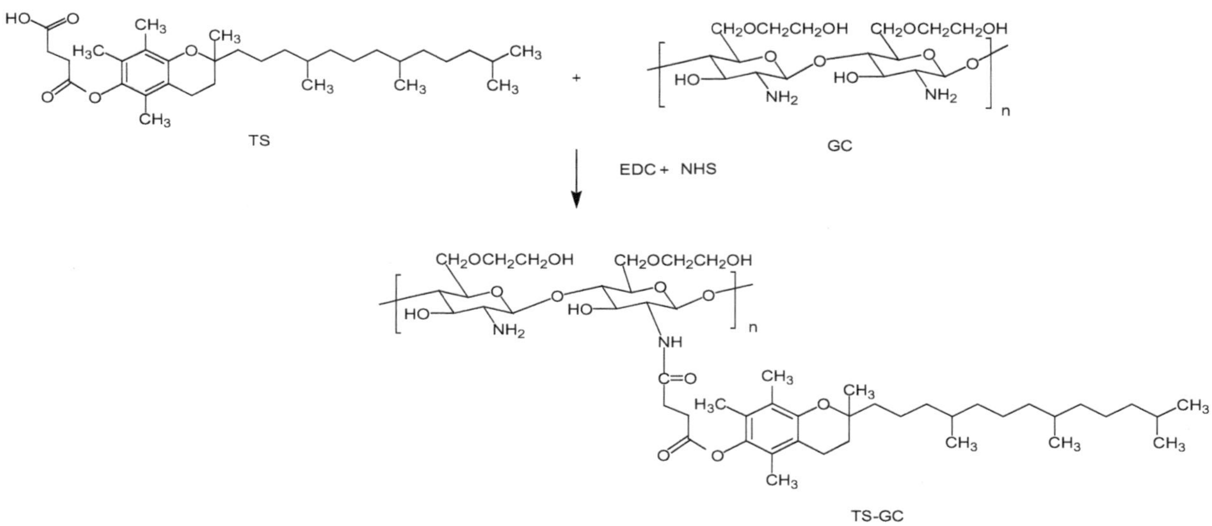

Figure 1. Synthesis of α-tocopherol succinate modified glycol chitosan (TS-GC).

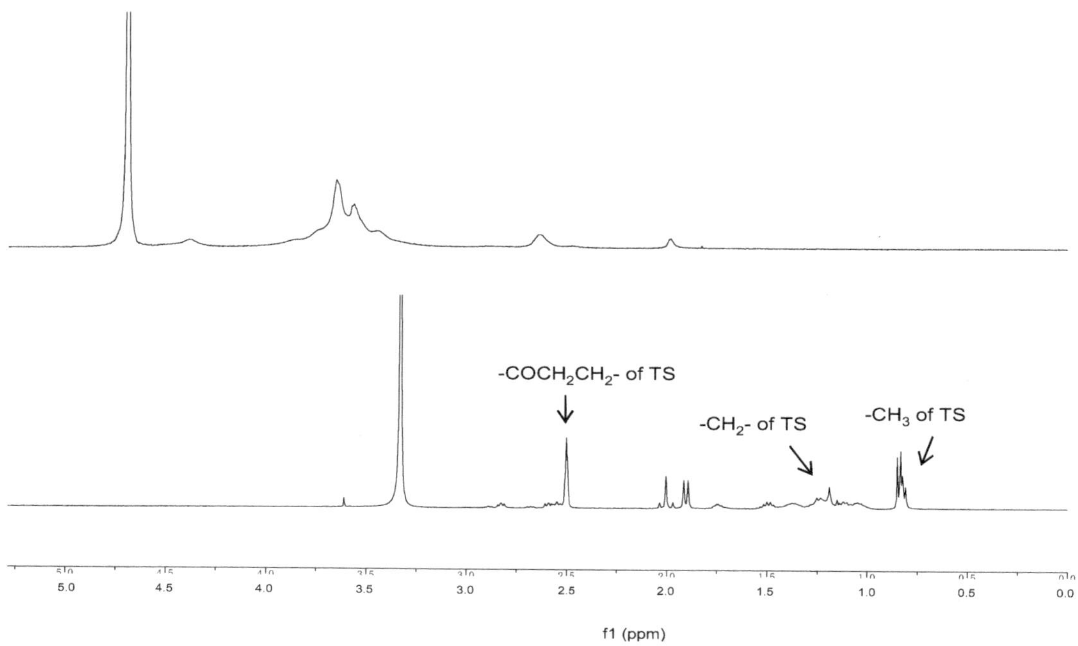

Figure 2. Proton nuclear magnetic resonance (^{1}H-NMR) spectra of (**a**) glycol chitosan (GC) and (**b**) TS-GC.

FT-IR was used to characterize the functional groups of GC before and after modification. The FT-IR spectra of GC (a), TS-GC (b), physical mixture of GC and TS (c) and TS (d) are shown in Figure 3. In curve a, GC showed characteristic signals at 3133 cm^{-1} (O–H stretch overlapped with N–H stretch), 1669 cm^{-1} (amide I band, C=O stretch of acetyl group), 1558 cm^{-1} (amide II band, N–H bending) and 1399 cm^{-1} (C–H bending), respectively. The spectrum of TS (curve d) showed characteristic peaks of C=O bond at 1754 cm^{-1} (the carbonyl of ester bond) and at 1715 cm^{-1} (carboxylic C=O). And the peak at 934 cm^{-1} was attributed to the out-of-plane bending vibrations of carboxylic C–OH. Compared with GC, the increased intensity of peaks at 1669 cm^{-1} and 1558 cm^{-1} in TS-GC (curve b) indicated the formation of amide bond. In addition, disappearance of carboxylic C=O signal at 1715 cm^{-1} and carboxylic C–OH signal at 934 cm^{-1} further confirmed full reaction of TS with GC. All the above indicated the successful introduction of TS.

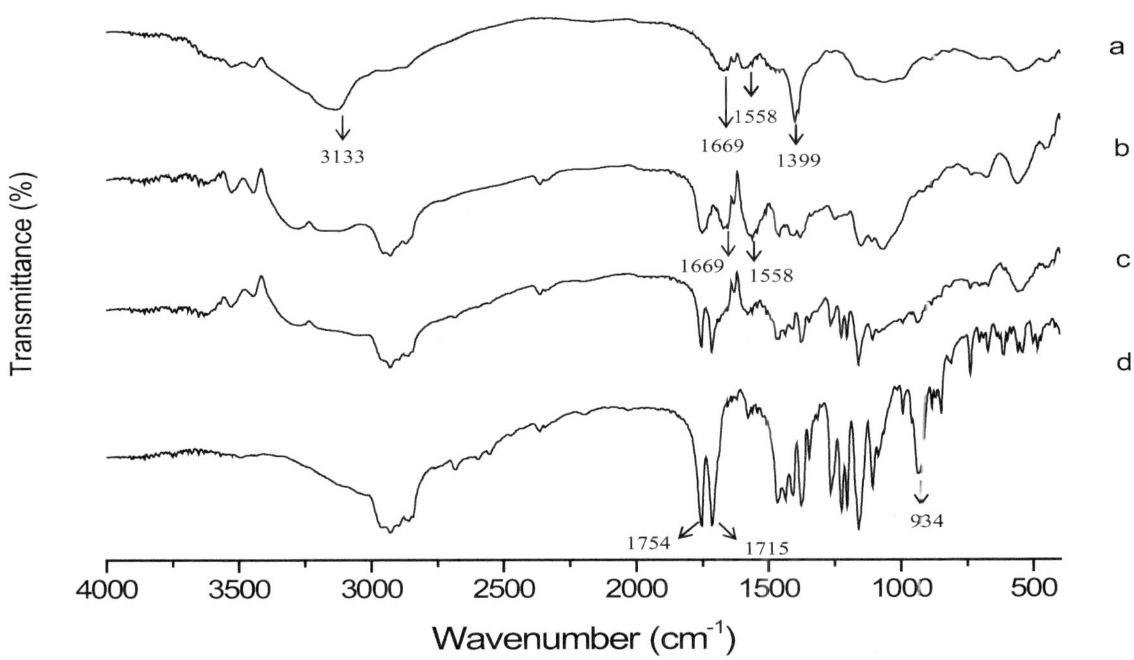

Figure 3. Fourier-transform infrared (FT-IR) spectra of (**a**) GC, (**b**) TS-GC, (**c**) physical mixture of GC and TS, and (**d**) TS.

The degree of amino substitution was calculated by measuring the amount of terminal amino groups of TS-GC with 2,4,6-trinitrobenzene sulphonic acid (TNBS reagent). The maximum absorbance of the yellow colour product was at 344 nm, and the absorbance was in proportion to the number of primary amino groups. From the calibration curve obtained with unmodified GC solution with different concentrations, the substitution degree of TS-GC in this experiment was calculated as 11.3%.

Critical aggregation concentration (CAC) plays an important role in maintaining the stability of micelles upon dilution. Only when the concentration of the polymer is higher than its CAC can micelles be formed. Polymeric micelles are generally more stable because of their markedly lower CAC. For the determination of the CAC of TS-GC, the pyrene fluorescence was employed to monitor the properties of TS-GC in solution. This method utilized pyrene's sensitivity to the local polarity of the environment. Below the CAC, the polymers only exist as single chains, and pyrene is solubilized in water. When the concentration increases to reach the critical value called CAC, polymer chains start to associate to form micelles, and pyrene partitions preferentially toward the hydrophobic cores of micelles. This leads to the increase of fluorescence intensity, and the intensity of the third peak increased significantly compared to that of the first peak. So the CAC values were calculated from curves of I_3/I_1 versus polymer concentration. The CAC was defined as the intercept of the tangents to the curve before and after the point of inflection [27]. For TS-GC, the CAC was calculated to be 7.2×10^{-3} mg/mL, which was significantly lower than that of the low molecular weight surfactants in water. This implied the TS-GC micelles could be stable and not easily dissociate upon dilution.

2.2. Preparation of TS-GC Micelles

In general, the aggregates of amphiphilic polymers can be prepared through diafiltration or sonication methods. As for diafiltration, the polymer was dissolved in co-solvent and then dialyzed against water. For TS-GC, it was not easily dissolved in the mixture of organic solvent with water, so the probe-sonication was employed. Micelle formation is a delicate balance between the attractive force that leads to the association of molecules and the repulsive force that prevents unlimited growth of the micelles [28]. It was reported that once the micelle structure was formed completely, the drugs were hardly to be incorporated into the micelles [29]. So, in this study, in order to get high encapsulation efficiency, the PTX loading occurred simultaneously with self-assembly of TS-GC under the probe-sonication treatment as stated in Section 3.5. The PTX encapsulation efficiency reached to 71.8%, and the drug loading capacity was calculated as 8.0%.

2.3. Characterization of PTX-Loaded TS-GC Micelles

2.3.1. XRD Analysis

To confirm the existence form of PTX in PTX-loaded TS-GC micelles, XRD analysis was conducted for PTX, blank micelles, their physical mixture and PTX-loaded micelles. As illustrated in Figure 4, typical intense diffraction peaks of PTX were still observed with weak intensity in the pattern obtained from the physical mixture of PTX and blank micelles, which reflected the presence of PTX crystal in the mixture. While the spectrum of lyophilized PTX-loaded micelles was similar to that of the blank micelles, and there were no diffraction peaks for PTX. It can be concluded that PTX was encapsulated in the polymeric micelles in molecular or amorphous state and there was no free drug on the surface of micelles.

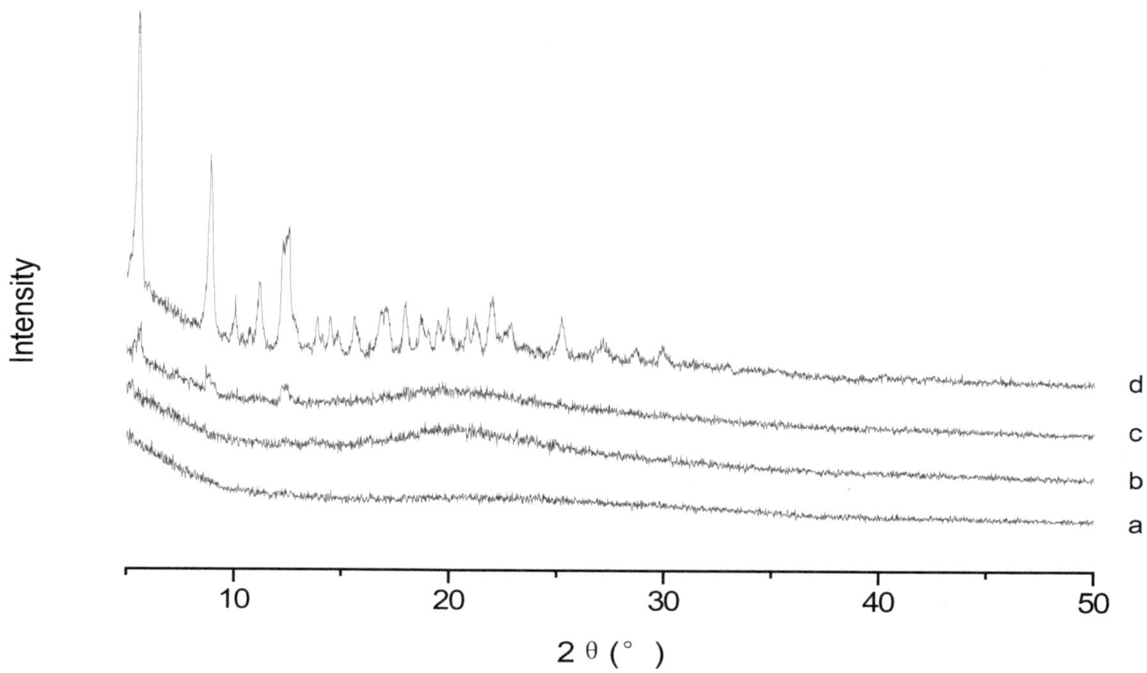

Figure 4. X-ray diffraction (XRD) spectra of (**a**) blank micelles; (**b**) paclitaxel (PTX)-loaded micelles; (**c**) physical mixture of PTX and blank micelles; and (**d**) PTX.

2.3.2. Particle Size and Zeta Potential

Size of the micelles is an important factor affecting the in vivo fate of the drug. In this study, the particle size and their distribution of the micelles were measured by dynamic light scattering method. From the result, the mean particle size of PTX-loaded micelles was 142 nm with polydispersity index (PDI) of 0.186, and it was larger than that of blank micelles (35 nm, PDI of 0.105), which indicated the encapsulation of PTX into the micelles. For both bare and PTX-loaded micelles, the size distribution was narrow. It was reasonably safe to assume an increasing accumulation of the drug in tumor tissue. Because it was reported that the small size of micelles (<200 nm) can reduce non-selective clearance by the reticuloendothelial system (RES) and show EPR effect for passive accumulation in certain tumor sites [30]. Moreover, the size-sieving may occur during the distribution process in the body, and the narrow size distribution may promote the selective accumulation at the target site. Transmission electron microscope (TEM) micrograph of PTX-loaded micelles is shown in Figure 5. It was obvious that the micelles were spherical in shape.

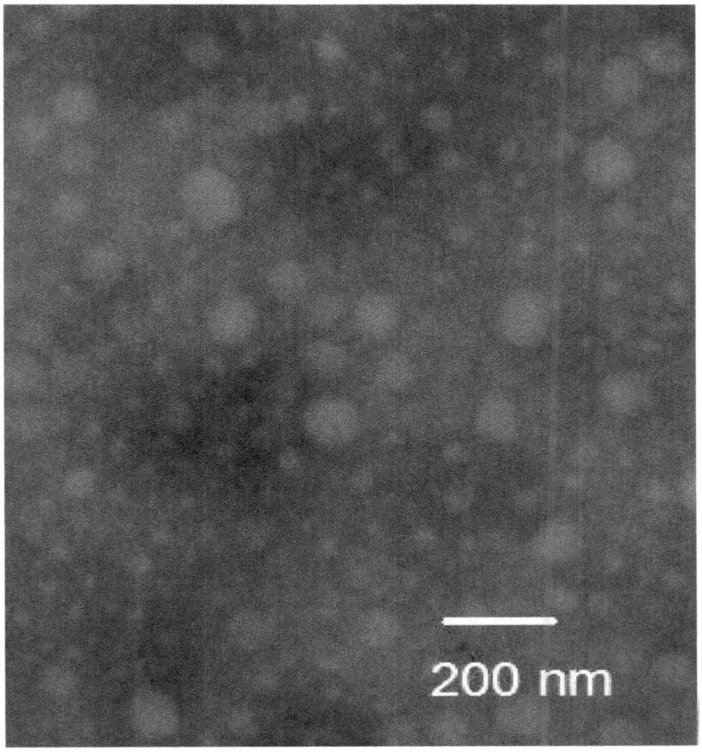

Figure 5. Transmission electron microscope (TEM) image of PTX-loaded micelles.

For a micellar solution, zeta potential can greatly influence the particle stability through the electrostatic repulsion. Relatively high surface charge could provide a repelling force between the particles, and thus increase the stability of the system. In this study, the blank micelles and PTX-loaded micelles in aqueous medium were positively charged with zeta potential of +19 mV and +16 mV, respectively, due to the presence of ionized amino groups of TS-GC distributing on the surface of micelles. It is reasonable to confirm the aqueous stability of the micelles.

2.4. *In Vitro Hemolytic Test*

As analogs of low molecular weight surfactants, amphiphilic polymers may solubilize lipids or be inserted into phospholipid membranes, and consequently lead to hemolysis of red blood cells following intravenous administration [31]. It is necessary to determine whether TS-GC induces hemolysis and is safe for intravenous injection. The hemolysis of PTX-loaded TS-GC micelles was compared with that of Cremophor EL-based formulation. It was observed that in the range of 10–200 μg/mL, the hemolysis of PTX-loaded TS-GC micelles was almost negligible, with only 4.2% at the concentration of 200 μg/mL, while the hemolysis induced by the Cremophor EL-based formulation increased from 0.04% to 10.9%. The results suggested that PTX-loaded micelles were not toxic to the erythrocytes.

2.5. *In Vitro Cytotoxicity Study*

The cytotoxicity study was conducted by 3-(4,5-dimethylthiazol-2-yl)-2,5-diphenyl tetrazolium bromide (MTT) method to assess the effectiveness of PTX-loaded TS-GC micelles. As shown in Figure 6, the cytotoxicity of PTX-loaded TS-GC micelles was similar to that of Cremophor EL-based formulation with equivalent doses in the concentration ranges used in this study. Furthermore, as the concentration and incubation time increased, PTX-loaded micelles and Cremophor EL-based formulation displayed increasing cytotoxicity. It was straightforward to understand that both concentration and incubation time played a major role in the in vitro cytotoxicity of PTX. For longer incubation periods, a larger number of cells enter the G2 and M cell cycle phases, during which PTX is more active. The findings suggested that PTX-loaded TS-GC micelles could be used as a potential PTX carrier.

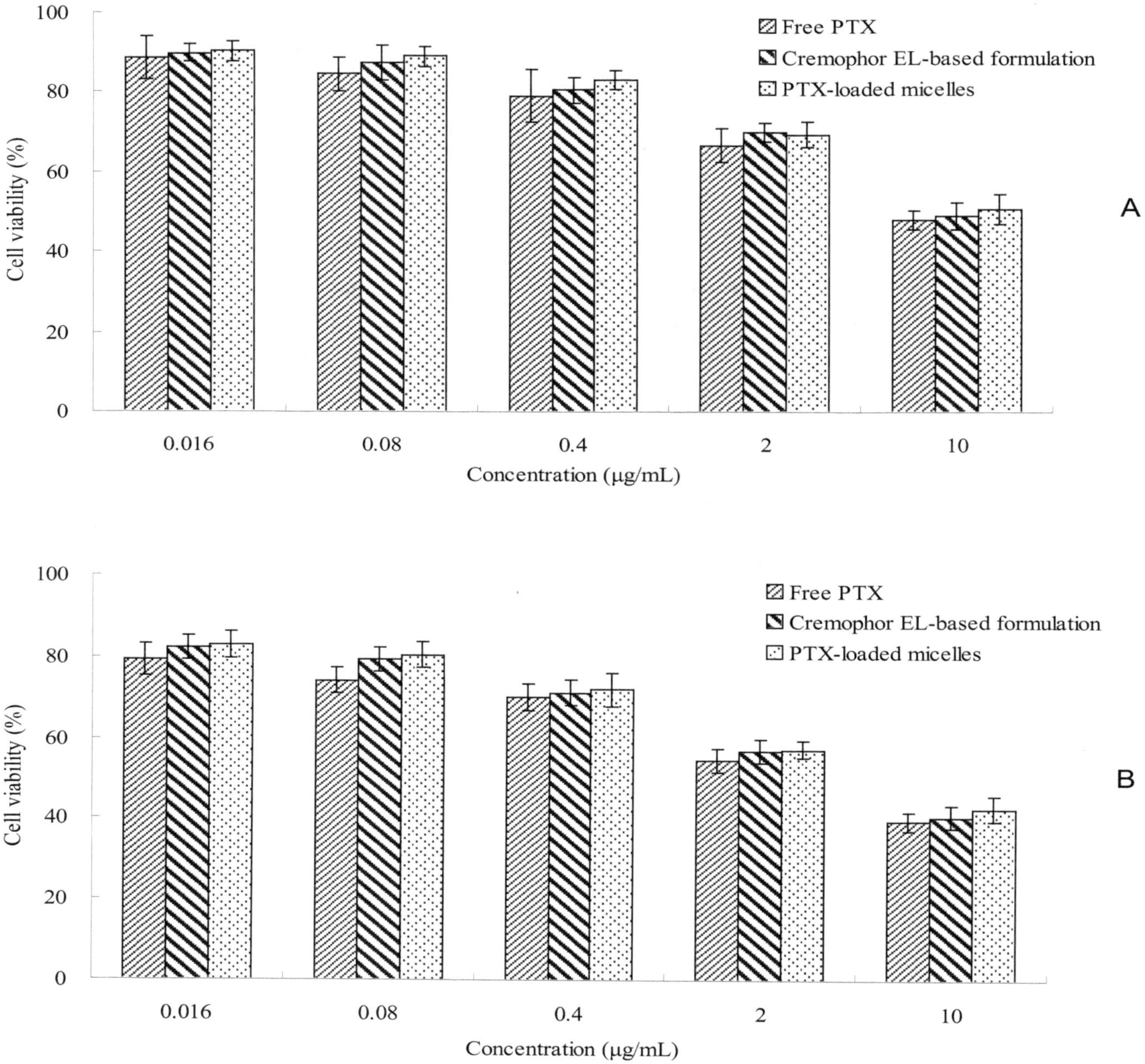

Figure 6. In vitro cytotoxicity of free PTX in dimethyl sulfoxide (DMSO), Cremophor EL-based formulation and PTX-loaded micelles against MCF-7 cells after 24 h (**A**) and 48 h (**B**) incubation (mean ± SD, $n = 3$).

2.6. In Vivo Antitumor Activity Study

The antitumor efficacy of PTX-loaded micelles in vivo was consistent well with the in vitro cell experiment. As shown in Figure 7, comparing with the saline control, a striking antitumor response was observed in all the treatment groups ($p < 0.05$). Furthermore, PTX-loaded micelles and Cremophor EL-based formulation had the similar antitumor efficacy at the same dose of 10 mg/kg, with tumor inhibition rate (TIR) of 72.3% and 69.6%, respectively, and the difference between the two groups was not statistically significant. The slight superior of the micelles might be explained by the increased concentration of PTX in tumor tissue due to the EPR effect, and the exact mechanism will be further studied.

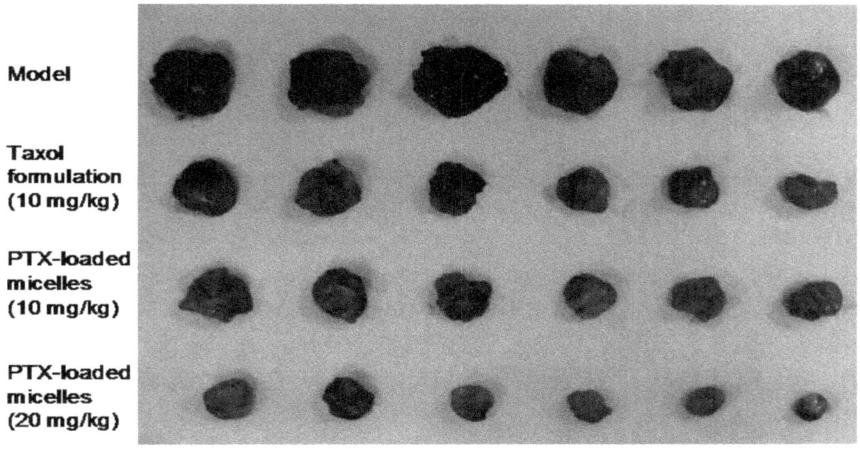

Figure 7. Photograph of tumors from each treatment group excised after intravenous injection treatment on day 13.

On the other hand, the toxic effect of PTX-loaded micelles was much less than that of the Cremophor EL-based formulation. At the high dose of 20 mg/mL, the intravenous administration of Cremophor EL-based formulation resulted in development of immediate ataxia and enhanced respiration, and 3 in 6 animals died although the injection speed was slowed down and the first aid treatment was given immediately after the symptoms appeared. In contrast, none of the mice treated with PTX-loaded micelles died throughout the study. This may be explained by the serious side effect of Cremophor EL. While for the micelles, all the components constructed TS-GC were nontoxic and biocompatible. It was obvious that the low toxicity of PTX-loaded micelles is of great value.

Taken together, these results implied that PTX-loaded TS-GC micelles not only had similar to better antitumor efficacy as Cremophor EL-based formulation, but also induced less systemic toxicity, especially at the higher dose. The PTX-loaded micelles might allow the administration of a higher dose of PTX.

3. Materials and Methods

3.1. Materials

Glycol chitosan (Mw 250 kDa, deacetylation degree of 82.7%), 2,4,6-trinitrobenzene sulfonic acid (TNBS), pyrene (purity > 99%) and 3-(4,5-dimethylthiazol-2-yl)-2,5-diphenyl tetrazolium bromide (MTT) were purchased from Sigma, St. Louis, MO, USA. 1-Ethyl-3-(3-dimethylaminopropyl) carbodiimide hydrochloride (EDC) and *N*-hydroxysuccinimide (NHS) were obtained from Shanghai Medpep Co., Ltd., Shanghai, China. α-Tocopherol succinate (TS) was kindly donated by Xinchang Pharmaceutical Co., Ltd., Shaoxing, China. Paclitaxel (PTX, purity of 99.9%) was supplied by Tianfeng Bioengineering Technology Co., Ltd., Shenyang, China. Cremophor EL was a kind gift from BASF Corp., Ludwigshafen, Germany. Dulbecco's modified Eagle's medium (DMEM), fetal bovine serum (FBS), and penicillin–streptomycin mixture were purchased from Gibco BRL, Carlsbad, CA, USA. All other chemicals and solvents were of analytical or chromatographic grade and used without further purification. Distilled water or Milli-Q water was used in all experiments.

3.2. Animals and Cell Line

The male New Zealand rabbit (weighing 2 kg) and specific pathogen-free female Kunming mice (5–6 weeks old, weighing 20–25 g) were acquired from Laboratory Animal Center of Harbin Medical University, Harbin, China. MCF-7 cells (human breast cancer cells) were obtained from American Type Culture Collection, Manassas, VA, USA. U14 cells (mouse uterine cervix carcinoma cells) were provided by the Cell Resource Center of Chinese Academy of Medical Sciences, Beijing, China. All animal procedures were performed in compliance with the animal care protocols approved by the Animal Ethics Committee of Harbin Medical University (18 May 2017, No. 201705180118).

3.3. *Synthesis of α-Tocopherol Succinate Modified Glycol Chitosan (TS-GC)*

α-Tocopherol succinate modified glycol chitosan (TS-GC) was synthesized by the coupling reaction between carboxyl group of TS with amine group of GC in the presence of EDC [32]. Briefly, GC (100 mg) was dissolved in 10 mL of distilled water and TS (64 mg) was dissolved in 10 mL of methanol. The TS solution was slowly added to the GC solution and followed by the addition of excessive EDC and NHS under gentle stirring at room temperature. After overnight reaction, the mixture was dialyzed against distilled water using a cellulose membrane (Viskase, Willowbrook, IL, USA, MWCO: 7000), and TS-GC was obtained by freeze-drying of the dialyzed solution.

3.4. *Characterization of TS-GC*

To confirm the formation of TS-GC, proton nuclear magnetic resonance (^{1}H-NMR) analysis was operated at 300 MHz using Bruker Avance spectrometer (AV-300, Bruker, Karlsruhe, Germany). GC and the conjugate were dissolved in D_2O and DMSO-d6 at the concentration of 1% (w/v), respectively.

In order to further investigate the structural changes during the synthesis process, Fourier-transform infrared (FT-IR) spectra of GC, TS, their physical mixture and the conjugate were obtained on FT-IR spectrometer (Tensor II, Bruker, Fällanden, Switzerland) in the range between 4000 and 400 cm^{-1} after compressed into KBr pellets.

The degree of substitution means the number of tocopherol groups per 100 anhydroglucose units (amino groups) of TS-GC. The amount of remaining terminal amino residues on TS-GC was measured using 2,4,6-trinitrobenzene sulfonic acid (TNBS reagent) according to TNBS method [33,34]. The numerical value was calculated from the calibration curve that obtained by GC solution.

The critical aggregation concentration (CAC) of TS-GC in aqueous milieu was measured using a fluorometer (F-2500 FL Spectrophotometer, Hitachi Ltd., Tokyo, Japan). The pyrene fluorescence was monitored to evaluate the micropolarity and hydrophobicity of the region in which it was solubilized, so as to prove the formation of the micelles [35].

3.5. *Preparation of PTX-Loaded TS-GC Micelles*

PTX-loaded TS-GC micelles were prepared by a probe-type ultrasonic method. Briefly, 10 mg of TS-GC was swollen in 10 mL of distilled water under gentle stirring overnight at room temperature to ensure complete dispersion. PTX solution was prepared by dissolving PTX in methanol at the concentration of 1 mg/mL. The PTX solution was added into the TS-GC solution, and the final mixture was sonicated for 10 min at 400 W in an ice bath by a ultrasonicator (JY92-II, Ningbo Scientz Biotechnology Co., Ltd., Ningbo, China). The sonication was carried out with the pulse function (turned on for 3 s and off for 2 s). The unloaded PTX was removed by centrifugation at 4000 rpm for 10 min. The resulting supernatant was lyophilized to obtain PTX-loaded TS-GC micelles. The blank micelles were prepared by the same procedure except no PTX was added.

3.6. *Characterization of PTX-Loaded TS-GC Micelles*

3.6.1. X-ray Diffraction (XRD) Analysis

X-ray diffraction diagrams were detected using an X-ray diffractometer (Geigerflex, Rigaku Co., Akishima, Japan) with Cu Kα radiation. Samples were scanned from 5 to 50° (2θ) at a scanning speed of 2°/min and step size of 0.02°. The X-ray system was operated at a potential of 30 kV and current of 30 mA.

3.6.2. Transmission Electron Microscopy (TEM) Observation

Morphology of the micelles was observed using a transmission electron microscope (TEM) (Jeol JEM1200EX, Tokyo, Japan) that operated at an accelerating voltage of 60 kV. For TEM, an aqueous droplet of micelles was immobilized on copper grids and negatively stained with phosphotungstate solution (2%, *w*/*v*), then dried at room temperature before observation.

3.6.3. Measurement of Particle Size and Zeta Potential

The number-weighted diameter, their distribution, and zeta potential of the micelles were measured using photon correlation spectroscopy with a Zetasizer Nano-ZS90 at 25 °C. The lyophilized samples were suspended in distilled water before measurement.

3.6.4. Determination of Drug Loading and Drug Encapsulation Efficiency

The drug loading and entrapment efficiency of the PTX-loaded micelles were determined as follows: 100 μL of PTX-loaded TS-GC micelles solution (concentration of TS-GC = 1.0 mg/mL) was centrifuged at 10,000 rpm for 10 min with ultrafilter (Vivaspin 500, MWCO 10 k, Sartorius Co., Göttingen, Germany). The unentrapped PTX amount in the ultrafiltrate (W_1) was determined using HPLC method. The HPLC system consisted of a mobile phase delivery pump (LC-10ATVP HPLC pump, Shimadzu, Kyoto, Japan) and a UV detector (SPD-10A UV/Vis detector, Shimadzu, Japan). For separation, a Diamonsil^{TM} C_{18} reverse-phase column (200 × 4.6 mm, 5 μm, Dikma technologies Inc., Beijing, China) was used. The mobile phase consisted of a mixture of acetonitrile and water (60:40, *v*/*v*). The detector wavelength, column temperature, and flow rate of the mobile phase were set at 227 nm, 30 °C and 1.0 mL/min, respectively. The injection volume of the test samples was 20 μL.

In order to measure the total PTX content (W_0) in the micelles, another 100 μL of identical PTX-loaded micelles solution was ultrasonicated in 10 mL methanol to extract PTX from the micelles. The encapsulation efficiency (EE%) and drug loading (DL%) of PTX were calculated using the following Equations (1) and (2):

$$\text{EE\%} = (W_0 - W_1)/W_0 \times 100\% \quad (1)$$

$$\text{DL\%} = (W_0 - W_1)/(W_0 - W_1 + 100) \times 100\% \quad (2)$$

where W_1 is the PTX content in the ultrafiltrate, and W_0 is the total PTX amount in the solution. The unit was microgram.

3.7. *In Vitro Hemolysis Test*

Hemolysis assessment of the TS-GC micelles was conducted as the method reported by Gong et al. [36]. The rabbit blood, freshly drawn from the ear vein, was centrifuged at 3000 rpm for 10 min to isolate the erythrocytes. The erythrocytes were washed with normal saline for three times to make the supernatant achromatic, and then suspended in normal saline to get a 2% (*v*/*v*) suspension. The lyophilized powder of PTX-loaded TS-GC micelles was dispersed in 0.9% NaCl, and different amounts of micelle solution were added into the tubes with 2.5 mL of 2% erythrocyte dispersion in each. Then adequate amounts of normal saline were added in every tube to obtain a final volume of 5 mL. After incubating at 37 °C for 4 h, the mixture was centrifuged at 3000 rpm for 10 min to remove intact red blood cells (RBC). The supernatant was analyzed for released hemoglobin at 540 nm using a spectrophotometer (UV-Vis Spectrophotometer Model 752, Shanghai Spectrum Instruments Co., Ltd., Shanghai, China). To obtain 0 and 100% hemolysis, 2.5 mL of saline and 2.5 mL of distilled water was added to 2.5 mL of RBC suspension, respectively. The degree of hemolysis was calculated by the following Equation (3).

$$\text{Hemolysis (\%)} = (A_{sample} - A_{0\%})/(A_{100\%} - A_{0\%}) \times 100\% \quad (3)$$

where A_{sample}, $A_{0\%}$, and $A_{100\%}$ are the absorbance of the samples, a solution of 0% hemolysis, and a solution of 100% hemolysis, respectively.

3.8. In Vitro Cytotoxicity

The in vitro cytotoxic activity of samples was evaluated by MTT method using MCF-7 cell line [37]. Briefly, 50 μL of MCF-7 cells growing in the logarithmic phase were cultured in Dulbecco's modified Eagle's medium (DMEM) containing 10% (*v*/*v*) fetal bovine serum, 100 IU/mL of penicillin G sodium and 100 μg/mL of streptomycin sulfate. The cells were seeded in a 96-well microtitre plate at the density of 1×10^4 cells per well and maintained in an incubator supplied with 5% CO_2 at 37 °C. After reaching 75% confluence, the cells were incubated with free drug in dimethyl sulfoxide (DMSO), Cremophor EL-based PTX formulation and PTX-loaded TS-GC micelles at the equivalent drug concentrations ranging from 0.016 to 10 μg/mL for 24 and 48 h. At designated time intervals, the medium was removed and the wells were washed with PBS for two times. Then, 10 μL of MTT (5 mg/mL in PBS) was added to the wells. After an additional 4 h of incubation, the MTT medium was aspirated off and 100 μL of DMSO was added to each well to dissolve the formazan crystals. The absorbance was measured at 570 nm with a microplate reader (Bio-Tek Instruments Inc., Vernusky, VT, USA). Untreated cells were taken as the control with 100% viability, and cells without addition of MTT were used as blank to calibrate the spectrophotometer to zero absorbance. The cytotoxicity was calculated as follows Equation (4):

$$\text{Cytotoxicity} = (\text{A}_{\text{culture medium}} - \text{A}_{\text{sample}})/\text{A}_{\text{culture medium}} \times 100\% \tag{4}$$

where $\text{A}_{\text{culture medium}}$ and A_{sample} are the absorbance of cells incubated with culture medium and absorbance of cells exposed to the sample, respectively.

3.9. In Vivo Antitumor Activity

Specific pathogen-free Kunming mice, 5 to 6 weeks old, weighing 20–25 g were used for this study. The animals were housed six per cage in standard size cages under standard laboratory conditions (21 ± 2 °C, 12-h light: 12-h dark cycle, relative humidity of 50–60%) and allowed to access sterilized food and water freely. The mice were left to acclimatize for a week prior to the experiment.

U14 cells of the third passage in vivo were used for tumor development. Animals were inoculated with 2.0×10^6 cells (0.2 mL/mouse) subcutaneously in the armpit of right anterior limb. Three days later, the tumors were palpable, and the tumor model was established. The animals were randomized and divided into different groups ($n = 6$): (1) normal saline group (negative control); (2) Cremophor EL-based PTX formulation groups (positive control, 10 mg/kg and 20 mg/kg); and (3) PTX-loaded TS-GC micelles groups (10 mg/kg and 20 mg/kg), and the treatments were initiated. All samples were injected intravenously via the tail vein every 3 days, four times in total. The day that mice received treatment was set as day 1. At the end of the experiment, on day 13, the animals were sacrificed by cervical dislocation, and the tumors were excised, weighed and imaged. The tumor inhibition rate (TIR) of each formulation was defined as follows Equation (5):

$$\text{TIR} = (\text{tumor weight of negative control group - tumor weight of treatment group})/\text{tumorweightofnegativecontrolgroup} \times 100\% \tag{5}$$

3.10. Statistical Analysis

Each experiment was performed in triplicate. Values were expressed as mean±standard deviation (SD). Statistical data analysis was performed using the Student's *t*-test with $p < 0.05$ as the level of significance.

4. Conclusions

In this study, a novel amphiphilic derivative of glycol chitosan was successfully synthesized by grafting α-tocopherol succinate onto the skeleton of glycol chitosan. In aqueous milieu, the conjugates provide stable self-aggregates above the CAC. Furthermore, the water-insoluble anticancer agent, PTX, was successfully encapsulated into the core of the micelles. The mean diameters of PTX-loaded micelles were about 142 nm, which was larger than blank ones. The spherical morphology of the micelles was visually confirmed by TEM. Moreover, the PTX-loaded TS-GC micelles possessed antitumor activities in vitro and in vivo, and had advantages over the commercially available Cremophor EL-based formulation in terms of low toxicity levels and increased tolerated dose. Therefore, this novel TS-GC polymer might be used as a potential carrier for PTX.

Author Contributions: N.L. and S.S. conceived and designed the experiments. N.L. performed the experiments, analyzed the data and wrote the paper; X.G. and Q.L. performed studies of in vitro cytotoxicity. F.C. and P.Y. gave us much useful advice and some pieces of guidance.

References

1. Abou-ElNaga, A.; Mutawa, G.; El-Sherbiny, I.; Abd-ElGhaffar, H.; Allam, A.; Ajarem, J.; Mousa, S. Novel nano-therapeutic approach actively targets human ovarian cancer stem cells after xenograft into nude mice. *Int. J. Mol. Sci.* **2017**, *18*, 813. [CrossRef] [PubMed]
2. Bernabeu, E.; Cagel, M.; Lagomarsino, E.; Moretton, M.; Chiappetta, D.A. Paclitaxel: What has been done and the challenges remain ahead. *Int. J. Pharm.* **2017**, *526*, 474–495. [CrossRef] [PubMed]
3. Gelderblom, H.; Verweij, J.; Nooter, K.; Sparreboom, A. Cremophor EL: The drawbacks and advantages of vehicle selection for drug formulation. *Eur. J. Cancer* **2001**, *37*, 1590–1598. [CrossRef]
4. Liu, Y.; Zhang, B.; Yan, B. Enabling anticancer therapeutics by nanoparticle carriers: The delivery of paclitaxel. *Int. J. Mol. Sci.* **2011**, *12*, 4395–4413. [CrossRef] [PubMed]
5. Ravar, F.; Saadat, E.; Gholami, M.; Dehghankelishadi, P.; Mahdavi, M.; Azami, S.; Dorkoosh, F.A. Hyaluronic acid-coated liposomes for targeted delivery of paclitaxel, in-vitro characterization and in-vivo evaluation. *J. Control. Release* **2016**, *229*, 10–22. [CrossRef] [PubMed]
6. Hou, J.; Sun, E.; Zhang, Z.H.; Wang, J.; Yang, L.; Cui, L.; Ke, Z.C.; Tan, X.B.; Jia, X.B.; Lv, H.X. Improved oral absorption and anti-lung cancer activity of paclitaxel-loaded mixed micelles. *Drug Deliv.* **2017**, *24*, 261–269. [CrossRef] [PubMed]
7. Alani, A.W.G.; Bae, Y.; Rao, D.A.; Kwon, G.S. Polymeric micelles for the pH-dependent controlled, continuous low dose release of paclitaxel. *Biomaterials* **2010**, *31*, 1765–1772. [CrossRef] [PubMed]
8. Erdoğar, N.; Esendağlı, G.; Nielsen, T.T.; Esendağlı-Yılmaz, G.; Yöyen-Ermiş, D.; Erdoğdu, B.; Sargon, M.F.; Eroğlu, H.; Bilensoy, E. Therapeutic efficacy of folate receptor-targeted amphiphilic cyclodextrin nanoparticles as a novel vehicle for paclitaxel delivery in breast cancer. *J. Drug Target* **2018**, *26*, 66–74. [CrossRef] [PubMed]
9. Nakamura, I.; Ichimura, E.; Goda, R.; Hayashi, H.; Mashiba, H.; Nagai, D.; Yokoyama, H.; Onda, T.; Masuda, A. An in vivo mechanism for the reduced peripheral neurotoxicity of NK105: A paclitaxel-incorporating polymeric micellar nanoparticle formulation. *Int. J. Nanomed.* **2017**, *12*, 1293–1304. [CrossRef] [PubMed]
10. Zhang, T.; Luo, J.; Fu, Y.; Li, H.; Ding, R.; Gong, T.; Zhang, Z. Novel oral administrated paclitaxel micelles with enhanced bioavailability and antitumor efficacy for resistant breast cancer. *Colloids Surf. B Biointerfaces* **2017**, *150*, 89–97. [CrossRef] [PubMed]
11. Torchilin, V.P. Micellar nanocarriers: Pharmaceutical perspectives. *Pharm. Res.* **2006**, *24*, 1–16. [CrossRef] [PubMed]
12. Deshmukh, A.S.; Chauhan, P.N.; Noolvi, M.N.; Chaturvedi, K.; Ganguly, K.; Shukla, S.S.; Nadagouda, M.N.; Aminabhavi, T.M. Polymeric micelles: Basic research to clinical practice. *Int. J. Pharm.* **2017**, *532*, 249–268. [CrossRef] [PubMed]
13. Rapoport, N. Physical stimuli-responsive polymeric micelles for anti-cancer drug delivery. *Prog. Polym. Sci.* **2007**, *32*, 962–990. [CrossRef]
14. Biswas, S.; Kumari, P.; Lakhani, P.M.; Ghosh, B. Recent advances in polymeric micelles for anti-cancer drug delivery. *Eur. J. Pharm. Sci.* **2016**, *83*, 184–202. [CrossRef] [PubMed]

15. Kwon, G.S.; Okano, T. Polymeric micelles as new drug carriers. *Adv. Drug Deliv. Rev.* **1996**, *21*, 107–116. [CrossRef]
16. Yang, Y.; Wang, S.; Wang, Y.; Wang, X.; Wang, Q.; Chen, M. Advances in self-assembled chitosan nanomaterials for drug delivery. *Biotechnol. Adv.* **2014**, *32*, 1301–1316. [CrossRef] [PubMed]
17. Prabaharan, M. Chitosan-based nanoparticles for tumor-targeted drug delivery. *Int. J. Biol. Macromol.* **2015**, *72*, 1313–1322. [CrossRef] [PubMed]
18. Fu, Y.-N.; Li, Y.; Li, G.; Yang, L.; Yuan, Q.; Tao, L.; Wang, X. Adaptive chitosan hollow microspheres as efficient drug carrier. *Biomacromolecules* **2017**, *18*, 2195–2204. [CrossRef] [PubMed]
19. Yu, J.; Liu, Y.; Zhang, L.; Zhao, J.; Ren, J.; Zhang, L.; Jin, Y. Self-aggregated nanoparticles of linoleic acid-modified glycol chitosan conjugate as delivery vehicles for paclitaxel: Preparation, characterization and evaluation. *J. Biomater. Sci. Polym. Ed.* **2015**, *26*, 1475–1489. [CrossRef] [PubMed]
20. Sharma, S.; Verma, A.; Teja, B.V.; Shukla, P.; Mishra, P.R. Development of stabilized paclitaxel nanocrystals: In-vitro and in-vivo efficacy studies. *Eur. J. Pharm. Sci.* **2015**, *69*, 51–60. [CrossRef] [PubMed]
21. Zhao, L.; Zhu, L.; Liu, F.; Liu, C.; Shan, D.; Wang, Q.; Zhang, C.; Li, J.; Liu, J.; Qu, X.; et al. pH triggered injectable amphiphilic hydrogel containing doxorubicin and paclitaxel. *Int. J. Pharm.* **2011**, *410*, 83–91. [CrossRef] [PubMed]
22. Park, J.S.; Han, T.H.; Lee, K.Y.; Han, S.S.; Hwang, J.J.; Moon, D.H.; Kim, S.Y.; Cho, Y.W. *N*-acetyl histidine-conjugated glycol chitosan self-assembled nanoparticles for intracytoplasmic delivery of drugs: Endocytosis, exocytosis and drug release. *J. Control. Release* **2006**, *115*, 37–45. [CrossRef] [PubMed]
23. Huo, M.; Fu, Y.; Liu, Y.; Chen, Q.; Mu, Y.; Zhou, J.; Li, L.; Xu, W.; Yin, T. *N*-mercapto acetyl-*N*′-octyl-*O,N*″-glycol chitosan as an efficiency oral delivery system of paclitaxel. *Carbohyd. Polym.* **2018**, *181*, 477–488. [CrossRef] [PubMed]
24. Nielsen, P.B.; Müllertz, A.; Norling, T.; Kristensen, H.G. The effect of α-tocopherol on the in vitro solubilisation of lipophilic drugs. *Int. J. Pharm.* **2001**, *222*, 217–224. [CrossRef]
25. Park, C.; Vo, C.L.-N.; Kang, T.; Oh, E.; Lee, B.-J. New method and characterization of self-assembled gelatin–oleic nanoparticles using a desolvation method via carbodiimide/*N*-hydroxysuccinimide (EDC/NHS) reaction. *Eur. J. Pharm. Biopharm.* **2015**, *89*, 365–373. [CrossRef] [PubMed]
26. Xu, X.; Li, L.; Zhou, J.; Lu, S.; Yang, J.; Yin, X.; Ren, J. Preparation and characterization of *N*-succinyl-*N*′-octyl chitosan micelles as doxorubicin carriers for effective anti-tumor activity. *Colloids Surf. B Biointerfaces* **2007**, *55*, 222–228.
27. Băran, A.; Stîngă, G.; Anghel, D.F.; Iovescu, A.; Tudose, M. Comparing the spectral properties of pyrene as free molecule, label and derivative in some colloidal systems. *Sens. Actuators B Chem.* **2014**, *197*, 193–199. [CrossRef]
28. Astafieva, I.; Zhong, X.F.; Eisenberg, A. Critical micellization phenomena in block polyelectrolyte solutions. *Macromolecules* **1993**, *26*, 7339–7352. [CrossRef]
29. Zhang, C.; Ping, Q.; Zhang, H. Self-assembly and characterization of paclitaxel-loaded *N*-octyl-*O*-sulfate chitosan micellar system. *Colloids Surf. B Biointerfaces* **2004**, *39*, 69–75. [CrossRef] [PubMed]
30. Danhier, F.; Danhier, P.; De Saedeleer, C.J.; Fruytier, A.-C.; Schleich, N.; Rieux, A.d.; Sonveaux, P.; Gallez, B.; Préat, V. Paclitaxel-loaded micelles enhance transvascular permeability and retention of nanomedicines in tumors. *Int. J. Pharm.* **2015**, *479*, 399–407. [CrossRef] [PubMed]
31. Huo, M.; Zhang, Y.; Zhou, J.; Zou, A.; Yu, D.; Wu, Y.; Li, J.; Li, H. Synthesis and characterization of low-toxic amphiphilic chitosan derivatives and their application as micelle carrier for antitumor drug. *Int. J. Pharm.* **2010**, *394*, 162–173. [CrossRef] [PubMed]
32. Liang, N.; Sun, S.; Li, X.; Piao, H.; Piao, H.; Cui, F.; Fang, L. α-Tocopherol succinate-modified chitosan as a micellar delivery system for paclitaxel: Preparation, characterization and in vitro/in vivo evaluations. *Int. J. Pharm.* **2012**, *423*, 480–488. [CrossRef] [PubMed]
33. Bernkop-Schnürch, A.; Krajicek, M.E. Mucoadhesive polymers as platforms for peroral peptide delivery and absorption: Synthesis and evaluation of different chitosan-EDTA conjugates. *J. Control. Release* **1998**, *50*, 215–223. [CrossRef]
34. Dosio, F.; Brusa, P.; Crosasso, P.; Arpicco, S.; Cattel, L. Preparation, characterization and properties in vitro and in vivo of a paclitaxel–albumin conjugate. *J. Control. Release* **1997**, *47*, 293–304. [CrossRef]
35. Yan, M.; Li, B.; Zhao, X. Determination of critical aggregation concentration and aggregation number of acid-soluble collagen from walleye pollock (*Theragra chalcogramma*) skin using the fluorescence probe pyrene. *Food Chem.* **2010**, *122*, 1333–1337. [CrossRef]

36. Gong, J.; Huo, M.; Zhou, J.; Zhang, Y.; Peng, X.; Yu, D.; Zhang, H.; Li, J. Synthesis, characterization, drug-loading capacity and safety of novel octyl modified serum albumin micelles. *Int. J. Pharm.* **2009**, *376*, 161–168. [CrossRef] [PubMed]
37. Laskar, P.; Samanta, S.; Ghosh, S.K.; Dey, J. In vitro evaluation of pH-sensitive cholesterol-containing stable polymeric micelles for delivery of camptothecin. *J. Colloid Interface Sci.* **2014**, *430*, 305–314. [CrossRef] [PubMed]

5

Thermal, Mechanical and Rheological Properties of Biocomposites made of Poly(Lactic Acid) and Potato Pulp Powder

Maria Cristina Righetti [1,*], Patrizia Cinelli [1,2,*], Norma Mallegni [1], Carlo Andrea Massa [1], Simona Bronco [1], Andreas Stäbler [3] and Andrea Lazzeri [1,2]

1. CNR-IPCF, National Research Council—Institute for Chemical and Physical Processes, Via Moruzzi 1, 56124 Pisa, Italy; norma.mallegni@pi.ipcf.cnr.it (N.M.); carlo.andrea.massa@pi.ipcf.cnr.it (C.A.M.); simona.bronco@pi.ipcf.cnr.it (S.B.); andrea.lazzeri@unipi.it (A.L.)
2. Department of Civil and Industrial Engineering, University of Pisa, Largo Lucio Lazzarino 1, 56122 Pisa, Italy
3. Fraunhofer Institute for Process Engineering and Packaging IVV, Giggenhauser Straße, 35, 85354 Freising, Germany; andreas.staebler@ivv.fraunhofer.de

* Correspondence: cristina.righetti@pi.ipcf.cnr.it (M.C.R.); patrizia.cinelli@unipi.it (P.C.)

Abstract: The thermal, mechanical, and rheological properties of biocomposites of poly(lactic acid) (PLA) with potato pulp powder were investigated in order to (1) quantify how the addition of this filler modifies the structure of the polymeric material and (2) to obtain information on the possible miscibility and compatibility between PLA and the potato pulp. The potato pulp powder utilized is a residue of the processing for the production and extraction of starch. The study was conducted by analyzing the effect of the potato pulp concentration on the thermal, mechanical, and rheological properties of the biocomposites. The results showed that the potato pulp powder does not act as reinforcement but as filler for the PLA polymeric matrix. A progressive decrease in elastic modulus, tensile strength, and elongation at break was observed with increasing the potato pulp percentage. This moderate loss of mechanical properties, however, still meets the technical requirements indicated for the production of rigid packaging items. The incorporation of potato pulp powder to PLA offers the possibility to reduce the cost of the final products and promotes a circular economy approach for the valorization of agro-food waste biomass.

Keywords: bio-based polymers; natural fibers; biomass; biocomposites; fiber/matrix adhesion

---

1. Introduction

Biodegradable bio-based polymers obtained from renewable resources represent an important alternative to petroleum-derived non-degradable polymers. For this reason, bio-based polymers—for example, poly(lactic acid) (PLA), polyhydroxylalkanoates (PHA), in particular polyhydroxylbutyrate (PHB) and its copolymers poly(hydroxylbutyrate-co-valerate) (PHBV), cellulose, and starch derived plastics, which are produced from renewable agricultural and biomass feedstock—have become a subject of crucial interest for academia and industry. Among them, PLA is the biodegradable polymer most present on the market and widely used for packaging purposes. Many properties of PLA, such as strength and stiffness, are comparable to those of traditional petroleum-based polymers. On the other hand, the drawbacks for its applications are the low toughness and the relatively low glass transition, which limits its use at a temperature above about 50 °C, because, due to its slow crystallization rate, processing of PLA generally results in amorphous products [1]. In addition, bio-based polymers are generally more expensive than conventional petroleum-derived polymers and often exhibit worse thermal and mechanical properties, which make them unacceptable for several applications.

In order to balance the cost and achieve suitable properties for different applications, a possible solution is offered by biocomposites. Biocomposites are a special class of composite materials obtained by blending natural fibers with bio-based polymers. Biocomposites represent an ecological and low-cost alternative to conventional petroleum-derived materials and, for this reason, are becoming progressively more utilized for a wide variety of uses. Natural fiber reinforced biocomposites have been reviewed in several articles [2–7].

Besides the benefit of being biodegradable and produced from renewable resources, natural fibers are also less abrasive to processing equipment than synthetic fibers like ceramic or glass fibers and traditional fillers like mica and glass, and they exhibit a lower density, which makes biocomposites lightweight and economical materials [8].

The chemical composition of the natural fibers, which varies depending on the plant from which they are derived, consists mainly of cellulose (50–70 wt%), hemicellulose (10–20 wt%), lignin (10–30 wt%), and pectin and waxes in smaller amounts [7]. The physico-mechanical properties of the natural fibers also depend on the original plants. The density, elastic modulus, tensile strength, and elongation at break are approximately in the ranges 0.8–1.5 g/cm^3, 5–20 GPa, 200–900 MPa, and 1.5–20%, respectively [7].

The mechanical properties of a biocomposite results from both the matrix and fiber properties [9,10] and strongly depend on the matrix/fiber interface [11]. The tensile strength is more sensitive to the matrix/fiber adhesion, whereas the modulus depends, in general, on both the matrix and the fiber properties. The mass or volume fraction of the fibers, the fiber aspect ratio (length to width ratio), the fiber–matrix adhesion, and the fiber orientation are crucial factors responsible for the final properties of natural fiber reinforced composites [12]. In a composite, the matrix plays the role of transferring the applied stress to the fibers. This occurs at the interface, and therefore a good matrix/fiber adhesion is necessary. Poor adhesion often characterizes biocomposites made of hydrophobic polymers and hydrophilic natural fibers, which can lead to scarce mechanical properties due to the tendency of the fibers to form aggregates during processing. In addition, the hydrophilic nature of the lignocellulosic fibers induces absorption of moisture in an amount that can vary up to about 10%, with the result that if the fibers are not efficiently dried previously, the process ability is worsened, and the formation of porous products is thus unluckily favored. The fiber aspect ratio strongly influences the tensile modulus and the fracture properties, and a critical fiber length is necessary to develop composites with high stiffness and strength. Fibers with low aspect ratio and irregular shape behave as fillers and not as reinforcement. Thus, multiple factors influence the mechanical properties of a composite, not only the intrinsic properties of the fibers but also their shape and dimensions and their dispersion in the polymeric matrix together with the interactions that are established at the matrix/filler interface.

In regard to PLA, several types of natural fibers have been used to produce biocomposites [13–18]. The effect of the processing conditions and, in some cases, surface chemical modification of the fibers on mechanical, rheological, and thermal properties of PLA/fibers biocomposites has been widely investigated [13–18].

Organic wastes have sometimes been utilized as reinforcements or additives for various polymers, examples being charcoal and lignin [19]. By considering that significant amounts of organic wastes from industry and agriculture are unutilized, it follows that the use of organic residue materials in biocomposites can represent an ecologically friendly method to produce materials for different applications. This can also allow the reduction of the cost of the products.

Potato wastes are biomasses rich in starch and lignocellulosic constituents. After extraction of the starch, the potato pulp accumulates in high amounts—approximately 0.75 tons of pulp arises per ton of purified starch. Within the European Union, about 140,000 tons of dried potato pulp are produced annually in the starch industry [20]. The original potato pulp contains water up to about

90%, but de-watering processes generally results in an increase in the dry matter up to about 90 wt%. Dried potato pulp, which consists mainly of lignocellulosic fibers, starch, and proteins, can be used as filler or for a reinforcement action on bio-based polymers. The cost of this raw material is low, which makes potato pulp even more interesting for industrial utilization [21].

In the present work, biocomposites made of poly(lactic acid) and potato pulp powder (PPP) have been produced by extrusion followed by injection molding and characterized in terms of process ability and thermal, mechanical, and rheological properties. It is worth pointing out that the composition of the potato pulp powder is expected to depend on the geographical origin of the agricultural product and on the climatic and harvesting conditions as well as on the industrial processing methods. For this reason, the present study provides only a general description of the properties of these PLA-based biocomposites, which could slightly change with varying the original materials. In order to make the processing of PLA with the potato pulp powder easier, a plasticizer, acetyl tributyl citrate (ATBC), was used. ATBC is an efficient plasticizer of PLA [22]. It is derived from naturally occurring citric acid, non-toxic, and accepted for food-contact [23]. In addition, calcium carbonate ($CaCO_3$) was used in low percentages as an inert filler to facilitate the removal of the injection-molded specimens from the mold. The thermal, mechanical, and rheological properties of the PLA-based matrix were investigated as preliminary step in order to better quantify the influence of the reinforcing fibers on the material properties.

The PLA-based samples investigated in the present study, with the relative composition, are listed in Table 1. All the samples were processed as described in Section 3.

Table 1. Composition of the poly(lactic acid) (PLA)-based matrix and biocomposites.

	Potato Pulp
PLA (100%)	-
[PLA (90 wt%) + ATBC (10 wt%)]	-
[PLA (85 wt%) + ATBC (10 wt%) + $CaCO_3$ (5 wt%)]	-
[PLA (85 wt%) + ATBC (10 wt%) + $CaCO_3$ (5 wt%)] (95 wt%)	PPP (5 wt%)
[PLA (85 wt%) + ATBC (10 wt%) + $CaCO_3$ (5 wt%)] (90 wt%)	PPP (10 wt%)
[PLA (85 wt%) + ATBC (10 wt%) + $CaCO_3$ (5 wt%)] (80 wt%)	PPP (20 wt%)

2. Results and Discussion

2.1. *Thermogravimetric Analysis of the Potato Pulp Powder, the PLA-Based Matrix, and the Biocomposite [PLA (85 wt%) + ATBC (10 wt%) + $CaCO_3$ (5 wt%)] + PPP (20 wt%)]*

The main drawback of the use of natural fibers as reinforcement or fillers in biocomposites is the relative low processing temperature required due to the thermal degradation that occurs at high temperatures which can irreversibly damage them. As all the components of the potato pulp can be subjected to thermal degradation during composite processing, it is of practical significance to investigate the thermal decomposition process of the potato pulp powder in order to estimate the temperature range in which the biocomposites can be processed.

The thermal stability of the potato pulp powder was determined by means of the thermogravimetric analysis under nitrogen flow, because the contact of the material with air is reduced in the extruder and molder. Figure 1 shows the thermogravimetric curve of the potato pulp powder, which reports the change in weight according to a fixed temperature program. The initial weight loss detected at temperatures lower than 150 °C is due to water vaporization. The water content in PPP is approximately 3 wt%. The weight residue that is observed at high temperature is due to the carbon deposit that remains in the presence of an inert atmosphere. The thermal degradation of natural fibers generally has multiple processes due to the different components [24,25]. Degradation of hemicellulose occurs mainly in the temperature range of 200–350 °C with the maximum mass loss rate at around 300 °C, whereas cellulose pyrolysis takes places at higher temperatures, between 250 and 400 °C. At high temperatures, the solid residue of pure hemicellulose and cellulose is about 20%. Lignin is the most difficult component to decompose; its degradation takes place slowly and continuously from about 250 °C, with a high solid residue that remains at high temperatures (around 50%). Starch degradation extends from approximately 300 to 350 °C, and its solid residue at 500 °C is approximately 20% [26]. Degradation of proteins also occurs in a similar temperature range, between 200 and 400 °C, with a maximum loss rate at around 300 °C and 20% residue at high temperatures [27]. From Figure 1, the potato pulp powder appears stable up to about 180 °C. This thermal stability assures that the potato pulp powder does not undergo substantial degradation during the processing of the PLA biocomposites at 180 °C, being that the residence time at this temperature is no longer than 1.5 min. It can therefore be concluded that the potato pulp can be confidently used for the production of PLA-based biocomposites.

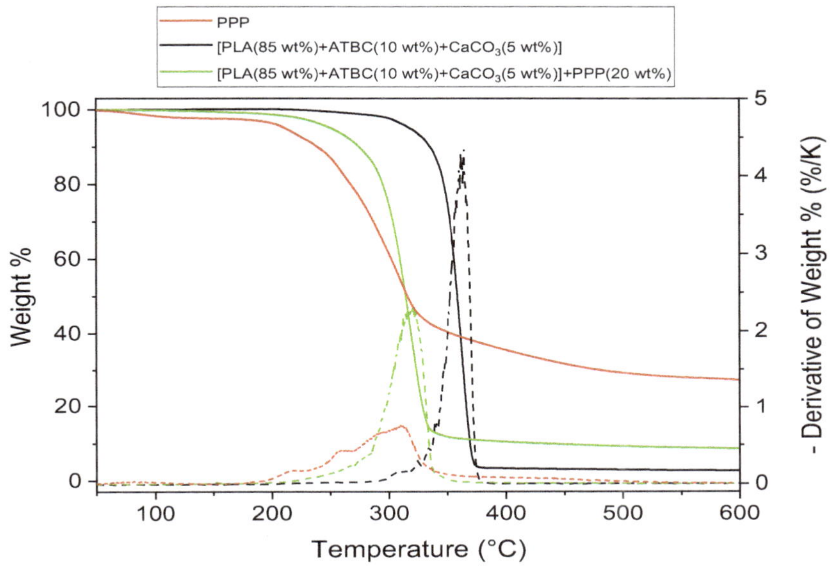

Figure 1. Thermogravimetric curves of the potato pulp powder (PPP), the poly(lactic acid) (PLA)-based matrix andbiocomposite [PLA (85 wt%) + acetyl tributyl citrate (ATBC; 10 wt%) + $CaCO_3$ (5 wt%)] + potato pulp powder (PPP; 20 wt%) at 10 K/min under nitrogen flow. The dotted lines are the derivatives of the weight % curves.

Figure 1 shows also the thermogravimetric curves of the PLA-based matrix and biocomposite with 20 wt% of PPP. Both these thermal degradations occur in a single step in a narrow temperature range. The initial degradation temperature of the PLA-based matrix is located at about 300 °C, whereas the maximum degradation rate is centered at about 365 °C, which is in excellent agreement with previous studies [28–30] in which the independence of the PLA thermal stability on the molar mass was also demonstrated [28]. The biocomposite with 20 wt% of potato pulp powder starts to degrade at about 180 °C, whereas the maximum degradation rate shifts to around 320 °C, thus approaching the thermal degradation of plain potato pulp powder. As expected, the residue of the biocomposite at 600 °C is higher compared to that of the PLA based matrix due to the presence of the lignocellulosic residue. In conclusion, Figure 1 reveals that the processing of the biocomposites at 180 °C does not substantially affect the potato pulp powder structure because the degradation of the biocomposite starts at the same temperature of the original unprocessed potato pulp powder.

2.2. *Scanning Electron Microscopy of the Potato Pulp Powder*

The morphologies of the potato pulp powder were investigated with scanning electron microscopy (SEM), and SEM images are shown in Figure 2. A homogeneous distribution of pulp fragments, which appear as small aggregates, can be observed. The aggregates are relatively large (200 μm and more).

The round-shaped particles detected at 1200× magnification in the potato pulp powder are either starch or pectin because they disappear after treatment with amylase and pectinase.

50× 500×

1200× 2000×

Figure 2. Scanning electron microscopy (SEM) images of the potato pulp powder (PPP) at the magnifications indicated.

2.3. Thermal, Mechanical, and Rheological Properties of the PLA-Based Matrix

The thermal, mechanical, and rheological properties of the PLA-based matrix were investigated as a preliminary step in order to better quantify the influence of the reinforcing fibers on the material properties.

The specific heat capacity (c_p) curves of PLA and PLA mixed with (1) the plasticizer ATBC and (2) with ATBC and the mineral filler $CaCO_3$, measured at 10 °C/min, are shown in Figure 3. As described in Section 3, the samples were injection-molded for 1 min at 90 °C. The glass transition, which occurs in the proximity of 60 °C, which is in agreement with literature data [31], is overlapped by an enthalpy recovery peak due to the permanence of the samples at room temperatures for one day. Before the melting endotherm, all the curves displayed an intense cold crystallization peak located approximately in the interval between 90 and 130 °C. In the temperature range between 100 and 120 °C, both the crystalline α'- and α-forms grew [32,33]. Thus the melting behavior that extends from approximately 130 to 160 °C results from the fusion of both the α- and the α'-crystals. At the heating rate of 10 K/min, the α'-crystals transform into the more ordered α-phase via melting and almost simultaneous recrystallization, with the result that a multiple melting behavior, independent of the molar mass, is commonly observed [34,35]. Reorganization and recrystallization events overlap the entire fusion process, which generally takes place in semi-crystalline polymers at a relatively low heating rate [36,37].

Table 2 lists the T_g values together with the enthalpy of cold crystallization (Δh_c) and the enthalpy of fusion (Δh_m) of the samples investigated in this section, calculated from the c_p curves shown in Figure 3. The Δh_c and Δh_m values collected in Table 2 are normalized to the PLA content. From these experimental values, an estimation of the crystalline weight fraction growing during the cold crystallization process (w_{Cc}) and disappearing during the melting process (w_{Cm}) was obtained by dividing Δh_c and Δh_m by the enthalpy of fusion of 100% crystalline PLA phase ($\Delta h_m{}^{\circ}$) at the crystallization and melting peak temperatures, respectively. As both the α'- and α-forms grow during cold crystallization and disappear during the melting process, the average values between the enthalpy of fusion of the α'- and α-forms were utilized [38], i.e., $\Delta h_m{}^\circ$ = 101 J/g for the cold crystallization centered at about 110 °C, $\Delta h_m{}^\circ$ = 96 J/g for the cold crystallization centered at about 100 °C, and $\Delta h_m{}^\circ$ = 119 J/g for the melting process centered approximately at 150 °C. The w_{Cc} and w_{Cm} values listed in Table 2 reveal that the PLA and the PLA samples with the addition of (1) only the plasticizer ATBC and (2) the plasticizer ATBC and the mineral filler $CaCO_3$, after processing for 1 min at 90 °C, are completely amorphous. This can be explained by considering the short times that characterize a typical injection molding and the slow crystallization kinetics of PLA [39]. In the presence of ATBC, the crystallinity that grows during the heating run is higher with respect to PLA, which can be ascribed to the higher mobility achieved by the PLA chains. The plasticizing effect of ATBC on PLA is evidenced by the decrease in the T_g value, the reduced peak temperature of the cold crystallization, and the higher crystallinity that develops upon heating, whereas the inertia of $CaCO_3$ towards the PLA phase transitions is proven by the unchanged thermal parameters.

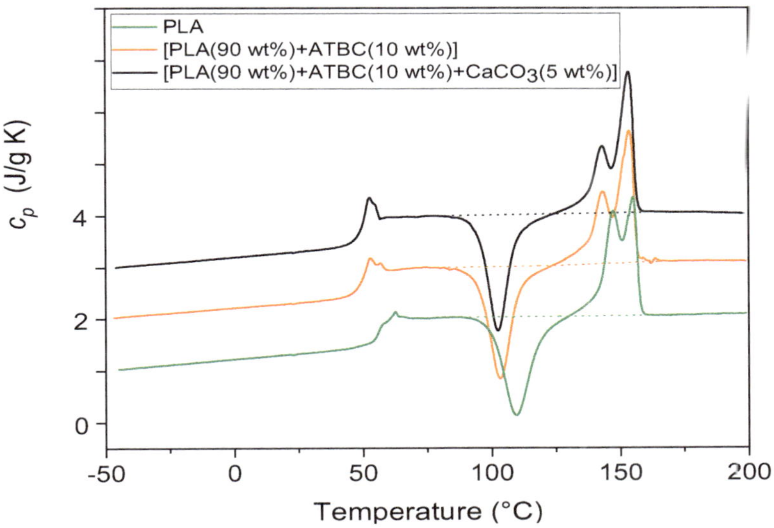

Figure 3. Specific heat capacity (c_p) of PLA and PLA mixed with (1) ATBC and (2) ATBC and the mineral filler $CaCO_3$ at the concentrations indicated. The curves were obtained upon heating at 10 K/min after previous fast cooling to −50 °C. The dotted lines are the baselines used for the calculation of the Δh_c and Δh_m values. The ordinate values refer only to the bottom curve. All the other curves are shifted vertically for the sake of clearness.

Table 2. Glass transition temperatures (T_g), enthalpy of cold crystallization (Δh_c), enthalpy of fusion (Δh_m), and crystalline weight fraction growing during cold crystallization (w_{Cc}) and disappearing during fusion (w_{Cm}) for PLA and PLA mixed with (1) ATBC and (2) ATBC and the mineral filler $CaCO_3$ (estimated errors: ± 0.5 °C for T_g; ± 1 J/g for Δh_c and Δh_m, and ± 0.02 for w_{Cc} and w_{Cm}).

	T_g (°C)	Δh_c (J/g)	w_{Cc}	Δh_m (J/g)	w_{Cm}
PLA	58	24	0.24	28	0.24
[PLA (90 wt%) + ATBC (10 wt%)]	50	27	0.28	33	0.28
[PLA (85 wt%) + ATBC (10 wt%) + $CaCO_3$ (5 wt%)]	50	27	0.28	33	0.28

The mechanical properties of PLA and of PLA after processing in the presence of (1) the plasticizer ATBC and (2) the plasticizer ATBC with the mineral filler $CaCO_3$ are summarized in Figure 4. PLA is a quite brittle polymer with a high elastic modulus and a high tensile strength. The addition of the plasticizer ATBC, as expected, modifies the mechanical properties: The elastic modulus does not substantially change, the tensile strength decreases, and conversely, the elongation at break greatly increases. In the presence of the plasticizer ATBC, the intermolecular forces between the PLA chains decrease, the mobility of the polymeric chains enhances, and an increase in flexibility and ductility is produced. A similar behavior has been reported for other PLA plasticized systems [40,41].

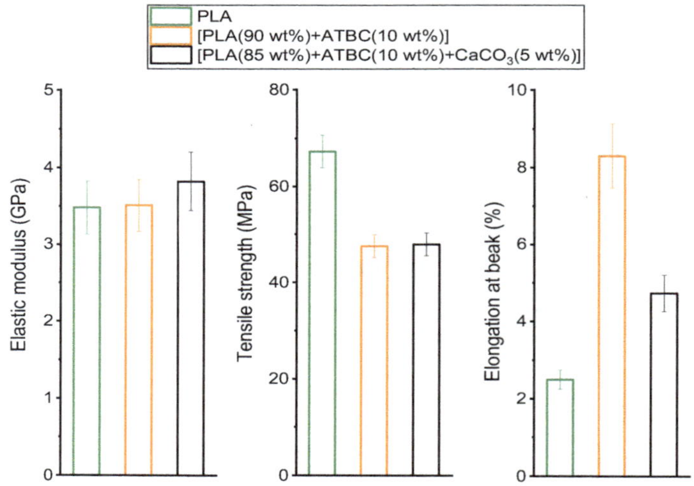

Figure 4. Elastic modulus, tensile strength and elongation at break of PLA and PLA mixed with (1) ATBC and (2) ATBC and the mineral filler $CaCO_3$ at the concentrations indicated.

The addition of the mineral filler $CaCO_3$ produces a slight increase in the elastic modulus and a strong decrease in the elongation at break, whereas it has negligible effect on the tensile strength. This behavior is in agreement with literature data. The modulus increase is the result of the reinforcement effect of $CaCO_3$ particles. However, if the mineral filler is present in small percentages, the influence on the elastic modulus is small because it can be assumed that the perturbed polymer fraction around each mineral particle is low compared with the unperturbed one [42,43]. The tensile strength remains unaffected by the presence of the mineral filler, which means that the mobility of the polymer chains is not influenced by the small percentage of $CaCO_3$ as also proven by the constant T_g value. Conversely, the elongation at break strongly decreases because the presence of micro-sized mineral particles modifies the fracture mechanism and hinders the elongation of the material. Mineral particles generally act as stress concentrators capable of initiating cracking and favoring specific fracture mechanisms.

The study of rheological properties of PLA is crucial for gaining a fundamental understanding of the processability of PLA materials, especially for injection molding and extrusion processes. PLA behaves like a pseudo-plastic, a non-Newtonian fluid, and a typical shear thinning fluid in which, at high shear rates, the macromolecules orient and decrease the entanglement number [44,45].

Figure 5 shows the dependence of the modulus of the complex viscosity on the deformation frequency for PLA and PLA with the addition of (1) the plasticizer ATBC, and (2) the plasticizer ATBC and the mineral filler $CaCO_3$. The rheological measurements were performed from low to high frequencies. All the curves indicate a decrease in complex viscosity with increasing rotational frequency. In the terminal zone, which defines the zero-shear viscosity η_o, a small decrease in viscosity is observed, probably due an original entangled structure not completely destroyed before the beginning of the rheological test. The η_o value for PLA at 175 °C is about 1.8×10^6 mPa·s. This value is in the range typical of commercial polymers (M_w = 140–160 kg/mol), i.e., 10^5–10^7 mPa·s [46]. The incorporation of ATBC reduces the viscosity due to its plasticizing effect, which leads to a decrease in the intermolecular forces and an increase in the mobility of the polymeric chains [47,48]. The addition of $CaCO_3$ causes a further small reduction of viscosity, which is in excellent agreement with literature data [49]. A low $CaCO_3$ content is insufficient to form a filler network. As a consequence, the incorporation of the mineral filler weakens the interaction between PLA chains segments, thus producing a viscosity reduction.

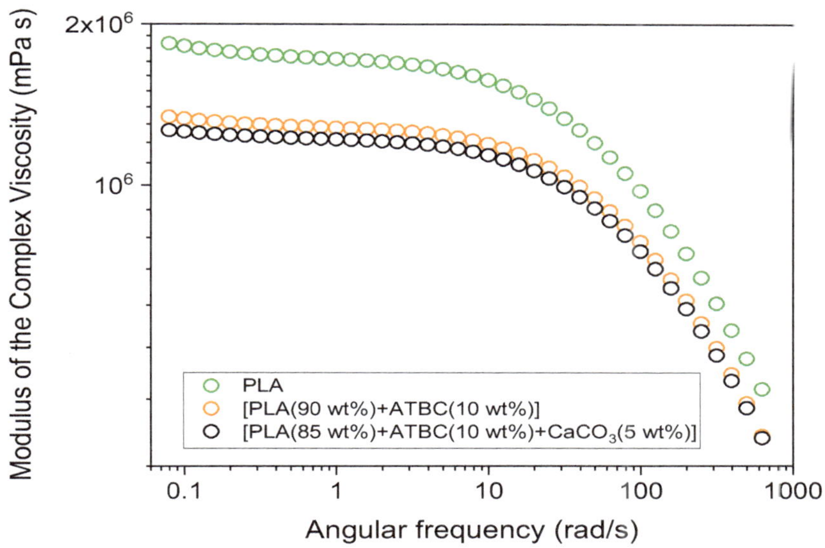

Figure 5. Modulus of the complex viscosity vs. angular frequency at 175 °C for PLA and PLA with the addition of (1) the plasticizer ATBC, and (2) the plasticizer ATBC and the mineral filler $CaCO_3$.

2.4. *Thermal, Mechanical and Rheological Properties of the PLA-Based Biocomposites*

The thermal, mechanical, and rheological properties of the biocomposites of PLA with potato pulp powder were investigated in order to quantify how the addition of these fibers modifies the structure of the polymeric material. These data together provide information on the possible miscibility and compatibility between PLA and potato pulp. The PLA-based biocomposites with potato pulp were studied by analyzing the effect of potato pulp concentration.

The specific heat capacity (c_p) curves of the [PLA (85 wt%) + ATBC (10 wt%) + $CaCO_3$ (5 wt%)] matrix with the addition of an increasing percentage of PPP (5 wt%, 10 wt%, and 20 wt%), measured at 10 K/min, are shown in Figure 6.

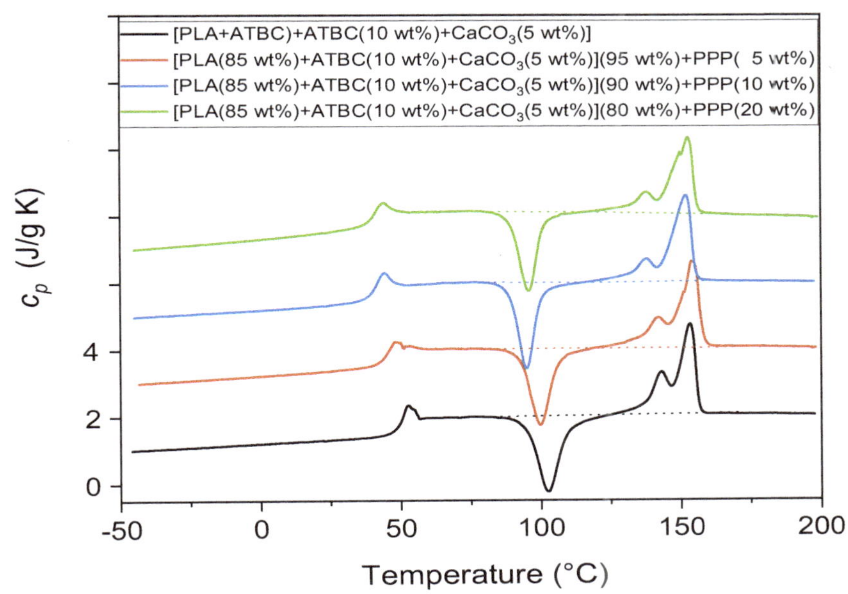

Figure 6. Specific heat capacity (c_p) of the PLA-based matrix and biocomposites indicated in the legend. The curves were obtained upon heating at 10 K/min after previously fast cooling to -50 °C. The dotted lines are the baselines used for the calculation of the Δh_c and Δh_m values. The ordinate values refer only to the bottom curve. All the other curves are shifted vertically for the sake of clarity.

Table 3 lists the T_g values together with the enthalpy of cold crystallization (Δh_c) and the enthalpy of fusion (Δh_m) of the samples investigated in this section as calculated from the c_p curves shown in Figure 6. The addition of PPP affects the glass transition temperature, which decreases with increasing the PPP percentage. This reveals that the polymer chain mobility increases after addition of the potato pulp powder. This can be ascribed to the development of weak intermolecular interactions between PLA and the potato pulp particles, which could cause the formation of outspread free volume at the matrix/fiber interface.

Table 3. Glass transition temperatures (T_g), enthalpy of cold crystallization (Δh_c), enthalpy of fusion (Δh_m), and crystalline weight fraction growing during cold crystallization (w_{Cc}) and disappearing during fusion (w_{Cm}) for the PLA-based matrix and biocomposites (estimated errors: $\pm$ 0.5 °C for T_g; $\pm$ 1 J/g for Δh_c and Δh_m, and $\pm$ 0.02 for w_{Cc} and w_{Cm}).

	T_g (°C)	Δh_c (J/g)	w_{Cc}	Δh_m (J/g)	w_{Cm}
[PLA (85 wt%) + ATBC (10 wt%) + $CaCO_3$ (5 wt%)]	50	27	0.28	33	0.28
[PLA (85 wt%) + ATBC (10 wt%) + $CaCO_3$ (5 wt%)] (95 wt%) + PPP (5 wt%)	45	26	0.27	34	0.28
[PLA (85 wt%) + ATBC (10 wt%) + $CaCO_3$ (5 wt%)] (90 wt%)+PPP (10 wt%)	40	26	0.32	35	0.32
[PLA (85 wt%) + ATBC (10 wt%) + $CaCO_3$ (5 wt%)] (80 wt%)+PPP (20 wt%)	40	28	0.34	37	0.34

The peak temperature of the cold crystallization process decreases in the biocomposites, proving that PPP plays a nucleating role during the heating run, which accelerates the PLA cold crystallization process. This a general behavior of the natural fibers, as also found in other PLA/natural fibers composites [15,50], which has been explained as due to the formation of a transcrystalline layer around the fibers. The occurrence of transcrystallinity was reported for a large combination of semi-crystalline thermoplastic matrices and fibers [51,52]. For biocomposites containing 10 and 20 wt% of PPP, cold crystallization occurs almost completely at temperatures lower than 100 °C, which means that only the α'-form develops [32,33]. At the heating rate of 10 K/min, however, these α'-crystals are expected to transform into the more ordered α-phase [34]. For biocomposites with 10 and 20 wt% of PPP, w_{Cc} was calculated by dividing the measured Δh_c by the enthalpy of fusion of the α'-form at about 95 °C ($\Delta h_m{}^\circ$ = 81 J/g), whereas w_{Cm} was obtained from the ratio between Δh_m and the enthalpy of fusion of the α'-form at about 150 °C ($\Delta h_m{}^\circ$ = 108 J/g) [38], as the enthalpy of crystallization and the enthalpy of fusion of the α-crystals cancel out. For the biocomposite with 5 wt% of PPP, for which cold crystallization extends up to above 110 °C, the average values between the enthalpy of fusion of the α'- and α-forms were utilized, i.e., $\Delta h_m{}^\circ$ = 96 J/g for the cold crystallization centered at about 100 °C, and $\Delta h_m{}^\circ$ = 119 J/g for the melting process centered approximately at 150 °C [38].

The w_{Cc} and w_{Cm} values listed in Table 3, as derived from the measured Δh_c and Δh_m values, respectively, reveal that the PLA-based biocomposites containing PPP, processed for 1 min at 90 °C, are completely amorphous. Despite the nucleating action of PPP, which leads to a progressively and slightly higher crystallinity to develop upon heating, the residence time of the materials in the mold is too short to allow the crystallization of PLA.

The mechanical properties of the [PLA (85 wt%) + ATBC (10 wt%) + CaCO3 (5 wt%)] matrix and the biocomposites with potato pulp powder are summarized in Figure 7.

As a general rule, the modulus of a composite depends on the modulus of both the matrix and the reinforcement and filler, and the modulus of the natural fibers is generally higher than the modulus of the polymeric matrices [12]. Figure 7 reveals that the elastic modulus decreases by increasing the PPP percentage, which could mean that PPP acts for PLA matrix as filler and not as reinforcement, or it could be ascribed to PLA degradation during the processing step induced by fiber moisture [53]; although, both PLA and the PPP powder were previously dried for a long time, and the water content of the potato pulp powder was about 3 wt% (Figure 1). The number-average molar mass

(M_n) and the mass-average molar mass (M_w) of PLA of both the matrix and the biocomposite with 20 wt% of potato pulp powder processed under identical conditions were measured by size exclusion chromatography (SEC). For the PLA-based matrix, M_n = 92,000 g/mol and M_w = 170,000 g/mol, whereas for the biocomposite [PLA (85 wt%) + ATBC (10 wt%) + $CaCO_3$ (5 wt%)] (80 wt%) + PPP (20 wt%), M_n = 70,000 g/mol and M_w = 130,000 g/mol were obtained. The observed decrease of about 20% in the PLA molar mass can thus be ascribed to the PLA degradation occurring during processing in the presence of the moisture introduced by the potato pulp powder. This could explain the loss in the mechanical properties. However, for M_w higher than about 100,000 g/mol, the mechanical properties of PLA were not strongly influenced by the molar mass [54]. Thus, besides the possible effect of PLA degradation induced by the fiber moisture, the reduction in the elastic modulus of the PLA-based biocomposites with potato pulp powder has to be ascribed also to poor adhesion between the matrix and the filler. Potato pulp powder is composed mainly of lignocellulosic fibers and starch, which are highly hydrophilic, so poor interactions are expected at the interface with the less hydrophilic PLA. Poor adhesion between the potato pulp powder and the polymeric matrix is also evidenced by the tensile strength, which decreases with PPP percentage. The elongation at break is reduced drastically because the dispersed filler particles act as stress concentrators at the polymer/filler interface by inhibiting the deformation, which leads to reduced ductility of the material.

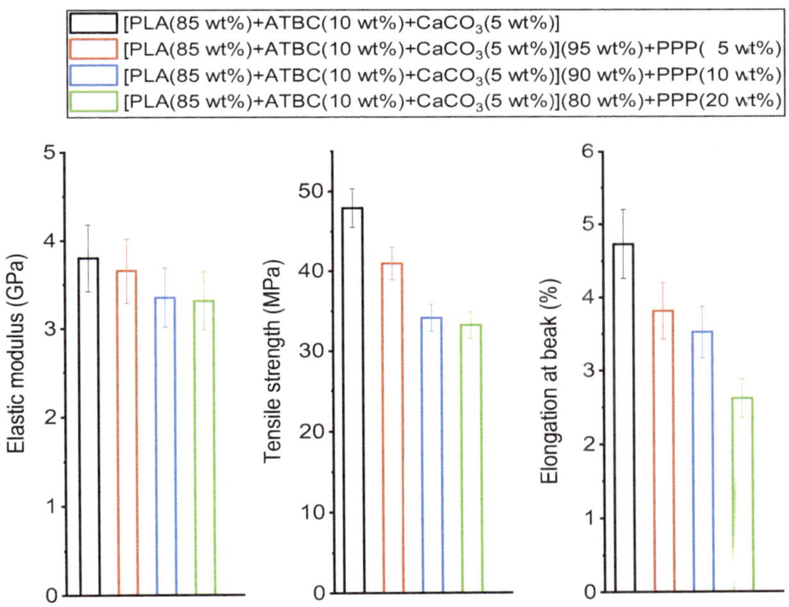

Figure 7. Elastic modulus, tensile strength, and elongation at break of the PLA-based matrix and biocomposites indicated in the legend.

Figure 8 shows the dependence of the modulus of the complex viscosity on the deformation frequency at 175 °C for the PLA-based biocomposites containing potato pulp powder. The experiments were performed from high to low frequencies in a narrow deformation frequency range outside but not far from the Newtonian plateau in order to considerably reduce the measurement times, and consequently, fiber degradation. This also allows the comparison between the viscoelastic behaviors of the different biocomposites. Figure 8 indicates that the viscosity of the biocomposites is smaller than that of the polymeric matrix and decreases with an increase in PPP percentage. For $M_w > 100{,}000$ g/mol, the zero shear viscosity η_o, referring to the Newtonian plateau, can be described by the relationship: $\log \eta_o = \log K + 3.4 \log M_w$ [55,56]. Assuming for the PLA-based matrix that $\log \eta_o \approx 6.1$ (Figure 5), the molar mass reduction detected for the [PLA (85 wt%) + ATBC (10 wt%) + $CaCO_3$ (5 wt%)] (80 wt%) + PPP (20 wt%) biocomposite should produce a decrease in $\log \eta_o$ of about 7%. The decrease in the modulus of the complex viscosity for the biocomposite with 20 wt% of PPP is actually about 16% in the investigated deformation frequency range, which is close to the Newtonian plateau. This finding further confirms that the presence of free volume at the polymer/fiber interfaces due to poor adhesion between potato pulp powder and PLA reduces the viscosity in the biocomposites and favors the flow of the PLA chains.

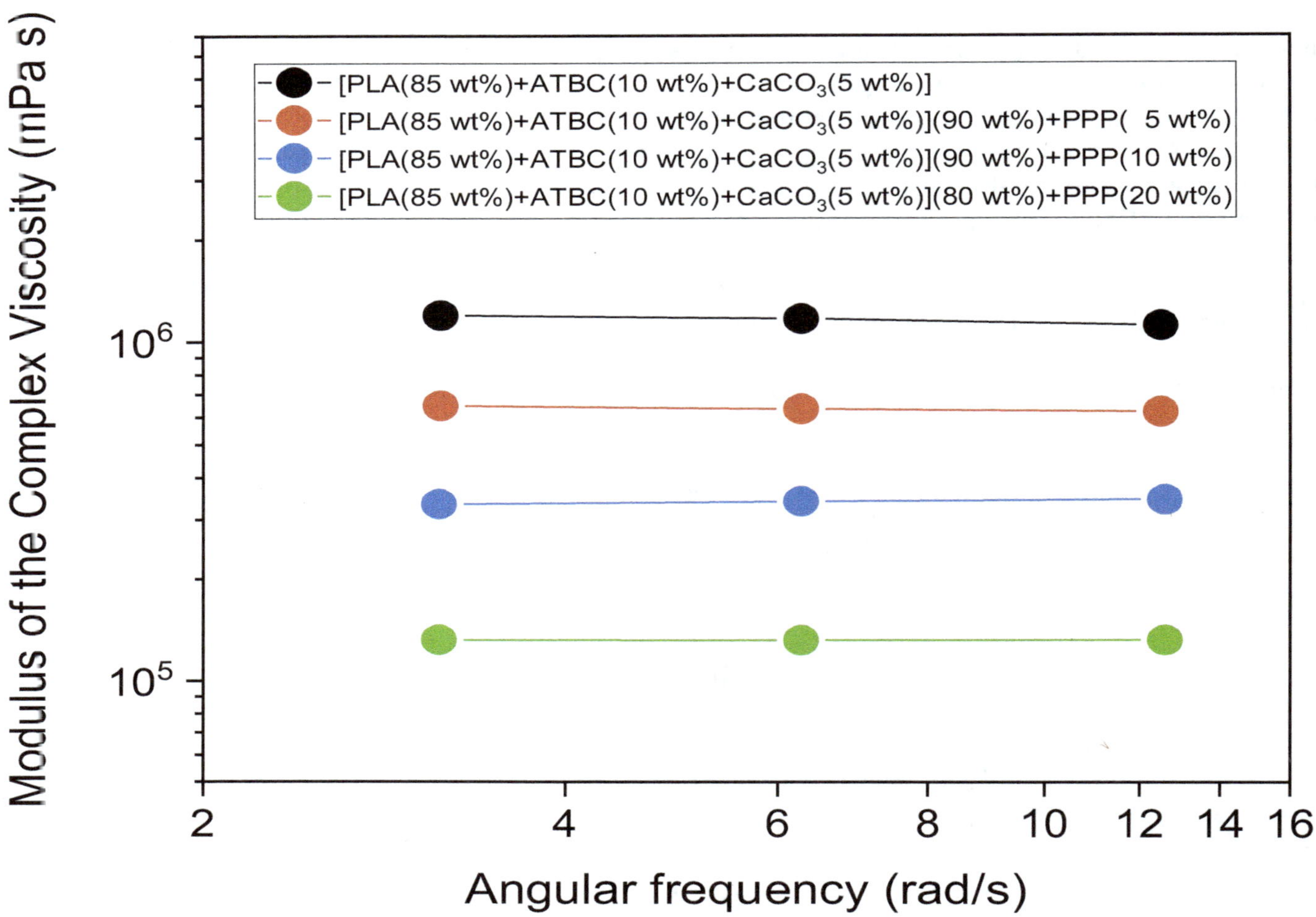

Figure 8. Modulus of the complex viscosity vs. angular frequency at 175 °C for PLA-based matrix and biocomposites indicated in the legend.

2.5. Morphological Properties of the PLA-Based Biocomposites

The morphology of fracture surfaces of the PLA-based biocomposites with PPP from dog-bone specimens was studied by scanning electron microscopy in order to investigate the dispersion of the potato pulp particles and the compatibility between the matrices and the fibers.

Figure 9 shows that the potato pulp particles are well dispersed within the matrix and their distribution is uniform, which means that they were satisfactorily separated during the extrusion process. The micrographs clearly show that the interfacial adhesion between the PLA-based matrix and the potato pulp powder is quite poor because a gap is well evident between the polymeric matrix and the potato pulp particles as indicated by the black arrows. Poor adhesion leads to brittle materials as also proven by the fiber pullout indicated by the red arrow, which illustrates how, in these biocomposites, failure occurs at the matrix/fiber interface. The finding is in agreement with the decrease in the tensile strength discussed above (Figure 7).

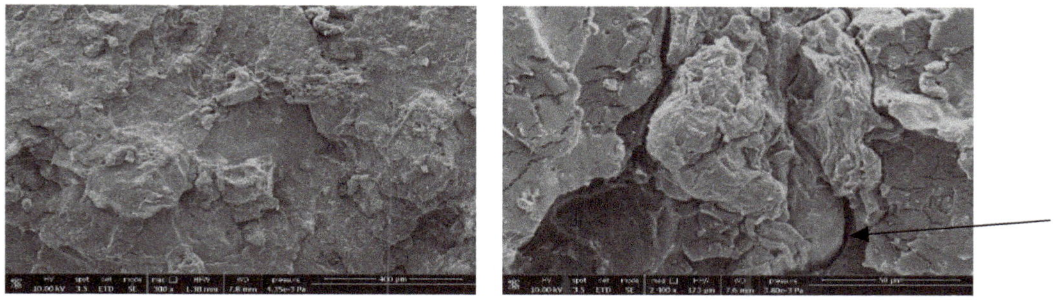

300× 2400×

[PLA (85 wt%) + ATBC (10 wt%) + $CaCO_3$ (5 wt%)] (90 wt%) + PPP (10 wt%)

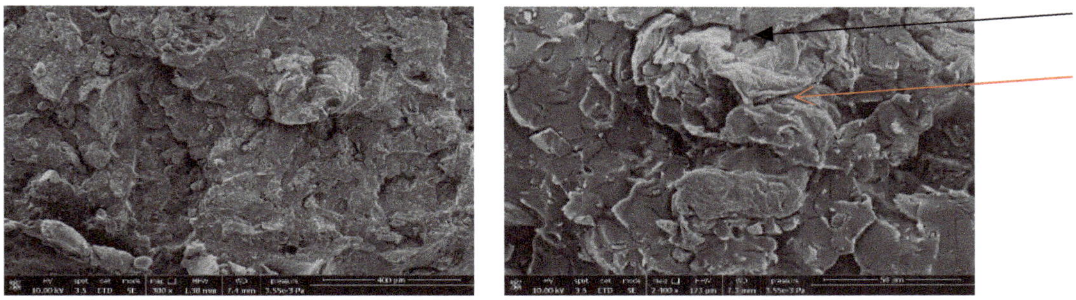

300× 2400×

[PLA (85 wt%) + ATBC (10 wt%) + $CaCO_3$ (5 wt%)] (80 wt%) + PPP (20 wt%)

Figure 9. SEM images of the [PLA (85 wt%) + ATBC (10 wt%) + $CaCO_3$ (5 wt%)] (90 wt%) + PPP (10 wt%) and [PLA (85 wt%) + ATBC (10 wt%) + $CaCO_3$ (5 wt%)] (80 wt%) + PPP (20 wt%) biocomposite at the indicated magnification.

3. Materials and Methods

3.1. Materials

Poly(lactic acid) (PLA), derived from natural resources, was obtained from 2003D NatureWorks (Minnetonka, MN, USA). The product grade was suitable for thermoforming and extrusion processes and contained 3% D-lactic acid units (melt flow index (MFI): 6 g/10 min (210 °C, 2.16 kg); nominal average molar mass: 200,000 g/mol).

The plasticizer acetyl tributyl citrate (ATBC) was purchased from Sigma Aldrich S.R.L (Milan, Italy).

The calcium carbonate ($CaCO_3$) OMYACARB® was an inert filler supplied by the company OMYA (Oftringen, Switzerland). The powder had fine granulometry with particle size distribution centered at 12 μm.

Dried potato pulp powder (PPP) was produced by the company SüdStärke (Schrobenhausen, Germany). The moisture content was about 3 wt%, and the composition of the dry matter was as follows: proteins 7 wt%, starch 25 wt%, cellulose 16 wt%, hemicellulose 7 wt%, lignin 20 wt%, pectin 17 wt%, and ash 5 wt%.

3.2. Composite Preparation

Biocomposites of PLA with potato pulp powder were prepared by adding the filler PPP in different percentages to the polymeric matrix constituted by the bio-based PLA (with a concentration of 85 wt%), the plasticizer ATBC (with a concentration of 10 wt%), and $CaCO_3$ (with a concentration of 5 wt%). For comparison, pure PLA and PLA mixed only with the plasticizer ATBC were also processed in the same way.

Before processing, the PLA and potato pulp powder were dried at a temperature of 60 °C for at least 24 h. The PLA-based matrix and biocomposites were prepared by using a MiniLab II HAAKE Rheomex CTW 5—a co-rotating conical twin-screw extruder. The molten materials were transferred from the mini extruder through a preheated cylinder to a mini injection molder (Thermo Scientific HAAKE MiniJet II), which allows the preparation of dog-bone tensile bar specimens to be used for thermal, mechanical, and rheological characterization. The dimensions of the dog-bone tensile bars were as follows: Width in the larger section—10 mm, width in the narrow section—4.8 mm, thickness—1.35 mm, and length—90 mm. The extruder operating conditions adopted for all the formulations are reported in Table 4.

Table 4. Operating condition use for the extrusion and injection molding process.

Extrusion Temperature (°C)	Screw Speed (rpm)	Cycle Time (s)	Injection Temperature (°C)	Injection Pressure (bar)	Molding Time (s)	Mold Temperature (°C)
180	100	90	180	500	60	90

After preparation, all the samples were stored in a desiccator and analyzed the day after in order to avoid physical ageing effects on the physical properties investigated.

3.3. Composite Characterization

The thermal stability of the potato pulp powder and selected samples was investigated by thermogravimetric analysis (TGA) carried out on about 10 mg of sample by using a SII TG/DTA 7200 EXSTAR Seiko (Minato, Tokyo, Japan) under nitrogen flow (200 mL/min) and at a heating speed of 10 K/min from 50 to 600 °C.

Number-average molar mass (M_n) and weight-average molar mass (M_w) were determined using size exclusion chromatography (SEC) with the Agilent Technologies 1200 Series (Santa Clara, CA, USA) and calculated with the Agilent ChemStation Software. The instrument was equipped with an Agilent degasser, an isocratic HPLC pump, two PLgel 5 μm MiniMIX-D columns conditioned at 35 °C, and an Agilent refractive index (RI) detector. The mobile phase was chloroform ($CHCl_3$) at a flow rate of 0.3 mL/min. The system was calibrated with polystyrene standards in a range from 500 to 3×10^5 g/mol. Samples were dissolved in $CHCl_3$ (2 mg/mL) and filtered through a 0.20 μm syringe filter before analysis.

The morphology of the PPP powder was investigated by scanning electron microscopy (SEM) with an FEI-Quanta 450 ESEM instrument (ThermoFisher, Waltham, MA, USA). The micrographs of samples fractured with liquid nitrogen and etched with gold were collected. Backscattered electrons generated the images, the resolution for which was provided by beam deceleration with a landing energy of 2 kV.

Differential scanning calorimetry (DSC) measurements were performed with a Calorimeter DSC 8500 Perkin Elmer (Waltham, MA, USA) equipped with an IntraCooler III as a refrigerating system. The instrument was calibrated to temperature with high purity standards (indium, naphthalene, cyclohexane) according to the procedure for standard DSC [57]. Energy calibration was performed with indium. Dry nitrogen was used as a purge gas at a rate of 30 mL/min. The as prepared samples were analyzed from −50 to 200 °C at the heating rate of 10 K/min and with a single thermal scan.

Tensile tests on the samples prepared with the injection molder were performed at room temperature at a crosshead speed of 10 mm/min and by means of an INSTRON 5500 R universal testing machine (Instron Corp., Canton, MA, USA) equipped with a 10 kN load cell and interfaced with a computer running the Testworks 4.0 software (MTS Systems Corporation, Eden Prairie, MN, USA). At least five specimens were tested for each sample in accordance to ASTM D 638, and the average values were reported.

Oscillatory shear measurements were performed by means of a Rheometer Anton Paar MCR 302 (Graz, Austria) equipped with parallel plates of 25 mm diameter under nitrogen flow to minimize oxidation and to maintain a dry environment. Frequency sweep experiments were performed at 175 °C at a fixed strain (3%) with a gap of 1 mm, in the linear regime, in order to measure the modulus of the complex viscosity. For the analysis of the PLA-based matrix, the angular frequencies were swept from 0.068 to 628 rad/s with five points per decade. For the analysis of the biocomposites, a narrower angular frequency was covered to reduce the time measurement and avoid degradation of the potato pulp.

4. Conclusions

Potato pulp powder, which is an organic waste from the production and extraction of starch, can be suitable for processing in the melt with the bio-based and biodegradable PLA, presenting no main problems in processing. Due to the low aspect ratio of the potato pulp powder, there is no reinforcing effect on the polymeric matrix, but there is a moderate loss in mechanical properties. The potato pulp power utilized to produce PLA-based biocomposites acts as a filler and not as a reinforcement for the polymeric matrix. The loss in mechanical properties is, however, not dramatically high, because the elastic modulus decreases by only 13% when the PPP content is 20 wt%. The adhesion between the potato pulp powder and the polymeric matrices is poor, as attested by the tensile strength, which appears progressively smaller for the biocomposites with respect to the PLA matrix. The ductility of the biocomposites is also smaller with respect to that of the PLA matrix because the dispersed potato pulp particles act as stress concentrators and promote microcrack formation. Conversely, the lower viscosity of the biocomposites containing PPP is a factor that favors the processing and the molding of these materials.

Notwithstanding, the present PLA-based biocomposites still present properties valuable for practical applications and still meet the technical requirements indicated for rigid packaging production. In addition, the presence of natural fibers promotes and accelerates the biodegradation of the biocomposite material [58].

Finally, the incorporation of about 20 wt% of potato pulp powder, no-food competition biomass, to PLA offers the possibility to markedly reduce the cost of the final products when considering the relatively high cost of PLA. This opens an innovative option for the valorization of an abundant agro-food biomass such as potato pulp and thus meeting circular economy expectations.

Author Contributions: M.C.R. contributed to funding acquisition, supervised the study, and wrote and revised the paper. P.C. contributed to funding acquisition, and co-supervised the study and revised the paper. N.M. prepared the biocomposites, and performed the mechanical and morphological characterization. C.A.M. performed the rheological measurements. S.B. performed the thermal characterization and the SEC measurements. A.S. provided the potato pulp powder and performed the composition analysis. A.L. contributed to funding acquisition and revised the paper.

References

1. Hamad, K.; Kaseem, M.; Ayyoob, M.; Joo, J.; Deri, F. Polylactic acid blends: The future of green, light and tough. *Prog. Polym. Sci.* **2018**, *85*, 83–127. [CrossRef]
2. Bledzki, A.K.; Gassan, J. Composites reinforced with cellulose based fibres. *Prog. Polym. Sci.* **1999**, *24*, 221–274. [CrossRef]
3. Mohanty, A.K.; Misra, M.; Hinrichsen, G. Biofibers, biodegradable polymers and biocomposites: An overview. *Macromol. Mater. Eng.* **2000**, *276–277*, 1–24. [CrossRef]
4. Yu, L.; Dean, K.; Lin, L. Polymer blends and composites from renewable resources. *Prog. Polym. Sci.* **2006**, *31*, 576–602. [CrossRef]
5. Maya Jacob, J.; Sabu, T. Review—Biofibres and biocomposites. *Carbohydr. Polym.* **2008**, *71*, 343–364. [CrossRef]
6. Satyanarayana, K.G.; Arizaga, G.G.C.; Wypych, F. Biodegradable composites based on lignocellulosic fibers—An overview. *Prog. Polym. Sci.* **2009**, *34*, 982–1021. [CrossRef]

7. Faruk, O.; Bledzki, A.K.; Fink, H.P.; Sain, M. Biocomposites reinforced with natural fibers: 2000–2010. *Prog. Polym. Sci.* **2012**, *37*, 1552–1596. [CrossRef]
8. Pandey, J.K.; Ahn, S.H.; Lee, C.S.; Mohanty, A.K.; Misra, M. Recent Advances in the Application of Natural Fiber Based Composites. *Macromol. Mater. Eng.* **2010**, *295*, 975–989. [CrossRef]
9. Nielsen, L.E.; Landel, R.F. *Mechanical Properties of Polymers and Composites*, 2nd ed.; Faulkner, L.L., Ed.; CRC Press: New York, NY, USA, 1994; ISBN 0824761837.
10. Ku, H.; Wang, H.; Pattarachaiyakoop, N.; Trada, M. A review on the tensile properties of natural fiber reinforced polymer composites. *Compos. Part B* **2011**, *42*, 856–873. [CrossRef]
11. Signori, F.; Pelagaggi, M.; Bronco, S.; Righetti, M.C. Amorphous/Crystal and Polymer/Filler Interphases in Biocomposites from Poly(butylene succinate). *Thermochim. Acta* **2012**, *543*, 74–81. [CrossRef]
12. Saheb, D.N.; Jog, J.P. Natural Fiber polymer composites: A Review. *Adv. Polym. Technol.* **1999**, *18*, 351–363. [CrossRef]
13. Li, J.; He, Y.; Inoue, Y. Thermal and mechanical properties of biodegradable blends of poly(L-lactic acid) and lignin. *Polym. Int.* **2003**, *52*, 949–955. [CrossRef]
14. Porras, A.; Maranon, A. Development and characterization of a laminate composite material from polylactic acid (PLA) and woven bamboo fabric. *Compos. Part B* **2012**, *43*, 2782–2788. [CrossRef]
15. Plackett, D.; Andersen, T.L.; Pedersen, W.B.; Nielsen, L. Biodegradable composites based on L-polylactide and jute fibres. *Compos. Part A* **2003**, *63*, 1287–1296. [CrossRef]
16. Masirek, R.; Kulinski, Z.; Chionna, D.; Pirokowska, E.; Pracella, M. Composites of poly(L-lactide) with hemp fibers: Morphology and thermal and mechanical properties. *J. Appl. Polym. Sci.* **2007**, *105*, 255–268. [CrossRef]
17. Johari, A.P.; Mohanty, S.; Kurmvanshi, S.K.; Nayak, S.K. Influence of different treated cellulose fibers on the mechanical and thermal properties of poly(lactic acid). *ACS Sustain. Chem. Eng.* **2016**, *4*, 1619–1629. [CrossRef]
18. Dong, Y.; Ghataura, A.; Takagi, H.; Haroosh, H.J.; Nakagaito, A.N.; Lau, K.T. Polylacticacid (PLA) biocomposites reinforced with coir fibres: Evaluation of mechanical performance and multifunctional properties. *Compos. Part A* **2014**, *63*, 76–84. [CrossRef]
19. Väisänen, T.; Haapala, A.; Lappalainen, R.; Tomppo, L. Utilization of agricultural and forest industry waste and residues in natural fiber-polymer composites: A review. *Waste Manag.* **2016**, *54*, 62–73. [CrossRef]
20. Fritsch, C.; Staebler, A.; Happel, A.; Cubero Márquez, M.A.; Aguiló-Aguayo, I.; Abadias, M.; Gallur, M.; Cigognini, I.M.; Montanari, A.; Jose López, M.; et al. Processing, Valorization and Application of Bio-Waste Derived Compounds from Potato, Tomato, Olive and Cereals: A Review. *Sustainability* **2017**, *9*, 1492. [CrossRef]
21. Schmid, M.; Herbst, C.; Mueller, K.; Staebler, A.; Schlemmer, D.; Coltelli, M.B.; Lazzeri, A. Effect of potato pulp filler on the mechanical properties and water vapor transmission rate of thermoplastic WPI/PBS blends. *Polym. Plast. Technol. Eng.* **2016**, *55*, 510–517. [CrossRef]
22. Mekonnen, T.; Mussone, P.; Khalil, H.; Bressler, D. Progress in bio-based plastics and plasticizing modifications. *J. Mater. Chem. A* **2013**, *1*, 13379–13398. [CrossRef]
23. European Food Safety Authority. Scientific Opinion on Flavouring Group Evaluation 10, Revision 3 (FGE.10Rev3): Aliphatic primary and secondary saturated and unsaturated alcohols, aldehydes, acetals, carboxylic acids and esters containing an additional oxygenated functional group and lactones from chemical groups 9, 13 and 30. *EFSA J.* **2012**, *10*, 25–63. [CrossRef]
24. Yang, H.; Yan, R.; Chen, H.; Lee, D.H.; Zheng, C. Characteristic of hemicellulose, cellulose and lignin pyrolysis. *Fuel* **2007**, *86*, 1781–1788. [CrossRef]
25. Garcia-Perez, M.; Chaala, A.; Yang, J.; Roy, C. Co-pyrolysis of sugarcane bagasse with petroleum residue. Part I: Thermogravimetric analysis. *Fuel* **2001**, *80*, 1245–1258. [CrossRef]
26. Aggarwal, P.; Dollimore, D.; Heon, K. Comparative thermal analysis study of two biopolymers, starch and cellulose. *J. Therm. Anal.* **1997**, *50*, 7–17. [CrossRef]
27. Kamur, P.; Sandeep, K.P.; Alavi, S.; Truong, V.D.; Gorga, R.E. Preparation and characterization of bio-nanocomposite films based on soy protein isolate and montmorillonite using melt extrusion. *J. Food Eng.* **2010**, *100*, 480–489. [CrossRef]
28. McNeill, I.C.; Leiper, H.A. Degradation Studies of Some Polyesters and Polycarbonates-1. Polylactide: General Features of the Degradation Under Programmed Heating Conditions. *Polym. Degrad. Stab.* **1985**, *1*, 264–285. [CrossRef]

29. Awal, A.; Rana, M.; Sain, M. Thermorheological and mechanical properties of cellulose reinforced PLA bio-composites. *Mech. Mater.* **2015**, *80*, 87–95. [CrossRef]
30. Zhao, Y.-Q.; Cheung, H.-Y.; Lau, K.-T.; Xu, C.-L.; Zhao, D.-D.; Li, H.-L. Silkworm silk/poly(lactic acid) biocomposites: Dynamic mechanical, thermal and biodegradable properties. *Polym. Degrad. Stab.* **2010**, *95*, 1978–1987. [CrossRef]
31. Pyda, M.; Bopp, R.C.; Wunderlich, B. Heat capacity of poly(lactic acid). *J. Chem. Thermodyn.* **2004**, *36*, 731–742. [CrossRef]
32. Pan, P.; Inoue, Y. Polymorphism and isomorphism in biodegradable polyesters. *Prog. Polym. Sci.* **2009**, *34*, 605–640. [CrossRef]
33. Di Lorenzo, M.L.; Rubino, P.; Luijkx, R.; Helou, M. Influence of chain structure on crystal polymorphism of poly(lactic acid). Part 1: Effect of optical purity of the monomer. *Colloid Polym. Sci.* **2014**, *292*, 399–409. [CrossRef]
34. Androsch, R.; Schick, C.; Di Lorenzo, M.L. Melting of conformationally disordered crystals α′-phase of poly(L-lactic acid). *Macromol. Chem. Phys.* **2014**, *215*, 1134–1139. [CrossRef]
35. Pan, P.; Kai, W.; Zhu, B.; Dong, T.; Inoue, Y. Polymorphs Crystallization and Multiple Melting Behavior of Poly(L-lactide): Molecular Weight Dependence. *Macromolecules* **2007**, *40*, 6898–6905. [CrossRef]
36. Righetti, M.C.; Laus, M.; Di Lorenzo, M.L. Temperature dependence of the rigid amorphous fraction in poly(ethylene terephthalate). *Eur. Polym. J.* **2014**, *58*, 60–68. [CrossRef]
37. Righetti, M.C.; Di Lorenzo, M.L. Rigid amorphous fraction and multiple melting behavior in poly(butylene terephthalate) and isotactic polystyrene. *J. Therm. Anal Calorim.* **2016**, *126*, 521–530. [CrossRef]
38. Righetti, M.C.; Gazzano, M.; Di Lorenzo, M.L.; Androsch, R. Enthalpy of melting of α′ and α-crystals of poly(L-lactic acid). *Eur. Polym. J.* **2015**, *70*, 215–220. [CrossRef]
39. Saeidlou, S.; Huneault, M.A.; Li, H.; Park, C.B. Poly(lactic acid) crystallization. *Prog. Polym Sci.* **2012**, *37*, 1657–1677. [CrossRef]
40. Baiardo, M.; Frisoni, G.; Scandola, M.; Rimelen, M.; Lips, D.; Ruffieux, K.; Wintermantel, E. Thermal and mechanical properties of plasticized poly(l-lactic acid). *J. Appl. Polym. Sci.* **2003**, *90*, 1731–1738. [CrossRef]
41. Jacobsen, S.; Fritz, H.G. Plasticizing polylactide—The effect of different plasticizers on the mechanical properties. *Polym. Eng. Sci.* **1999**, *39*, 1303–1310. [CrossRef]
42. Kim, H.S.; Park, B.H.; Choi, J.H.; Yoon, J.S. Mechanical Properties and Thermal Stability of Poly(L-lactide)/Calcium Carbonate Composites. *J. Appl. Polym. Sci.* **2008**, *109*, 3087–3092. [CrossRef]
43. Molnár, K.; Móczó1, J.; Murariu, M.; Dubois, P.; Pukánszky, B. Factors affecting the properties of PLA/CaSO4 composites: Homogeneity and interactions. *eXPRESS Polym. Lett.* **2009**, *3*, 49–61. [CrossRef]
44. Lim, L.T.; Auras, R.; Rubino, M. Processing technologies for poly(lactic acid). *Prog. Polym. Sci.* **2008**, *33*, 820–852. [CrossRef]
45. Auras, R.A.; Lim, L.T.; Selke, S.E.M.; Tsuji, H. *Poly(Lactid Acid) Synthesis, Structures, Properties, Processing, and Applications*; John Wiley & Sons: Hoboken, NJ, USA, 2010; ISBN 978-0-470-29366-9.
46. Domenek, S.; Fernandes-Nassar, S.; Ducruet, V. Rheology, Mechanical Properties, and Barrier Properties of Poly(lactic acid). *Adv. Polym. Sci.* **2017**, *279*, 303–342. [CrossRef]
47. Wang, N.; Zhang, X.; Ma, X.; Fang, J. Influence of carbon black on the properties of plasticized poly(lactic acid) composites. *Polym. Degrad. Stab.* **2008**, *93*, 1044–1052. [CrossRef]
48. Xu, Y.Q.; Qu, J.P. Mechanical and Rheological Properties of Epoxidized Soybean Oil Plasticized Poly(lactic acid). *J. Appl. Polym. Sci.* **2009**, *112*, 3185–3191. [CrossRef]
49. Gu, S.Y.; Zou, C.Y.; Zhou, K.; Ren, J. Structure-Rheology Responses of Polylactide/Calcium, Carbonate Composites. *J. Appl. Polym. Sci.* **2009**, *114*, 1648–1655. [CrossRef]
50. Shibata, M. Poly(lactic acid)/Cellulosic Fiber composites. In *Biodegradable Polymer Blends and Composites from Renewable Resources*; John Wiley & Sons, Inc.: Hoboken, NJ, USA, 2009. [CrossRef]
51. Zafeiropoulos, N.E.; Baillie, C.A.; Matthews, F.L. A study of transcrystallinity and its effect on the interface in flax fibre reinforced composite materials. *Composits Part A* **2001**, *32*, 525–543. [CrossRef]
52. Quan, H.; Li, Z.M.; Yang, M.B.; Huang, R. On transcrystallinity in semi-crystalline polymer composites. *Compos. Sci. Technol.* **2005**, *65*, 999–1102. [CrossRef]
53. Middleton, J.C.; Tipton, A.J. Synthetic biodegradable polymers as orthopedic devices. *Biomaterials* **2000**, *21*, 2335–2346. [CrossRef]
54. Maharana, T.; Mohanty, B.; Negi, Y.S. Melt-solid polycondensation of lactic acid and its biodegradability. *Prog. Polym. Sci.* **2009**, *34*, 99–124. [CrossRef]

55. Palade, L.-I.; Lehermeier, H.J.; Dorgan, J.R. Melt Rheology of High L-Content Poly(lactic acid). *Macromolecules* **2001**, *34*, 1384–1390. [CrossRef]
56. Othman, N.; Acosta-Ramirez, A.; Mehrkhodavandi, P.; Dorgan, J.R.; Hatzikiriakos, S.G. Solution and melt viscoelastic properties of controlled microstructure poly(lactide). *J. Rheol.* **2011**, *55*, 987–1004. [CrossRef]
57. Sarge, S.M.; Hemminger, W.; Gmelin, E.; Höhne, G.W.H.; Cammenga, H.K.; Eysel, W. Metrologically based procedures for the temperature, heat and heat flow rate calibration of DSC. *J. Therm. Anal.* **1997**, *49*, 1125–1134. [CrossRef]
58. John, M.J. Environmental degradation in biocomposites. In *Biocomposites for High-Performance Applications*; Woodhead Publishing: Cambridge, UK, 2017. [CrossRef]

6

Lignin: A Biopolymer from Forestry Biomass for Biocomposites and 3D Printing

Mihaela Tanase-Opedal [1], Eduardo Espinosa [2], Alejandro Rodríguez [2] and Gary Chinga-Carrasco [1,*]

1 RISE PFI, Høgskoleringen 6b, 7491 Trondheim, Norway; mihaela.tanase@rise-pfi.no
2 Chemical Engineering Department, Faculty of Science, Universidad de Córdoba, Building Marie-Curie, Campus de Rabanales, 14014 Córdoba, Spain; eduardo.espinosa@uco.es (E.E.); a.rodriguez@uco.es (A.R.)
* Correspondence: gary.chinga.carrasco@rise-pfi.no

Abstract: Biopolymers from forestry biomass are promising for the sustainable development of new biobased materials. As such, lignin and fiber-based biocomposites are plausible renewable alternatives to petrochemical-based products. In this study, we have obtained lignin from Spruce biomass through a soda pulping process. The lignin was used for manufacturing biocomposite filaments containing 20% and 40% lignin and using polylactic acid (PLA) as matrix material. Dogbones for mechanical testing were 3D printed by fused deposition modelling. The lignin and the corresponding biocomposites were characterized in detail, including thermo-gravimetric analysis (TGA), differential scanning calorimetry (DSC), Fourier transform infrared (FTIR) spectroscopy, X-ray diffraction analysis (XRD), antioxidant capacity, mechanical properties, and scanning electron microscopy (SEM). Although lignin led to a reduction of the tensile strength and modulus, the reduction could be counteracted to some extent by adjusting the 3D printing temperature. The results showed that lignin acted as a nucleating agent and thus led to further crystallization of PLA. The radical scavenging activity of the biocomposites increased to roughly 50% antioxidant potential/cm^2, for the biocomposite containing 40 wt % lignin. The results demonstrate the potential of lignin as a component in biocomposite materials, which we show are adequate for 3D printing operations.

Keywords: lignin; polylactic acid (PLA); 3D printing; biocomposites; biopolymers

---

1. Introduction

Environmental pollution and the increasing awareness of limited resources have been a major driver in finding renewable alternatives to replace traditional fossil-based plastics, with bio-based materials (for ex., biocomposites) derived from carbon-neutral feedstocks [1]. Natural fibres have various advantages, including good mechanical properties, no emission of toxic substances, and reduction in cost [2]. Additionally, natural fiber-reinforced biocomposites have attracted increasing attention due to several beneficial properties, e.g., low-cost, good mechanical properties, and lightweight.

Poly (lactic) acid (PLA) is a promising biopolymer which has been introduced commercially as a renewable alternative to fossil-based polymers [3,4]. A number of promising PLA-based products are presently found commercially, such as automotive parts [5]. PLA provides good mechanical properties and relatively easy melt-processability. However, PLA is relatively expensive and some properties (e.g., brittleness) limit the utilization of PLA in some applications [6]. Therefore, additional polymers can be used in PLA blends in order to tailor the properties of the final products [7]. Moreover, some alternatives have been suggested to improve e.g. the mechanical and thermal properties, including the addition of fibre or filler materials [8] and cellulose nanofibres [9].

Lignin as the second most abundant renewable bio-resource, next to cellulose, is considered as a waste product in several industrial processes. Attempts for lignin valorisation have been published in a vast number of papers and reviews over the last years [10–12].

The notable properties of lignin, such as highly abundance, low-cost and biodegradability, high carbon content, high aromaticity, and reinforcing capability make it a good candidate as a potential component for biocomposites [13]. Each year, over 50 million tons of lignin are produced worldwide as a side product from biorefineries, of which 98% are burned to generate energy. Only 2% of the lignin has been used for other purposes, mainly in applications such as dispersants, adhesives, and fillers [14–17]. The commercial lignin is mainly lignosulfonates from sulphite pulp mills (about 1 mill. tons/year) and less than 100,000 tons/year of kraft lignin [18]. Since lignin has lower energy content than coal and because lignin-rich side streams are wet, the energy value is limited to 50 US Dollar/dry ton [19]. Thus, cost-efficient valorisation of lignin into value-added products offers a significant opportunity to enhance operational efficiency and generates additional revenue so that the production of bioethanol or other products from the hydrolysed carbohydrates becomes more competitive. However, lignin properties (for ex., high heterogeneity and complex structure), make it difficult to predict how the lignin loading will affect the properties of a given biocomposite. Recent reviews summarized the research done on lignin-reinforced thermoplastic biocomposites [20,21].

Lignin can be used as a component in biocomposites, with or without modification, depending on the target application. Gordobil et al. [22] acetylated kraft lignin to improve the affinity with PLA. However, incorporation of kraft lignin decreased the tensile strength properties of the PLA with increasing lignin loading of 10 wt % and 20 wt %. Furthermore, when acetylated lignin was blended with a thermoplastic the tensile strength was improved [21,22]. Without modification, lignin can be directly incorporated into a polymeric matrix, such as UV-light stabilizer, antioxidant, flame retardant, plasticizer, and flow enhancer to reduce production cost, reduce plastic, and potentially improve material properties [23–26].

Lignin can also be used as a coupling agent in natural fibre biocomposites. Lignin can act as a compatibilizer between the hydrophilic fibres and hydrophobic matrix polymer, thus strengthening the fibre matrix interface [27–29]. Lignin treatments of hemp fibres [28] and flax fibres [27] have been shown to improve compatibility between fibres and the thermoset matrix, thus also improving the mechanical properties of the biocomposites. Graupner [29] reported increased tensile properties of compression-moulded PLA-cotton composites when the fibres were treated with lignin.

In our study, lignin was extracted by a soda process. Soda pulping pre-treatment is similar to that of alkaline pulping process, which uses alkali (e.g., NaOH, $Ca(OH)_2$) to solubilize or depolymerize lignin, and make lignin extractable from biomass matrix [30]. Soda pre-treatment disrupts the lignin structure and breaks the aryl-ether, ester, and C-C linkages among the lignin and hemicelluloses and hence open up the lignocellulose structure [31]. This process has mainly been studied on herbaceous biomass and to some extent hardwood [12,32,33]. The main difference of this process compared to the chemical pulping process is the moderate treatment severity, separating lignin with low condensation structure [19]. The soda process for obtaining lignin has some essential advantages from the environmental point of view in comparison with the sulphate process, namely, the soda cooking liquor has a lower content of low molecular products of wood degradation, which get to waste water, and the formation of the unpleasant sulphur organic compound odorants is prevented [34]. Soda lignin has found application in multiple areas: the production of phenolic resins [35,36], animal nutrition [37], dispersants [38], and synthesis of polymers [39]. A major difference between soda lignin compared with kraft lignin and lignosulphonates is that soda lignin is sulphur-free, without odour and with a chemical composition closed to pure lignin [38,39]. Additionally, soda lignin derived from non-wood plants contains a larger fraction of carboxyl groups and p-hydroxyl units [39].

Three-dimensional (3D) printing has evolved rapidly in recent years. This technology allows the production of unique, complex and customized structures by digitized and computer assisted processes, reducing production time and costs [40]. In addition, less waste in production and lower chemical consumption is required by this technique in comparison with the traditional processing manufacturing. Fused deposition modelling (FDM) is one of the most used 3D printing technologies, which consists in the melting of thermoplastic materials at high temperature, which are solidified when cooling. Currently, there is great interest in the use of biomass and biomass components for use in 3D printing by FDM. However, unlike petroleum-derived thermoplastic compounds, lignocellulosic components are difficult to melt for the extrusion and injection moulding processes. Therefore, the development of new materials from biomass suitable for 3D printing is a challenge [41–43]. The production of pure lignin composites is limited by its high thermal transition temperature and high flow resistance [40]. For this reason, lignin is mixed with other polymers that favour its melting behaviour and flow. Recently, organosolv hardwood lignin [40] and kraft softwood lignin [44] have been applied to manufacture filaments for FDM, based on acrylonitrile-butadiene-styrene and PLA polymers, respectively.

The purpose of this study was to demonstrate the suitability of PLA/soda lignin biocomposite filament for 3D printing. A motivation for selecting soda lignin is that it is sulphur-free. Soda lignin was thus expected to reduce the typical smell that is experienced when melt-processing biocompounds containing kraft lignin or lignosulfonates. Thus, samples with varying PLA/soda lignin weight ratios were manufactured and the mechanical (tensile testing), thermal (TGA, DSC analysis), morphological (SEM), FTIR, X-ray diffraction, and antioxidant properties were assessed.

2. Materials and Methods

2.1. *Raw Materials*

The lignin used in this study is a softwood lignin extracted from cooking liquor, using a soda cooking process of Norway spruce chips, collected from Norske Skog, Skogn, Norway. The PLA used in this study was a commercial grade for 3D printing (Ingeo PLA 3D850, NatureWorks LCC, Minnetonka, MN, USA).

2.2. *Lignin Extraction*

Lignin was extracted from Norway spruce by soda process using a MK circulation reactor. In this method, 400 g dried chips were cooked with cooking liquor which comprised 30% NaOH (30 g NaOH/100 g dried chips). The liquid wood ratio was 7.5:1. The temperature was increased slowly in order to get a good impregnation of the chips with the cooking liquor. Once the reactor reached the operating temperature of 180 °C, it was maintained for 100 min. After cooling, the cooking liquor was removed from the bottom of the reactor.

2.3. *Lignin Precipitation*

Sulphuric acid was slowly added to the cooking liquor under agitation until final pH of 2.5. A color change from black to brown was observed at pH 5.5. In addition, a viscosity change was observed, the liquid was more viscose at lower pH. These changes occur due to the initial stage of the lignin precipitation. The mixture was then centrifuged (3500 rpm, 10 min) to recover the lignin. The lignin was repeatedly washed with water, then dried in an oven at 105 °C overnight. Volatile organic compounds (VOC) were removed after drying in a 2L Parr reactor (N4622, Parr Instruments, Illinois, USA).

2.4. *Filaments and 3D Printing*

Neat PLA (100%) and blends of PLA and lignin (20 wt % and 40 wt %) were extruded through a Noztek Xcalibur (Table 1). The temperatures in the three heating chambers were 200 °C, 205 °C, and 210 °C. The speed of the screw extruder was set to 15 mm/s. The target diameter of the filaments was 1.75 mm. The filaments were extruded twice in order to improve the mixing of the PLA and lignin.

Table 1. Mechanical properties of PLA and PLA/Lignin biocomposites.

3D Printed Sample	E_t (MPA)	σ_M (MPa)	ε_M %
PLA	2890 ± 14.14	58.45 ± 0.55	2.45 ± 0.10
PLA + 20%Lignin	2460 ± 155.56	39.35 ± 1.05	1.8 ± 0.10
PLA + 40%Lignin (205 °C)	1955 ± 19.92	32 ± 2.10	1.8 ± 0.20
PLA + 40%Lignin (215 °C)	2695 ± 148.49	45.65 ± 0.05	1.9 ± 0.08
PLA + 40%Lignin (230 °C)	1930 ± 183.85	29.25 ± 1.35	1.65 ± 0.10

E_t (tensile elastic modulus); σ_M (tensile strength); ε_M (elongation at tensile).

An original Prusa i3 MK3S was used for 3D printing by fused deposition modeling (FDM), using a 0.4 mm nozzle and a printing speed of 35 mm/s. Dogbone samples (length 63 mm, width 3 mm) were printed for mechanical testing. Three sets of dogbones were printed with varying printing temperatures of 205, 215, and 230 °C. The printing direction was 45°. In addition, a smartphone protective case was printed as a demonstration.

2.5. Characterisation

The lignin and carbohydrates analyses were performed according to the standard procedures of NREL (NREL/TP–510–42623). The lignin samples were hydrolysed in two steps using sulfuric acid: first step hydrolysis was done with 72 wt % H_2SO_4, 30 °C for 60 min, followed by the second step with 4 wt % H_2SO_4 at 121 °C for 1 h). The resulting supernatant was filtered (0.2uL) and the filtrates were analysed by High Performance Anion Exchange Chromatography (HPAEC, ICS-5000, Dionex, CarboPac PA1 4 × 250 mm column), for carbohydrates. Klasson lignin (acid insoluble lignin) contents were determined gravimetrically, while acid soluble lignin was determined by using TAPPI method (TAPPI UM 250). The absorbance was measured with UV-VIS spectrometry at 205 nm, ε of 110. Carbohydrates were detected on a pulsed amperometry detector (PAD). All carbohydrate contents are reported as anhydrosugars. Purity of lignin samples was calculated from the sum of the ash and sugar results.

2.5.1. Thermo-Gravimetric Analysis (TGA) and Differential Scanning Calorimetry (DSC)

TGA was used to determine the thermal stability, decomposition temperature, and char yield for soda lignin, PLA, and different blends of lignin/PLA. The analyses were performed using a Netzsch Jupiter F3 equipment, operating in nitrogen environment. Samples for TGA for each measurement were maintained at 14 ± 5 mg and scans were preformed from 30 °C to 800 °C with the heating rate of 10 °C/min to observe thermal degradation and stability of lignin, PLA, and the corresponding biocomposites. DSC was performed to measure the glass transition temperature (Tg) of soda lignin.

2.5.2. Scanning Electron Microscopy (SEM)

The fracture surface of the dogbones after mechanical testing was sputtered with a layer of gold to make it conductive under the electron beam. Scanning electron microscopy (SEM, Hitachi SU3500, Hitachi High-Technologies Co., Tokyo, Japan) was performed in secondary electron imaging mode using an acceleration voltage of 5 kV and a working distance of 5–6 mm.

2.5.3. Mechanical Testing

Test specimens obtained from the 3D printing process were used for tensile testing of the biocomposites.

The 3D printed dogbones were mechanically tested with a Zwick Roell Proline (Zwick GmbH & Co. KG, Ulm, Germany) and a load cell of 2.5 kN. Four specimens of each series were tested. The speed and the grip distance were 20 mm/min and 50 mm, respectively.

2.5.4. Fourier Transform Infrared (FTIR) Spectroscopy

FTIR spectra were collected using a spectrometer FTIR-ATR Perkin Elmer Spectrum (Perkin Elmer, Waltham, United States). Two single spectra were collected in the wavelength range from 4000 to 450 cm^{-1} with a resolution of 4 cm^{-1} and a total of 40 scans.

2.5.5. X-ray Diffraction Analysis (XRD)

X-ray diffraction analysis of the test specimens was carried out in a Bruker D8 Discover Instrument (Bruker Corporation, Karlsruhe, Germany) with a monochromatic source (CuKα1) over an angular range of 5–50° at a scan speed of 1.56°/min.

2.5.6. Antioxidant Activity

The 2,2′-Azino-Bis-3-Ethylbenzothiazoline-6-Sulfonic Acid (ABTS) assay was used to measure the antioxidant activity of the biocomposites. A radical solution (7 mM ABTS and 2.45 mM potassium persulphate) was prepared and left in dark during 14–16 h before testing. The radical solution was adjusted to an absorbance of 0.70 ± 0.02 at 734 nm diluting with ethanol. A specimen of 0.5–1 cm^2 of the biocomposites was added to a 4 mL of radical solution and the absorbance of the solution was measured at 734 nm using ethanol as blank. The absorbance was measured after 6 min. The antioxidant activity was determined according to the following equation [45]:

$$AOP\ (\%) = \frac{A_{734,ABTS6'} - A_{734,sample6'}}{A_{734,ABTS0'}} \times 100 \quad (1)$$

where $A_{734,ABTS6'}$ is the absorbance at 734 nm of the radical solution after 6 min, $A_{734,sample6'}$ the absorbance at 734 nm of the sample after 6 min, and $A_{734,ABTS0'}$ the absorbance at 734 nm of the radical solution before the 6 min. Because the radical scavenging activity was performed in the surface of the biocomposites, the antioxidant activity was related to the surface exposed to the radical solution, expressed as antioxidant potential, (AOP%/cm^2).

3. Results and Discussion

3.1. Lignin Composition

The yield of lignin extraction was estimated to be 25%, based on Klasson lignin. The chemical composition shows that lignin is composed of 82% acid insoluble lignin, 6% acid soluble lignin, 1.6% carbohydrates, and ash. Surprisingly, glucomannan, the predominant constituent of softwood hemicellulose, was not found in the soda lignin. However, xylan was found in relatively high amounts. This shows that cleavage of the lignin–carbohydrates complexes is less complete for the soda process applied in this study. These carbohydrates may originate from lignin–carbohydrates complexes or from carbohydrates that are trapped during the lignin precipitation step which end up non-covalently bonds in the lignin after drying [46]. Furthermore, the high ash content in soda lignin may partly originate from the salts resulted after neutralization of cooking liquor during precipitation. These results are in accordance with literature results [47]. The chemical composition of lignin and ash content depends on the feedstock as well as on the isolation process. In general, alkaline lignin contains more residual sugar than other type of lignins [47].

The moisture content of isolated lignin was less than 2.5%. Sameni et al. [48] reported that lignin with higher impurities, containing hydrophilic compounds, led to more moisture content.

3.2. TGA

TGA curves reveal the residual mass of materials with respect to the temperature of thermal degradation and was used to assess the thermal stability of the both polymers PLA and Lignin but also of the biocomposites, PLA/20%Lignin and PLA/40%Lignin. The different chemical bonds present in the lignin molecular structure, leads to a range of degradation temperatures, extending from 100 to 800 °C [49]. The results show that 42 wt % of lignin sample still remained at 800 °C. This is due to the formation of highly condensed aromatic structures which have the ability to form char (Figure 1). Degradation of lignin sample can be divided into three stages [36]. Firstly, an initial weight loss between 50–120 °C, due to water evaporation. Secondly, above 220 °C degradation of carbohydrates occurs, which are converted to gasses such as CO, CO_2, and CH_4 [50]. The last stage occurs to around 420 °C, and then continues to lose mass at a slower rate, leading to the formation of gaseous products and phenolics, alcohols and aldehyde acids [50]. PLA, PLA/20%Lignin, and PLA/40%Lignin present a similar thermal degradation behavior. They show a faster thermal degradation between 340–400 °C. However, the residual mass at 800 °C is higher for biocomposites with lignin due to the carbonaceous composition of the lignin. With respect to the starting temperature of decomposition, the lignin shows a faster degradation (326 °C) compared to neat PLA (348 °C). As expected, the biocomposites resulting from the mixtures PLA/Lignin present an intermediate degradation start temperature (347 °C and 339 °C for 20% and 40% lignin content, respectively).

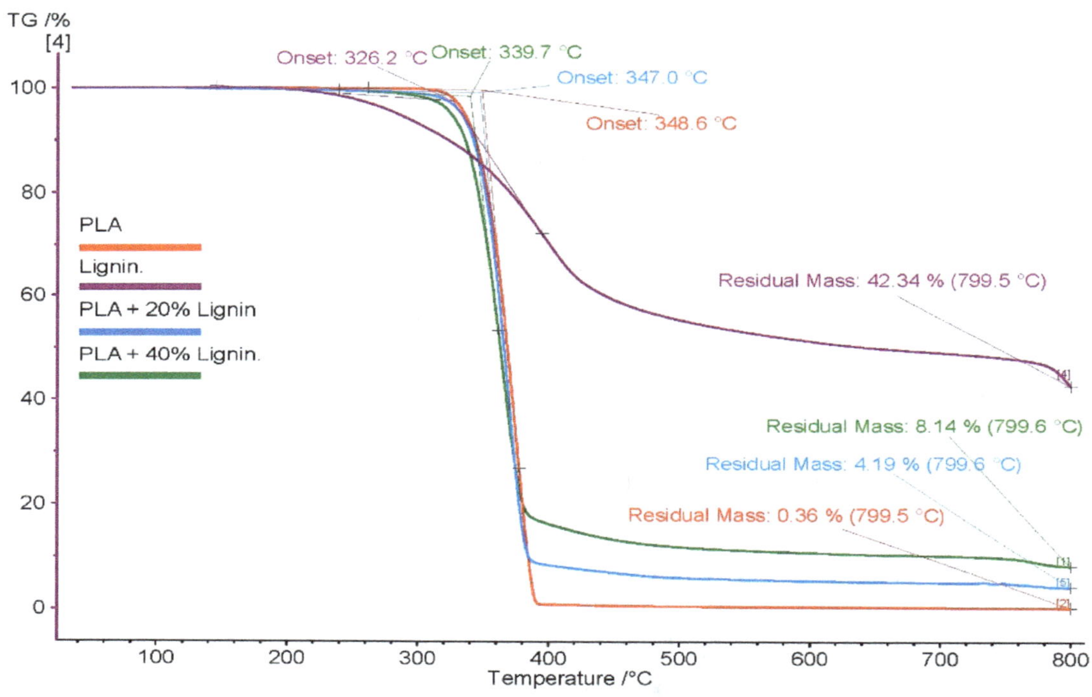

Figure 1. Thermo-gravimetric analysis (TGA) plot of Lignin, polylactic acid (PLA), and PLA/Lignin biocomposites.

3.3. DSC

The reaction energy was measured on soda lignin, polymer and blends of lignin and polymer to assess the thermal degradation. Enthalpy measurements were similar for PLA/40%Lignin (27 J/g) compared with neat PLA (31 J/g). At higher a temperature of 376 °C (PLA) the enthalpy of 628 J/g was similar with PLA/20%Lignin and higher than PLA/40%Lignin (362 °C, 409 J/g), see Figure 2. No significant variations in Tm (temperature maximum degradation) were found with the addition of lignin to PLA matrix. The temperature of maximum degradation occurred between 360–400 °C for all biocomposite samples. Decomposition of aromatic rings is expected above 500 °C [36]. The glass transition temperature (Tg) for soda lignin in our study was found to be at 109 °C. Additionally, the Tg value can be correlated with the molecular weight of lignin [51]. This value is in accordance with the values reported in the literature. The Tg for neat PLA is found to be at 71 °C in this study. The glass transition temperature of the PLA/Lignin biocomposites showed significant shift of the Tg from 71 °C towards lower temperature, Tg of 59 °C for PLA/Lignin 20%. This can be explained by different molecular factors such as interchain hydrogen bonding, crosslinking density, rigid phenyl groups, and molecular mass [52].

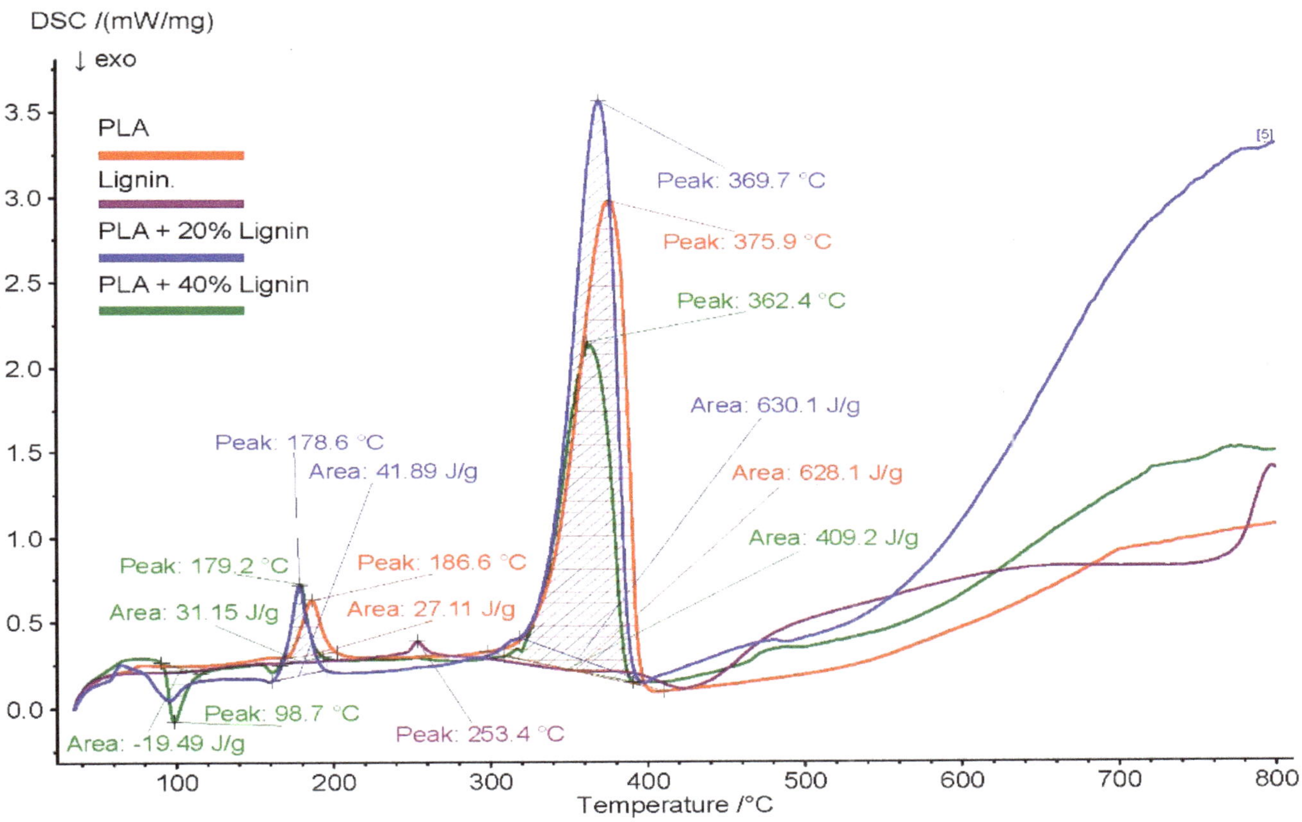

Figure 2. Differential scanning calorimetry (DSC) thermographs of Lignin, PLA, and PLA/lignin biocomposites.

3.4. Mechanical Properties

The filaments (PLA, PLA/20%Lignin, and PLA/40%Lignin) were used to 3D print dogbones for further characterization (Figure 3). The stress–strain curves of the biocomposites with unmodified lignin showed that the mechanical strength decreased when the lignin content increased (Figure 4), which is in accordance with previous studies [21,22]. This is most probably due to a low fusing between the printing layers, especially in biocomposites with 40 wt % of lignin. However, when the printing temperature was increased to 215 °C, the biocomposite revealed a relative increase of the mechanical properties. This was presumptively due to an improved adhesion of the 3D printed layers. The strength–strain curves also show that lignin led to a more fragile biocomposite that resists less deformation before rupture. Consequently, the elastic modulus decreased by 25%–32%, compared to the neat PLA sample. The addition of low content of lignin in PLA biocomposites presented an improvement of the ductility. However, lignin content above 10% has caused a decrease in the plasticity of biocomposites, either for acetylated or unmodified lignin [53].

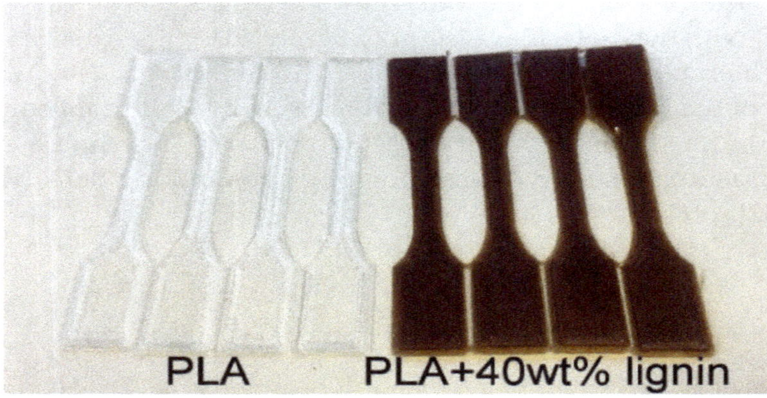

Figure 3. 3D printed dogbones for mechanical testing.

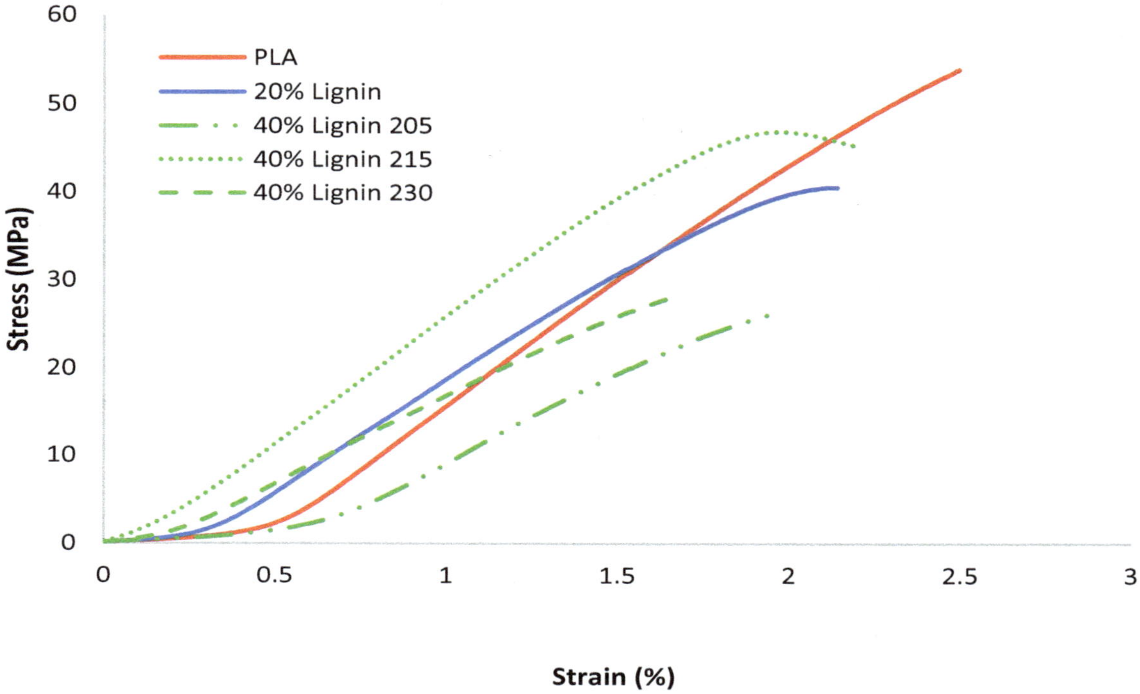

Figure 4. Stress–strain curves for the different biocomposites

3.5. Scanning Electron Microscopy (SEM)

The SEM images (Figure 5) of the fractured surface of the biocomposites provided an insight into the bonding interaction between lignin and PLA. The analysis of the PLA/40% Lignin sample revealed a 3D printed structure where the printed threads are clearly visible, thus confirming a poor inter-layer adhesion and a corresponding low mechanical performance (Table 1). In order to find a suitable printing temperature of the lignin-containing biocomposite and thus increase the inter-layer bonding, two additional temperatures were tested during the 3D printing process (215 °C and 230 °C). The results revealed that a suitable printing temperature for the biocomposite filaments developed in this study was 215 °C (Table 1 and Figure 4). The SEM pictures showed the improvement of the bonding of the printed layers, which favored the mechanical properties of the samples. However, increasing the printing temperature to 230 °C led to a decrease of the tensile strength and modulus. This was potentially caused by the degradation of the carbohydrates in the lignin fraction, which usually occurs over 220 °C, becoming volatile gases and presumptively creating microstructures within the polymeric matrix. Our results are in accordance with literature results of Thakur et al. [20] and Watkins et al. [50].

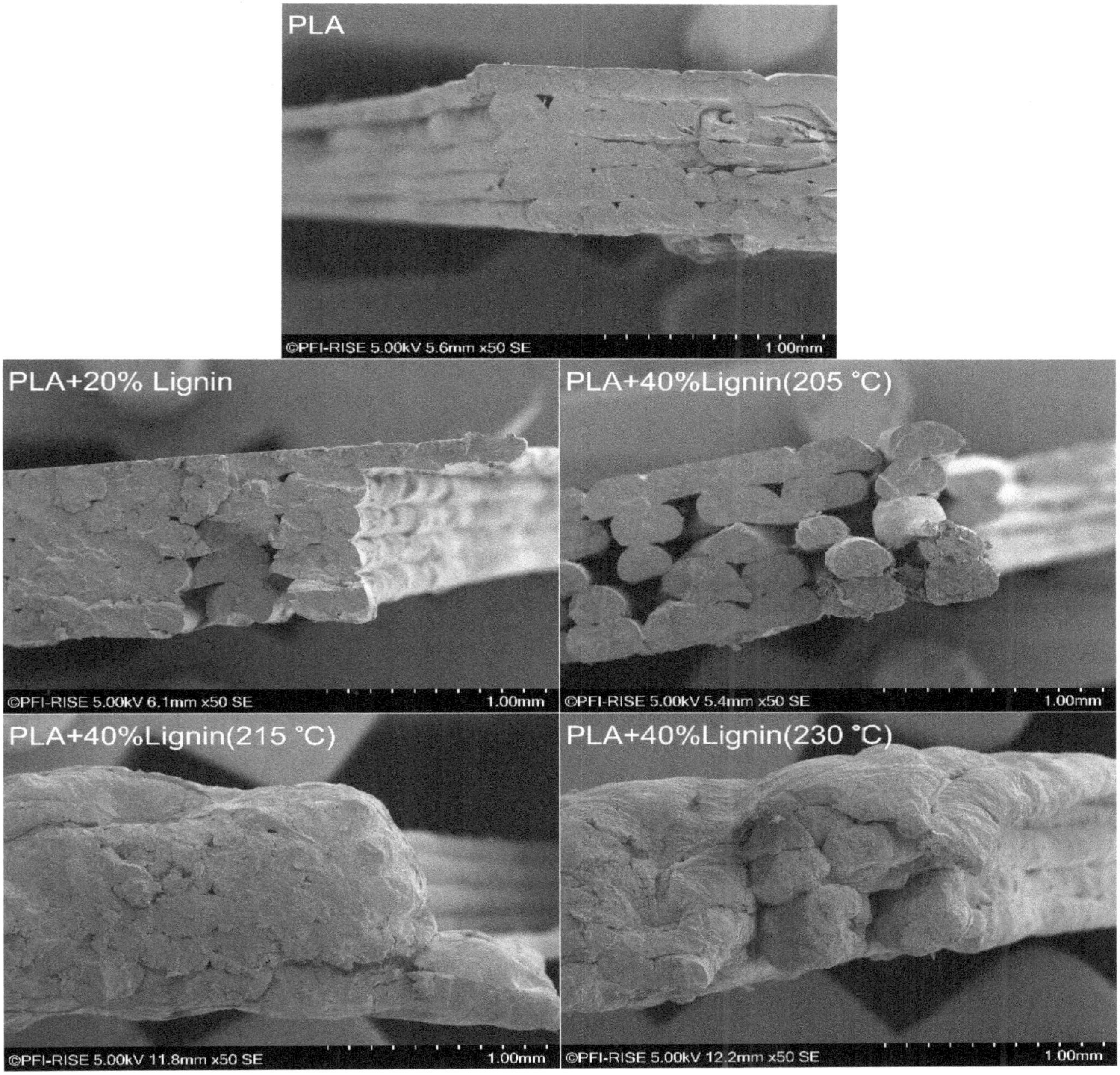

Figure 5. Scanning electron microscopy (SEM) analysis of the fracture surface of tensile tested dogbones.

3.6. FT-IR Spectra and XRD Analysis

FT-IR spectroscopy was applied to assess the functional groups of PLA/Lignin biocomposites at different temperatures. Generally, the curves of PLA/lignin showed similar bands with increased emissivity compared to neat PLA (Figure 6). Neat PLA showed peaks around 2995 cm^{-1} and 2930 cm^{-1}, which are associated to the asymmetric and symmetric stretching vibration of CH_3 group. The intense peak at 1749 cm^{-1} is attributed to the C=O stretching vibration [54]. The peak at 1450 cm^{-1} corresponds to CH_3 anti-symmetric bending vibration. The peaks at 1385 cm^{-1}, 1360 cm^{-1}, 1316 cm^{-1}, and 1300 cm^{-1} are associated to the deformation, symmetric, and bending mode of the CH group, respectively. The peaks at 1182 cm^{-1}, 1084 cm^{-1}, and 1038 cm^{-1} are attributed to C-O-C stretching vibrations [55]. This peak is obviously higher in the PLA/40%Lignin curve, which indicates that addition of lignin led to a higher content of hydroxy groups. In addition, the biocomposites containing lignin showed a small peak at 1510 cm^{-1} due to the C=C groups of the aromatic rings of lignin.

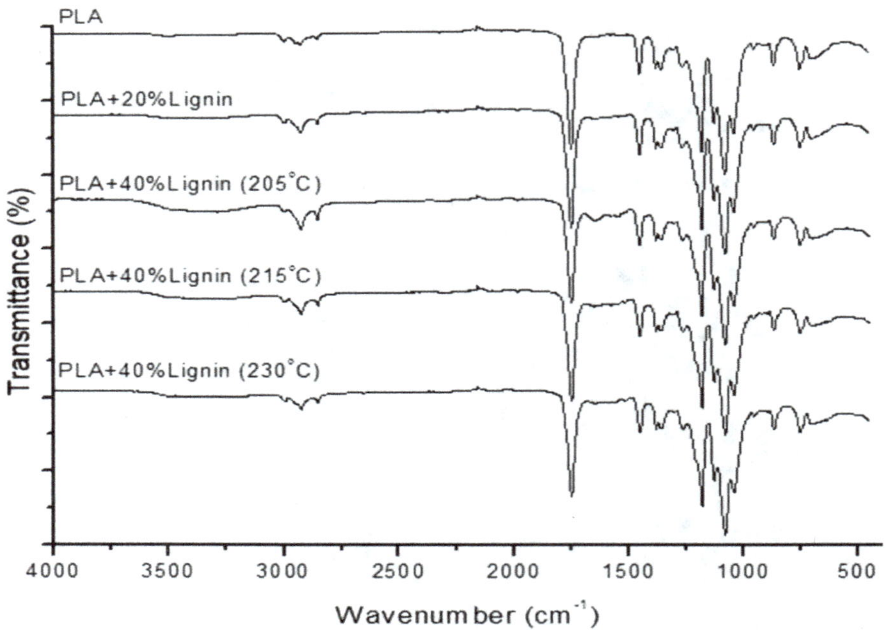

Figure 6. FT-IR spectra of PLA and PLA/lignin biocomposites.

X-ray analysis revealed a change in crystallinity as the lignin was included in the formulation (Figure 7). PLA exhibits a broad peak at 2θ degrees = 10°–25° associated with the semicrystalline nature of PLA. The appearance of peaks at 2θ = 32° and 34.5° in lignin-containing biocomposites indicated further crystallization of PLA, due to the action of lignin as nucleating agent [56].

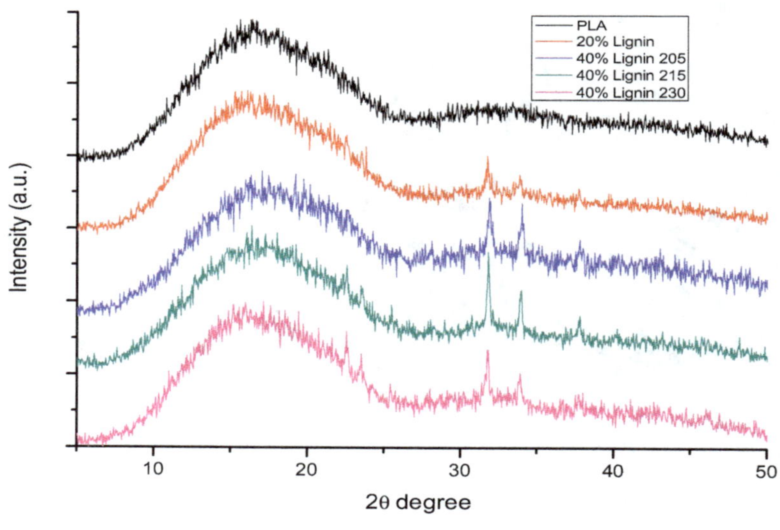

Figure 7. X-ray diffraction analysis (XRD) patterns of PLA and PLA/Lignin biocomposites.

3.7. Antioxidant Properties

Antioxidant capacity of the biocomposites was measured and expressed as radical scavenging activity (RSA). PLA shows a low antioxidant (activity only associated with the surface interaction with the radical ABTS) (Figure 8). However, the lignin-containing biocomposites show a significantly higher radical scavenging activity (~50%) due to the antioxidant activity of lignin. Although, the biocomposites containing 40% lignin have a slightly higher RSA compared to the PLA/20% Lignin, the differences are not significant. The reaction of ABTS with the biocomposite material occurs mostly on the surface of the specimens, which may explain this behavior. The antioxidant activity of the lignin has been reported previously in several studies [45,57,58], including the blending of low amounts of kraft lignin (0.5 wt %–3 wt %) in PLA for potential biomedical applications [44]. Materials with high antioxidant activity are highly demanded for application in food packaging and biomedicine. A wide variety of antioxidant compounds are described in literature (resveratrol, curcumin, ascorbic acid, carotenoids, etc.), however, these compounds are generally expensive. For this reason, the use of lignin in biocomposites is proposed as a low-cost option to produce materials with high antioxidant capacity.

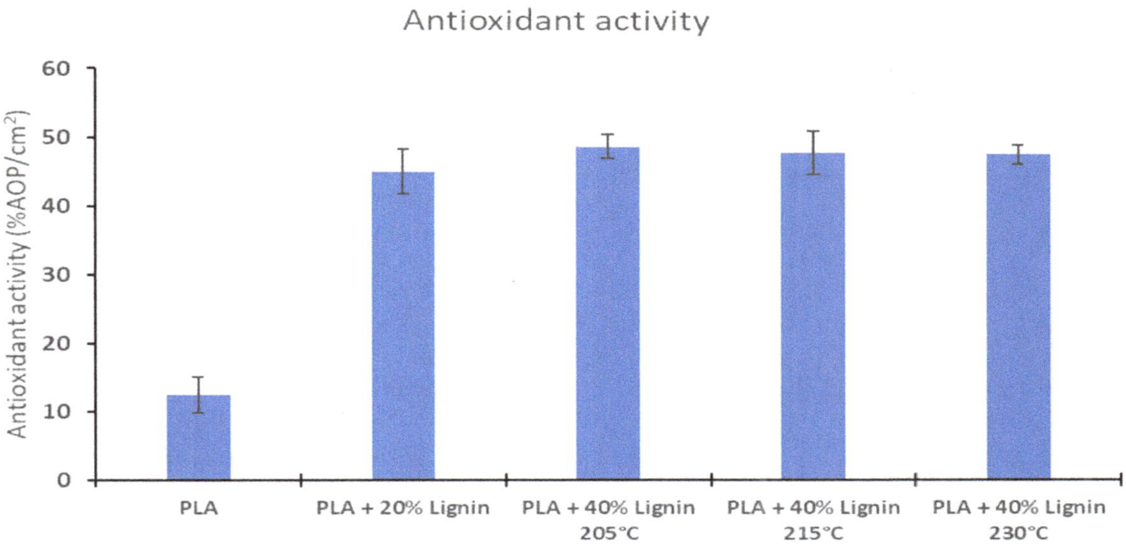

Figure 8. Antioxidant activity of PLA and biocomposites.

Keep in mind that lignin is a natural antioxidant, and lignin has been proposed to stabilize a given material against photo- and thermo-oxidation [14–17,23]. The antioxidant property of soda lignin has been confirmed in this study where the printed materials containing lignin has a significant larger antioxidant capability compared to PLA (Figure 7). This suggests that the biocomposites developed in this study are also suitable for additional materials, e.g., food packaging applications.

The suitability of the PLA/lignin biocomposite filament for 3D printing was also tested, by printing a smartphone protective case (Figure 9). The printing process revealed a good performance of the lignin-containing filament, and a functional protective case was effectively 3D printed. PLA/Lignin filaments are a plausible option for lignin utilization with potential in, e.g., rapid prototyping and consumer products [59]. It is worth to mention that the typical smell from some lignins was not detected during the extrusion of the filaments or during the printing process, which is an additional advantage of using soda lignin in PLA biocomposite materials.

Figure 9. 3D printing of a smartphone protective case with PLA/lignin biocomposite filament.

4. Conclusions

PLA/lignin biocomposites with different lignin loading were prepared and characterized in detail. Thermogravimetric analysis indicated that the thermal decomposition of lignin occurred over a wide temperature range. PLA/lignin biocomposites showed the highest antioxidant activity, due to the antioxidant activity of lignin. Biocomposites exhibited good extrudability and flowability with no observable agglomeration of the lignin. This suggests that lignin-containing biocomposites are plausible alternatives for 3D printing applications.

Author Contributions: Idea and supervision: G.C.-C.; Investigation: M.T.-O., E.E., G.C.-C., Draft preparation, reviewing and editing: M.T.-O., E.E., A.R., G.C.-C.

Acknowledgments: Johnny Kvakland Melbø, Kenneth Aasarød, Ingebjørg Leirset and Cornelis van der Wijst at RISE PFI are acknowledged for valuable assistance in the laboratory work.

References

1. Calvo-Flores, F.G.; Dobado, J.; Isac-García, J.; Martin-Martinez, F. Applications of Modified and Unmodified Lignins. In *Lignin and Lignans as Renewable Raw Material*; John Wiley & Sons Ltd.: New York, NY, USA, 2015; pp. 247–288.
2. Oksman, K.; Skrifvars, M.; Selin, J.F. Natural fibres as reinforcement in polylactic acid (PLA) composites. *Compos. Sci. Technol.* **2003**, *63*, 1317–1324. [CrossRef]
3. Lim, L.T.; Auras, R.; Rubino, M. Processing technologies for poly(lactic acid). *Prog. Polym. Sci.* **2008**, *33*, 820–852. [CrossRef]
4. Ray, S.S.; Yamada, K.; Okamoto, M.; Ueda, K. Polylactide-Layered Silicate Nanocomposite: A Novel Biodegradable Material. *Nano Lett.* **2002**, *2*, 1093–1096. [CrossRef]
5. Pang, X.; Zhuang, X.; Tang, Z.; Chen, X. Polylactic acid (PLA): Research, development and industrialization. *Biotechnol. J.* **2010**, *5*, 1125–1136. [CrossRef] [PubMed]
6. Cohn, D.; Salomon, A.H. Designing biodegradable multiblock PCL/PLA thermoplastic elastomers. *Biomaterials* **2005**, *26*, 2297–2305. [CrossRef] [PubMed]
7. Nofar, M.; Sacligil, D.; Carreau, P.J.; Kamal, M.R.; Heuzey, M.-C. Poly (lactic acid) blends: Processing, properties and applications. *Int. J. Biol. Macromol.* **2019**, *125*, 307–360. [CrossRef]
8. Ochi, S. Mechanical properties of kenaf fibers and kenaf/PLA composites. *Mech. Mater.* **2008**, *40*, 446–452. [CrossRef]
9. Vatansever, E.; Arslan, D.; Nofar, M. Polylactide cellulose-based nanocomposites. *Int. J. Biol. Macromol.* **2019**, *137*, 912–938. [CrossRef]
10. Rinaldi, R.; Jastrzebski, R.; Clough, M.T.; Ralph, J.; Kennema, M.; Bruijnincx, P.C.A.; Weckhuysen, B.M. Paving the Way for Lignin Valorisation: Recent Advances in Bioengineering, Biorefining and Catalysis. *Angew. Chem. Int. Ed.* **2016**, *55*, 8164–8215. [CrossRef]
11. Figueiredo, P.; Lintinen, K.; Hirvonen, J.T.; Kostiainen, M.A.; Santos, H.A. Properties and chemical modifications of lignin: Towards lignin-based nanomaterials for biomedical applications. *Prog. Mater. Sci.* **2018**, *93*, 233–269. [CrossRef]
12. Schutyser, W.; Renders, T.; Van den Bosch, S.; Koelewijn, S.F.; Beckham, G.T.; Sels, B.F. Chemicals from lignin: An interplay of lignocellulose fractionation, depolymerisation, and upgrading. *Chem. Soc. Rev.* **2018**, *47*, 852–908. [CrossRef] [PubMed]
13. Liu, L.; Qian, M.; Song, P.A.; Huang, G.; Yu, Y.; Fu, S. Fabrication of Green Lignin-based Flame Retardants for Enhancing the Thermal and Fire Retardancy Properties of Polypropylene/Wood Composites. *ACS Sustain. Chem. Eng.* **2016**, *4*, 2422–2431. [CrossRef]
14. Zhao, W.; Simmons, B.; Singh, S.; Ragauskas, A.; Cheng, G. From lignin association to nano-/micro-particle preparation: Extracting higher value of lignin. *Green Chem.* **2016**, *18*, 5693–5700. [CrossRef]
15. Grossman, A.; Vermerris, W. Lignin-based polymers and nanomaterials. *Curr. Opin. Biotechnol.* **2019**, *56*, 112–120. [CrossRef] [PubMed]
16. Gellerstedt, G.; Tomani, P.; Axegard, P.; Backlund, B. Lignin recovery and lignin-based products. In *Integrated Forest Biorefineries: Challenges and Opportunities*; RSC Green Chemistry: Cambridge, UK, 2013; Chapter 8; pp. 180–210.

17. Ragauskas, A.J.; Beckham, G.T.; Biddy, M.J.; Chandra, R.; Chen, F.; Davis, M.F.; Davison, B.H.; Dixon, R.A.; Gilna, P.; Keller, M.; et al. Lignin Valorization: Improving Lignin Processing in the Biorefinery. *Science* **2014**, *344*, 1246843. [CrossRef]
18. Christopher, L.P. Integrated Forest Biorefineries: Current State and Development Potential. In *Integrated Forest Biorefineries: Challenges and Opportunities*; RCS Green Chemistry: Cambridge, UK, 2013; Chapter 1; pp. 1–66.
19. Wang, H.; Pu, Y.; Ragauskas, A.; Yang, B. From lignin to valuable products–strategies, challenges, and prospects. *Bioresour. Technol.* **2019**, *271*, 449–461. [CrossRef] [PubMed]
20. Thakur, V.K.; Thakur, M.K.; Raghavan, P.; Kessler, M.R. Progress in Green Polymer Composites from Lignin for Multifunctional Applications: A Review. *ACS Sustain. Chem. Eng.* **2014**, *2*, 1072–1092. [CrossRef]
21. Kun, D.; Pukánszky, B. Polymer/lignin blends: Interactions, properties, applications. *Eur. Polym. J.* **2017**, *93*, 618–641. [CrossRef]
22. Gordobil, O.; Delucis, R.; Egüés, I.; Labidi, J. Kraft lignin as filler in PLA to improve ductility and thermal properties. *Ind. Crop. Prod.* **2015**, *72*, 46–53. [CrossRef]
23. Pouteau, C.; Dole, P.; Cathala, B.; Averous, L.; Boquillon, N. Antioxidant properties of lignin in polypropylene. *Polym. Degrad. Stab.* **2003**, *81*, 9–18. [CrossRef]
24. Domenek, S. Potential of Lignins as Antioxidant Additive in Active Biodegradable Packaging Materials. *J. Polym. Environ.* **2013**, *21*, 692–701. [CrossRef]
25. Mishra, S.B.; Mishra, A.K.; Kaushik, N.K.; Khan, M.A. Study of performance properties of lignin-based polyblends with polyvinyl chloride. *J. Mater. Process. Technol.* **2007**, *183*, 273–276. [CrossRef]
26. De Chirico, A.; Armanini, M.; Chini, P.; Cioccolo, G.; Provasoli, F.; Audisio, G. Flame retardants for polypropylene based on lignin. *Polym. Degrad. Stab.* **2003**, *79*, 139–145. [CrossRef]
27. Thielemans, W.; Can, E.; Morye, S.S.; Wool, R.P. Novel applications of lignin in composite materials. *J. Appl. Polym. Sci.* **2002**, *83*, 323–331. [CrossRef]
28. Thielemans, W.; Wool, R.P. Kraft lignin as fiber treatment for natural fiber-reinforced composites. *Polym. Compos.* **2005**, *26*, 695–705. [CrossRef]
29. Graupner, N. Application of lignin as natural adhesion promoter in cotton fibre-reinforced poly(lactic acid) (PLA) composites. *J. Mater. Sci.* **2008**, *43*, 5222–5229. [CrossRef]
30. Kim, J.S.; Lee, Y.Y.; Kim, T.H. A review on alkaline pretreatment technology for bioconversion of lignocellulosic biomass. *Bioresour. Technol.* **2016**, *199*, 42–48. [CrossRef] [PubMed]
31. Saratale, G.D.; Oh, M.-K. Improving alkaline pretreatment method for preparation of whole rice waste biomass feedstock and bioethanol production. *RSC Adv.* **2015**, *5*, 97171–97179. [CrossRef]
32. González-García, S.; Moreira, M.T.; Artal, G.; Maldonado, L.; Feijoo, G. Environmental impact assessment of non-wood based pulp production by soda-anthraquinone pulping process. *J. Clean. Prod.* **2010**, *18*, 137–145. [CrossRef]
33. Rodríguez, A.; Sánchez, R.; Requejo, A.; Ferrer, A. Feasibility of rice straw as a raw material for the production of soda cellulose pulp. *J. Clean. Prod.* **2010**, *18*, 1084–1091. [CrossRef]
34. Vishtal, A.G.; Kraslawski, A. Challenges in industrial applications of technical lignins. *BioResources* **2011**, *6*, 3547–3568.
35. Gosselink, R.J.A.; Abächerli, A.; Semke, H.; Malherbe, R.; Käuper, P.; Nadif, A.; van Dam, J.E.G. Analytical protocols for characterisation of sulphur-free lignin. *Ind. Crop. Prod.* **2004**, *19*, 271–281. [CrossRef]
36. Tejado, A.; Peña, C.; Labidi, J.; Echeverria, J.M.; Mondragon, I. Physico-chemical characterization of lignins from different sources for use in phenol–formaldehyde resin synthesis. *Bioresour. Technol.* **2007**, *98*, 1655–1663. [CrossRef] [PubMed]
37. Baurhoo, B.; Ruiz-Feria, C.A.; Zhao, X. Purified lignin: Nutritional and health impacts on farm animals—A review. *Anim. Feed Sci. Technol.* **2008**, *144*, 175–184. [CrossRef]
38. Nadif, A.; Hunkeler, D.; Käuper, P. Sulfur-free lignins from alkaline pulping tested in mortar for use as mortar additives. *Bioresour. Technol.* **2002**, *84*, 49–55. [CrossRef]
39. Wörmeyer, K.; Ingram, T.; Saake, B.; Brunner, G.; Smirnova, I. Comparison of different pretreatment methods for lignocellulosic materials. Part II: Influence of pretreatment on the properties of rye straw lignin. *Bioresour. Technol.* **2011**, *102*, 4157–4164. [CrossRef]
40. Nguyen, N.A.; Bowland, C.C.; Naskar, A.K. A general method to improve 3D-printability and inter-layer adhesion in lignin-based composites. *Appl. Mater. Today* **2018**, *12*, 138–152. [CrossRef]

41. Zhao, D.X.; Cai, X.; Shou, G.Z.; Gu, Y.Q.; Wang, P.X. Study on the Preparation of Bamboo Plastic Composite Intend for Additive Manufacturing. *Key Eng. Mater.* **2016**, *667*, 250–258. [CrossRef]
42. Henke, K.; Treml, S. Wood based bulk material in 3D printing processes for applications in construction. *Eur. J. Wood Wood Prod.* **2013**, *71*, 139–141. [CrossRef]
43. Le Duigou, A.; Castro, M.; Bevan, R.; Martin, N. 3D printing of wood fibre biocomposites: From mechanical to actuation functionality. *Mater. Des.* **2016**, *96*, 106–114. [CrossRef]
44. Domínguez-Robles, J.; Martin, N.K.; Fong, M.L.; Stewart, S.A.; Irwin, N.J.; Rial-Hermida, M.I.; Donnelly, R.F.; Larrañeta, E. Antioxidant PLA Composites Containing Lignin for 3D Printing Applications: A Potential Material for Healthcare Applications. *Pharmaceutics* **2019**, *11*, 165. [CrossRef] [PubMed]
45. García, A.; González Alriols, M.; Spigno, G.; Labidi, J. Lignin as natural radical scavenger. Effect of the obtaining and purification processes on the antioxidant behaviour of lignin. *Biochem. Eng. J.* **2012**, *67*, 173–185. [CrossRef]
46. Huijgen, W.J.J.; Telysheva, G.; Arshanitsa, A.; Gosselink, R.J.A.; de Wild, P.J. Characteristics of wheat straw lignins from ethanol-based organosolv treatment. *Ind. Crop. Prod.* **2014**, *59*, 85–95. [CrossRef]
47. Constant, S.; Wienk, H.L.J.; Frissen, A.E.; de Peinder, P.; Boelens, R.; van Es, D.S.; Grisel, R.J.H.; Weckhuysen, B.M.; Huijgen, W.J.J.; Gosselink, R.J.A.; et al. New insights into the structure and composition of technical lignins: A comparative characterisation study. *Green Chem.* **2016**, *18*, 2651–2665. [CrossRef]
48. Sameni, J.; Krigstin, S.; de Santos Rosa, D.; Leao, A.; Sain, M. Thermal Characteristics of Lignin Residue from Industrial Processes. *BioResources* **2013**, *9*, 725–737. [CrossRef]
49. Yang, H.; Yan, R.; Chen, H.; Lee, D.H.; Zheng, C. Characteristics of hemicellulose, cellulose and lignin pyrolysis. *Fuel* **2007**, *86*, 1781–1788. [CrossRef]
50. Watkins, D.; Nuruddin, M.; Hosur, M.; Tcherbi-Narteh, A.; Jeelani, S. Extraction and characterization of lignin from different biomass resources. *J. Mater. Res. Technol.* **2015**, *4*, 26–32. [CrossRef]
51. Schmidl, G. Molecular Weight Characterization and Rheology of Lignins for Carbon Fibers. Ph.D. Thesis, University of Florida, Gainesville, FL, USA, 1992.
52. Heitner, C.; Dimmel, D.; Schmidt, J.A. *Lignin and Lignans: Advances in Chemistry*; CRC Press, Tylor & Francis Group: Boca Raton, FL, USA, 2010.
53. Gordobil, O.; Egüés, I.; Llano-Ponte, R.; Labidi, J. Physicochemical properties of PLA lignin blends. *Polym. Degrad. Stab.* **2014**, *108*, 330–338. [CrossRef]
54. Al-Itry, R.; Lamnawar, K.; Maazouz, A. Reactive extrusion of PLA, PBAT with a multi-functional epoxide: Physico-chemical and rheological properties. *Eur. Polym. J.* **2014**, *58*, 90–102. [CrossRef]
55. Weng, Y.-X.; Jin, Y.-J.; Meng, Q.-Y.; Wang, L.; Zhang, M.; Wang, Y.-Z. Biodegradation behavior of poly(butylene adipate-co-terephthalate) (PBAT), poly(lactic acid) (PLA), and their blend under soil conditions. *Polym. Test.* **2013**, *32*, 918–926. [CrossRef]
56. Rahman, M.; Afrin, S.; Haque, P.; Islam, M.; Islam, M.S.; Gafur, M. Preparation and Characterization of Jute Cellulose Crystals-Reinforced Poly (L-lactic acid) Biocomposite for Biomedical Applications. *Int. J. Chem. Eng.* **2014**, *2014*, 842147. [CrossRef]
57. Lu, Q.; Liu, W.; Yang, L.; Zu, Y.; Zu, B.; Zhu, M.; Zhang, Y.; Zhang, X.; Zhang, R.; Sun, Z.; et al. Investigation of the effects of different organosolv pulping methods on antioxidant capacity and extraction efficiency of lignin. *Food Chem.* **2012**, *131*, 313–317. [CrossRef]
58. Mahmood, Z.; Yameen, M.; Jahangeer, M.; Riaz, M.; Ghaffar, A.; Javid, I. Lignin as Natural Antioxidant Capacity. In *Lignin-Trends and Applications*; IntechOpen: London, UK, 2018; pp. 181–205.
59. Gkartzou, E.; Koumoulos, E.P.; Charitidis, C.A. Production and 3D printing processing of bio-based thermoplastic filament. *Manuf. Rev.* **2017**, *4*. [CrossRef]

7

Chitin Nanofibrils in Poly(Lactic Acid)(PLA) Nanocomposites: Dispersion and Thermo-Mechanical Properties

Maria-Beatrice Coltelli [1,2,*], Patrizia Cinelli [1,2], Vito Gigante [1,2], Laura Aliotta [1,2], Pierfrancesco Morganti [3,4], Luca Panariello [1,2] and Andrea Lazzeri [1,2,*]

1. Department of Civil and Industrial Engineering, University of Pisa, Via Diotisalvi 2, 56122 Pisa, Italy; patrizia.cinelli@unipi.it (P.C.); vito.gigante@dici.unipi.it (V.G.); laura.aliotta@dici.unipi.it (L.A.); luca.panariello@ing.unipi.it (L.P.)
2. National InterUniversity Consortium of Materials Science and Technology (INSTM), Via Giusti 9, 50121 Florence, Italy
3. Skin Pharmacology and Dermatology Unit, Campania University "Luigi Vanvitelli", 80100 Naples, Italy; pierfrancesco.morganti@mavicosmetics.it
4. MAVI SUD, Aprilia (LT), 04011 Aprilia, Italy

* Correspondence: maria.beatrice.coltelli@unipi.it (M.-B.C.); andrea.lazzeri@unipi.it (A.L.)

Abstract: Chitin-nanofibrils are obtained in water suspension at low concentration, as nanoparticles normally are, to avoid their aggregation. The addition of the fibrils in molten PLA during extrusion is thus difficult and disadvantageous. In the present paper, the use of poly(ethylene glycol) (PEG) is proposed to prepare a solid pre-composite by water evaporation. The pre-composite is then added to PLA in the extruder to obtain transparent nanocomposites. The amount of PEG and chitin nanofibrils was varied in the nanocomposites to compare the reinforcement due to nanofibrils and plasticization due to the presence of PEG, as well as for extrapolating, where possible, the properties of reinforcement due to chitin nanofibrils exclusively. Thermal and morphological properties of nanocomposites were also investigated. This study concluded that chitin nanofibrils, added as reinforcing filler up to 12% by weight, do not alter the properties of the PLA based material; hence, this additive can be used in bioplastic items mainly exploiting its intrinsic anti-microbial and skin regenerating properties.

Keywords: chitin nanofibrils; poly(lactic acid); nanocomposites

---

1. Introduction

Chitin is the second most abundant biopolymer on earth, having global reserves of 100 billion tons [1]. Waste from the seafood industry is a great source of chitin, because it represents the matrix of the hierarchically structured, fiber-based composite constituting the exoskeleton of crustaceans [2]. On a global level, about 6 mega tons of crustacean shells are discarded per year [3]. For this reason, many researchers evidenced the possibility of exploiting this resource [4] to obtain valuable materials such as chitosan and its derivatives [5], chitin nanofibrils [6–8], inorganics such as calcium carbonate, and molecules such as acetic acid and pyrrole [9].

Chitin nanofibrils (CNs), consisting of colloidal nano-rods, represent the crystalline fraction of the chitin extracted from sea food waste; it was reported that they show anti-microbial properties and favor cells regeneration [10–12].

Poly(lactic acid) (PLA) is a fully renewable polymer compatible with the human body, plants and the environment, being both resorbable in the human body and biodegradable in composting plants [13]. It is reported to have biocidal activity thanks to its tendency to hydrolyze on the surface, producing lactic acid, which exerts a slight anti-microbial activity [14]. For all these reasons PLA is one of the best alternatives to petrol-based polymers in the packaging, agricultural, personal care, cosmetic, biomedical and tissue engineering sectors [15–17]. The modulation of PLA properties to make it suitable for several processing techniques and final applications is an attractive topic of current research about nanomaterials [18]. The possibility of reinforcing PLA using nanofibers was investigated by considering several nano-reinforcements [19] such as cellulose nanocrystals or nanofibers [20,21], sometimes also combined with nanofillers with different aspect ratios, like clays, to modulate barrier properties [22]. Nanocellulose did not show anti-microbial and cell regenerative properties.

In contrast, the combination of PLA and CNs to obtain bionanocomposites can represent a good opportunity for the preparation of bioplastic materials with improved structural and functional properties [23]. Moreover, it is reported that chitin nanofibers require less energy intensive treatments in their production, compared to cellulose nanofibers [24], and that the combination with cellulose nanofibers to obtained nanocomposites can be synergistic [25].

The dispersion of CNs at a nanometric level in PLA was attempted by several techniques. The preliminary modification of chitin nanofibrils was identified as a good strategy to improve its stability in suspension, as observed by Araki, that prepared sterically stabilized chitin nanowhiskers by surface grafting monomethoxy poly(ethylene glycol) (mPEG) [26]. mPEG grafting has been previously reported to effectively improve the dispersion stability of cellulose nanowhiskers [27,28]. By following this strategy based on CNs chemical modification, Zhang et al. [29] acetylated chitin nanofibrils to improve their dispersion in poly (lactic acid) (PLA), but the mechanical properties of composites prepared by solvent casting (not in the molten state) were not improved. Moreover, the chemical modification of diluted suspension can be complex, in view of an industrialization of the composite's production. On the other hand, extrusion is a conventional melt polymer-processing technique, and it would be preferentially adopted for nanocomposite processing for industrial applications. Rizvi et al. [30], working in a twin-screw laboratory extruder, used more traditional compatibilizing agents to improve the adhesion between nanofibrils and the PLA matrix, investigating the effect of different contents of chitin nano-fibrils in melt-blended PLA/chitin and PLA/CN composites. In particular, maleic anhydride (MA) was grafted to PLA at about 2% by weight, producing a modified polymer (PLA-g-MA). Tensile tests showed an effective increase in Young's modulus in nano-composites compared to pure PLA. In particular, it was observed that the Young's modulus underwent an increase of up to 5% chitin content, while higher quantities did not produce any further reinforcement effect. Herrera et al. noticed that only 1% CNs was enough to observe an increase in Young's Modulus. [31]. The tensile strength decreased by increasing the nano-chitin content, but in any case, lower values than those of pure PLA were observed. This decreasing trend was attributed to the hydrolysis of PLA during the preparation, since aqueous solutions of CNs were added in the melt mixer to avoid agglomeration, and therefore, as the amount of reinforcement used increased, the introduced moisture increased accordingly, leading to an increasing impact of hydrolysis. This paper showed the good potential of CNs to improve mechanical performance, but at the same time, the difficulty of dispersing them at the nano-metric level and homogeneously in the PLA matrix. Guan and Naguib [32] investigated PLA/CN nano-composites using MA as a compatibilizing agent and N,N-dimethylacetamide (DMAc) as a dispersing agent. The latter was added to improve the dispersion of nano-fibrils in the composite. The CNs were re-dispersed in DMAc by mechanical stirring to prepare a nano-structured suspension. Moreover, PLA was prepared with grafted anhydride groups (PLA-g-MA) using 2 wt% of MA and 0.5 wt% peroxide as a radical initiator to improve the fiber-matrix adhesion. Tensile test results demonstrated that the presence of both DMAc and PLA-g-MA, without CNs, led to a decrease of the Young's modulus and of tensile strength; however, DMAc caused a significant worsening of the mechanical properties, explained by the authors by considering the degradation generated in the PLA through chain scission during the extrusion. It is also important to note that as the amount of nano-filler increased, two contrasting effects occurred: a reinforcing effect of chitin, which increased stiffness and strength at the expense of ductility; and a negative impact on the mechanical properties which was attributable to the presence of DMAc, which caused the degradation of the matrix. This article evidenced the need to use a dispersing agent, looking for one that does not negatively impact the mechanical performance of PLA. The use of known plasticizers of PLA in PLA-based nanocomposites has been studied by Herrera et al. [31,33,34], who developed triethyl citrate (TEC) in water/alcohol suspensions, fed in the extruder, and investigated the properties of composites containing 1% of CNs. They reported that this amount was enough to modify some key properties, such as anti-fungal activity and antistatic behavior puncture strength, and to improve mechanical properties thanks to the very high surface-to-volume ratio of the nanofibrils. Interestingly, the authors investigated the thermal properties of nanocomposites by DSC, and concluded that the simultaneous presence of CNs and TEC in the explored composition range did not provoke significant changes, except for the evident decrease of the glass transition due to the plasticizing effect of TEC. The effect of the plasticizer content to determine the synergic effect of the plasticizer as a dispersing and toughening aid with a minimum impact on the properties of PLA was considered an interesting topic to further investigate by the authors [31]. This molecule, as well as other citrates, due to the presence of ester groups, showed a very high affinity for PLA matrix [35–38], thus strongly favoring the nanodispersion of CNs in the matrix and the formation of an extended CN/PLA interface.

Poly(ethylene glycol) (PEG) was also successfully used to disperse CNs [39–41] in PLA. More recently, Li et al. [42], aiming at obtaining rigid nanocomposites, used high molecular weight poly(ethylene oxide) (PEO) or PEG and investigated the flexural and impact properties of composites with NC in the range 10–40%, observing a reinforcing effect of CN.

In all these studies, the exigence of effectively dispersing the CNs in PLA led to the necessity of using both a plasticizer and chitin nanofibrils.

Interestingly, Nakagaito et al. [43] used only water as a dispersion medium, which was removed by filtration and drying. Although PLA is insoluble in water, aqueous suspensions can be obtained by using PLA short fibers or particles, that can be easily mixed with cellulose nanofibers in water suspension. After dewatering, the mixture forms paper-like sheets that can be laminated and compression molded. Hence, nanocomposites were obtained by compression molding the filtrates. Static tensile test and dynamic mechanical analysis were performed to evaluate the reinforcement as a function of nanofiber content. Chitin nanofibers delivered reinforcement similar to that of cellulose nanofibers, being especially effective at up to 70 wt% fiber load. The ultimate tensile modulus and strength reached 7.7 GPa and 110 MPa, respectively, at a nanofiber content of 70 wt%. This work evidenced the interesting potential of CNs as a reinforcing agent in a pure PLA matrix. This methodology, however, cannot claim to reach full nanoscaled homogeneity because of the anisotropy flows typical of compression molding process. Subsequently, Li [42] compared PEG and PEO dispersion with this methodology, using a laboratory twin screw extruder. In this case, the Modulus measured by DMTA resultedin 7.6 GPa for the water method (a value in agreement with the one reported by Nakagaito et al. [43]), and 6,5 and 6,0 GPa for PEG and PEO respectively. The lower value obtained by using PEG and PEO can be attributed to their plasticizing effect.

Interestingly, very recently, Shanshina et al. [44] used an ionic liquid-based approach to co-dissolve PLA and CNs and produce PLA fibers containing up to 27% by weight of CNs. The fibers showed improved strength with respect to the fibers obtained by pure PLA. However, processes considering ionic liquid technology, which may be promising for future applications in several fields, are not yet well diffused in the industry.

By an analysis of the literature, it is evident that the combination of reinforcement with the necessity of using dispersing agents is an interesting topic of current research. The use of PEG resulted an effective overall limit of CN agglomeration, but a systematic study about the possibility of modulating nanocomposite properties, considering low amount of CNs, was never carried out. Hence, in this attempt, after the preparation of nanocomposites, their tensile and thermal properties as a function of the content of CNs and PEG were measured and discussed.

2. Results

2.1. Dispersion of Chitin Nanofibrils in Plasticized Pla

In order to study chitin nanofibrils morphology, water suspensions at 2% by weight of chitin nanofibrils were diluted 1:100, and then one drop of diluted suspension was deposited onto a glass window. By using a field emission scanning electron microscope (FESEM), it was possible to determine the shape of chitin nanofibrils (Figure 1): average length of 20 μm and an average width of 90 nm.

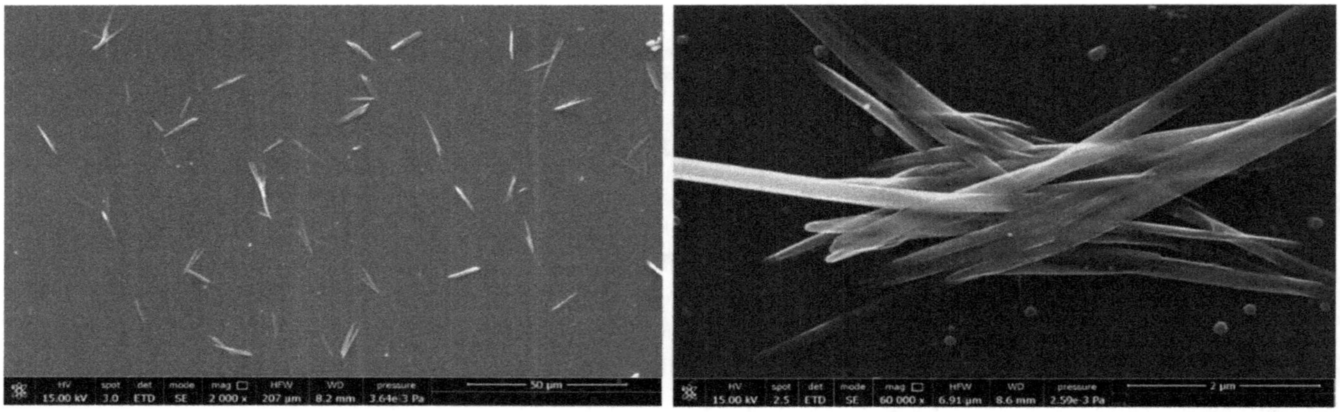

(a) (b)

Figure 1. FESEM (field emission scanning electron microscopy) micrographs obtained from 2% by weight water suspension of CN diluted 1:100 and deposited on glass. (**a**) 2000× magnification; (**b**) 60,000× magnification.

When the 2% by weight suspension was dried without any previous dilution, and characterized by Scanning Electron Microscopy (SEM), the chitin nanofibrils formed agglomerates (Figure 2). In Figure 2a, it is evident that the drying of nanosized chitin produced large flakes in which the fibers are agglomerated to form a compact structure like that of a sheet of paper. In fact, in between different nanofibrils, having a very high surface to volume ratio, the formation of a high number of hydrogen bonds is thermodynamically favored, and this reasonably results in this compact structure. The structure of a nano-fibrous disordered assembly is clearly observable at the edges of the flakes (Figure 2b).

With a concentration of 1:10 of the CN suspension, PEGs having 400, 1500, 4000, 6000 and 8000 as molecular weight were added in weight ratio 1:1 to CN to obtain five different pre-composites that were dried. The obtained materials were solid, except for the one obtained with PEG 400, that showed a pasty consistency. The solid pre-composites were characterized by SEM and showed, from a morphological point of view, a complex fibrous nano-structure (Figure 2c–f), indicating that PEG constituted the matrix in which chitin nano-fibrils were immersed. Hence, it seemed reasonable that a PEG polymer was present in between the CNs, thus avoiding the formation of compact agglomerates.

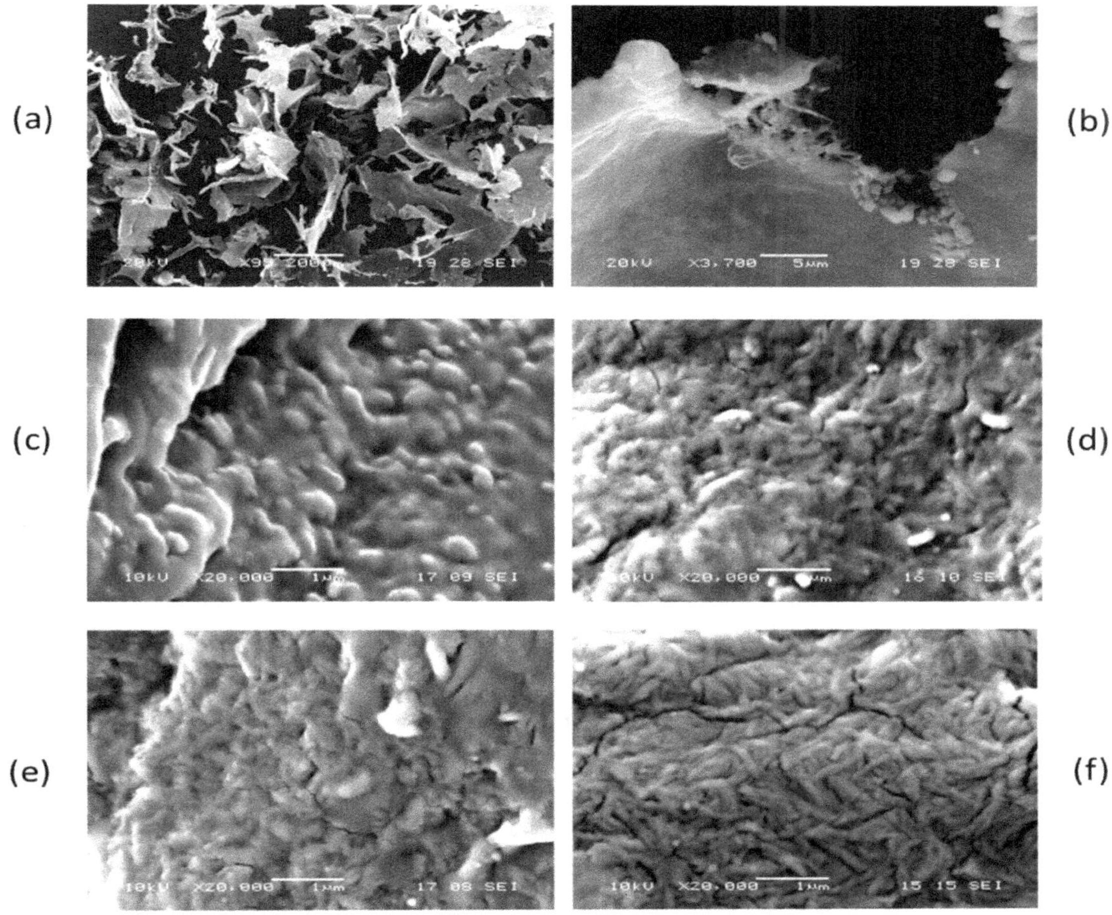

Figure 2. SEM micrographs of (**a**) dried CN suspension flakes; (**b**) magnification of flake edge; (**c**) pre-composite based on PEG 1500; (**d**) pre-composite based on PEG 4000; (**e**) pre-composite based on PEG 6000; (**f**) pre-composite based on PEG 8000.

With the purpose of investigating the interactions between PEG and chitin nanofibrils, some specific infrared characterizations and thermogravimetric tests were performed on the sample obtained by adding to chitin nanofibrils 2% of PEG 8000. The highest molecular weight was selected because, with respect to the same quantity of lower molecular weight PEG, it corresponded to a lower number of macromolecules, and consequently, demonstrated less efficient interactions with the CN. Thus, if the PEG 8000 is able to interact with the CNs, this effect will be stronger in samples with a lower molecular weight.

Regarding the infrared spectrum, CN showed characteristic amide I and Amide II bands at 1618 and 1550 cm^{-1} respectively. The Amide I band is typical of α-chitin [45], as it is split into two components at 1660 and 1630 cm^{-1}. This double band was attributed to the influence of hydrogen bonding or the presence of an enol form of the amide moiety [46–48]). Interestingly, it was found that the infrared bands typical of chitin resulted in the spectrum of the sample containing PEG8000, despite of only 2% of PEG8000 being present. This is particularly evident in the spectrum part where PEG 8000 (Figure 3) did not show any absorption bands, such as in the region 500–800 cm^{-1} and in the region 1500–1800 cm^{-1}. The presence of PEG, that probably at least partially interposes between CNs, thus favors the hypothesis of more complex interactions in between CNs.

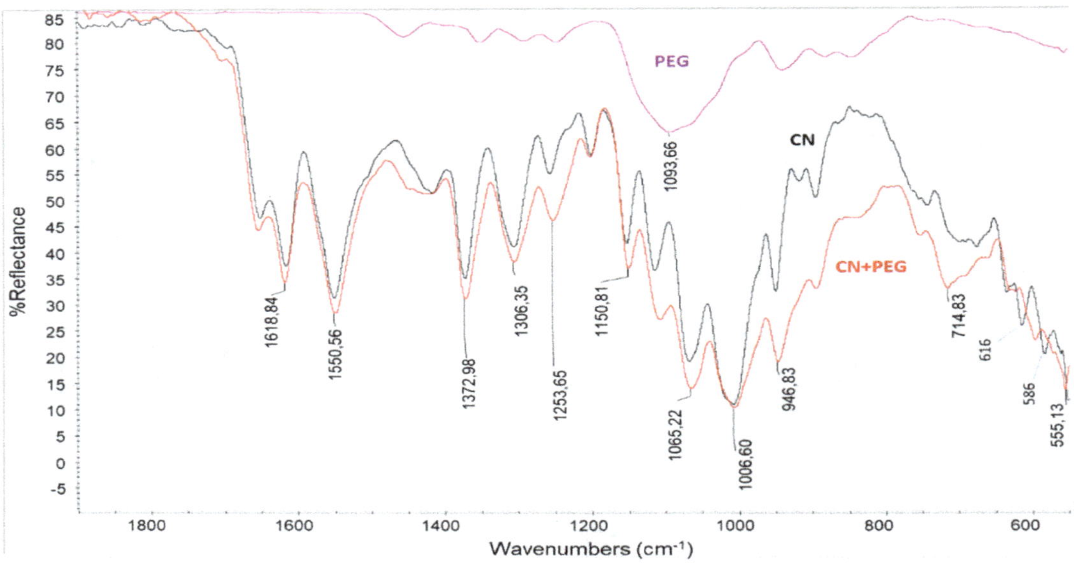

Figure 3. infrared ATR spectra of PEG 8000, CN and mixture CN/PEG8000 98:2. The spectrum of pure PEG is reported with a reduced reflectance intensity to allow a better visualization of the spectra.

The thermo-gravimetric analysis of pure CN in nitrogen flow showed that its main degradation step is at 349 °C (−21.2% by weight), and a second evident mass loss at 394 °C (−9.5%) is present. These values were calculated through the analysis of the derivative curve of the thermogram. A slight loss of mass can be observed also below 140 °C, but it accounts for a reduced loss of mass of the pure chitin (−2.6%). The mass loss observed below 100 °C was attributed to removal of humidity (−4%). The final residue was 58.1%. In the presence of 2% of PEG the thermogravimetry trend was completely changed (Figure 4). The main mass loss at 329 °C was −57.5%, and the final residue presented as 17.4% by weight. Hence, the presence of PEG induced a more efficient thermal degradation of CN, probably because it interposes between the CNs, avoiding the formation of compact agglomerations. This type of assembly is instead typical of pure CN, where inter-fibrils hydrogen bonding is predominant, resulting in the formation of a considerable quantity of carbonaceous residue after thermal degradation in nitrogen atmosphere. Similar results were obtained by Cheng et al., that considered the addition of PEG 1000 to cellulose nanofibrils [49].

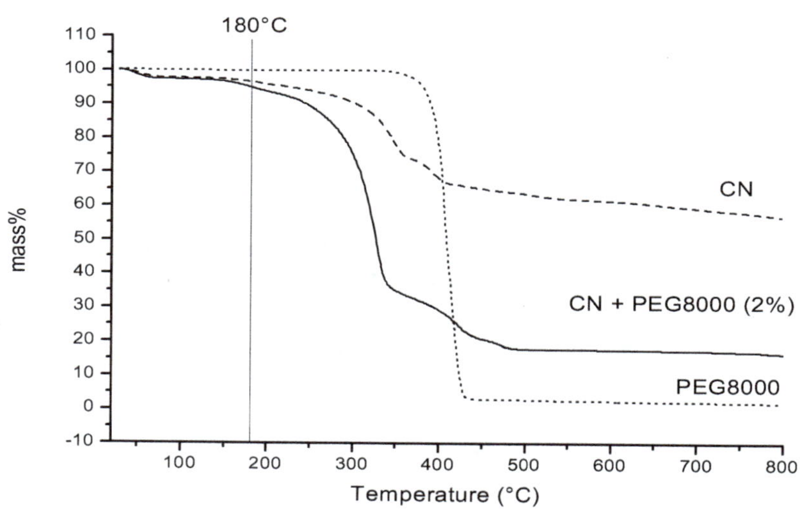

Figure 4. thermogravimetric curves of CN, PEG and CN + PEG (2%). The adopted extrusion temperature is marked with a dash line.

As the morphology of the different composites was very similar, only PEG400 liquid and PEG8000 solid (having the lowest and highest molecular weights respectively) -based samples were considered for preparing PLA based composites in a mini-extruder. The composition of the different composites, reported in Table 1, was selected by considering not only the two different molecular weights, but also with the aim of investigating the composite's properties as a function of PEG concentration and chitin nanofibrils concentration. As in the pre-composite, the weight ratio of PEG and CN was 1:1; for this reason, some PEG was added to obtain the desired content. For the P1Low2NC, a specific pre-composite was prepared with a PEG: CN weight ratio of 1:2.

Table 1. Composition of composites obtained by PLA, chitin nano-fibrils and PEG.

Samples	PEG (%)	MW of PEG	CN
P	—	—	—
P2CN* [a]	—	—	2
P10low	10	400	0
P10low2CN	10	400	2
P10high2CN	10	8000	2
P1low2CN	1	400	2
P5low2CN	5	400	2
P10low5CN	10	400	5
P10low12CN	10	400	12

[a] P2CN* is a sample of Chitin nanofibrils powder obtained by simply drying the CN suspension, without using PEG.

The extrusion was followed by the injection molding of specimens, that resulted in a transparent com pound, which was in agreement with the achievement of a nano-scaled dispersion [50,51]. As shown in Figure 5, P10low2CN and P10high 2CN resulted in a transparent compound, and also P5low2CN, but with the presence of some visible particles.

Figure 5. specimens obtained by injection moulding: comparison regarding transparency.

P10low5CN resulted in a transparent but brownish compound, whereas the sample containing 12% of CN (P10low12CN), resulted in a brown compound with a reduced transparency. In general, high transparency and colorlessness was achieved by decreasing the content of CN and increasing the content of PEG. In fact, both these characteristics positively affect the CN nano dispersion. The role of PEG as a dispersing agent is evident. In fact, the P2CN* composite was prepared by adding dried chitin nanofibrils without using PEG, and in this case, the resulting specimens were darker and not transparent because of the presence of visible particles. In this case, the CN formed aggregates that could not be well dispersed in the melt PLA during extrusion.

The samples were then characterized by SEM microscopy by preparing cryofractured surfaces from tensile specimens. From the micrographs in Figure 6, it is possible to see that the PLA containing only PEG400 (P10low sample) present a very high homogeneity apart from the presence of some round domains attributable to the presence of a second phase of PEG. The presence of very big aggregates, i.e., with diameters higher than 10 microns, could be observed in the sample obtained without preparing a pre-composite (P2CN*). In other composites where the chitin nanofibrils were dispersed by the addition of the pre-composite, it was not possible to reveal the presence of chitin nanofibrils aggregates, although several cryofractured surfaces were examined. The micrographs results were similar to those of the sample obtained in absence chitin nanofibrils, with submicrometric droplets being attributable to PEG. This result indicated a very good dispersion of chitin nanofibrils in the composite.

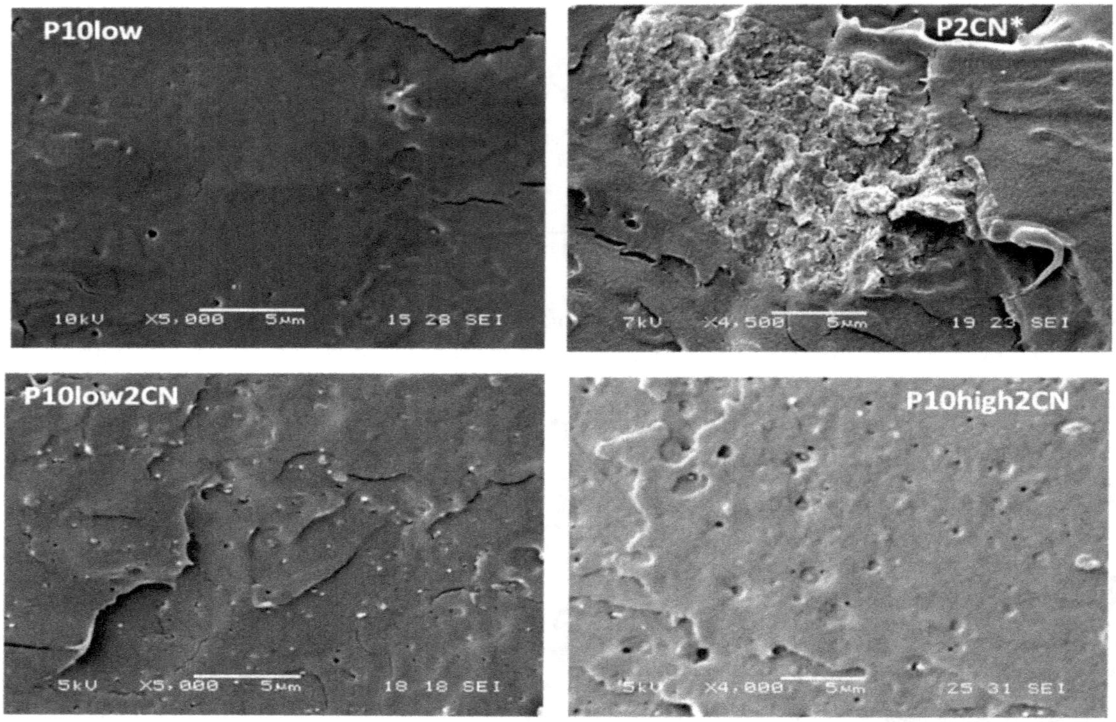

Figure 6. SEM micrographs obtained on cryofractured surfaces obtained from tensile specimens of different PLA/Chitin nanofibrils composites.

Interestingly, it was found that the dimensions of the PEG domains seemed to decrease by increasing the content of CN (Figure 7), reasonably indicating not negligible interactions between chitin nanofibrils and PEG 400, leading to a better dispersion of PEG. The interactions may be responsible for this decrease, considering that chitin nanofibrils are present both in the plasticized PLA matrix and in the PEG domains, thus acting as interfacial stabilizers. Moreover, micrometric PEG-based domains consist, during the melt extrusion of PLA nanocomposites, of liquid pools with a viscosity lower than that of the matrix, and these pools can contain CNs. The viscosity of such a liquid is increased because of the presence of chitin nanofibrils. In fact, in general, the viscosity of a suspension increases as a function of nanofibrils content [52,53], because of the tendency of gelling of the nanofibrils. The dimension of domains in immiscible polymer blends is reported to be dependent on the viscosity ratio, and in general it decreases when the ratio of the viscosity of the matrix and the dispersed phase is close to 1 [54,55]. The decrease of the PEG domain dimensions can be thus additionally ascribed to the viscosity variations in the two phases due to the presence in both of PEG and chitin nanofibrils.

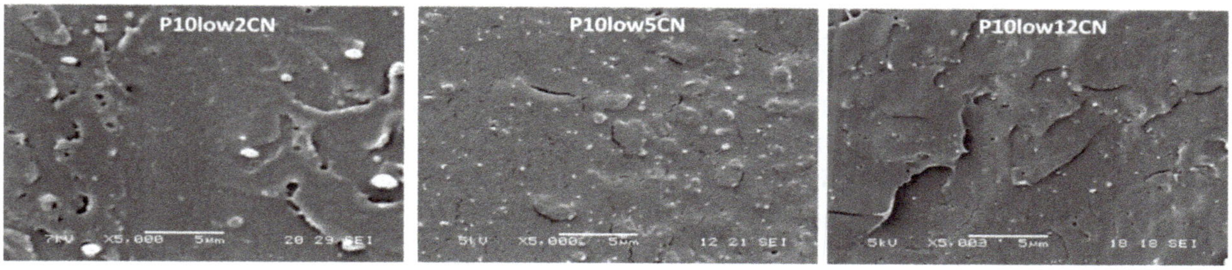

Figure 7. SEM micrographs obtained on cryofractured surfaces obtained from tensile specimens of different PLA/Chitin nanofibrils composites. From left to right the content of CN increases.

The composites P10low2CN was characterized by STEM microscopy (Figure 8) that revealed the presence of CNs as single fibers and bundles, dispersed at the nanometric scale. A good correspondence with the dimensions observed in FESEM micrographs of CNs (Figure 1) is evident.

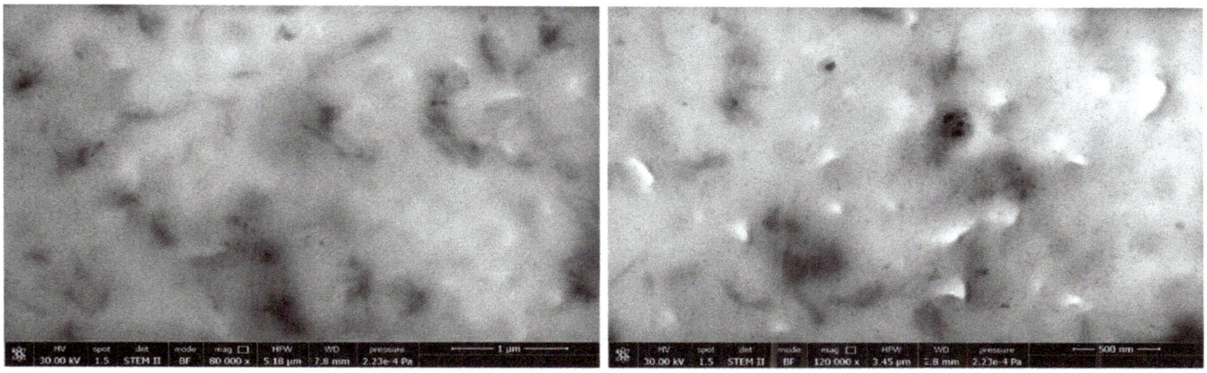

Figure 8. TEM micrographs obtained on the P10low2CN sample at different magnifications.

Infrared ATR spectroscopy was applied (Figure 9) to investigate the distribution of chitin nanofibrils on the surface of injection molded specimens. The ATR technique allowed us to record the vibrational spectrum of a material present on a surface at up to a few microns. In particular, the Smart itX ATR diamond plate allows a depth of penetration 2.03 micrometers at 1000 cm^{-1}. By overlaying the spectra of P10low, as a reference, with the spectra of the samples containing increasing amounts of CNs (2%, 5% and 12% by weight), it was found that some bands attributable to CNs are revealed. These bands are extremely weak in P10low2CN sample, whereas they are well evident in P10low5CN and P10low12CN. It is thus evident that a significant number of CNs are present on the specimen surface when the content of CNs is above 5% by weight. If the content is lower, the presence of CNs can not be revealed by this technique.

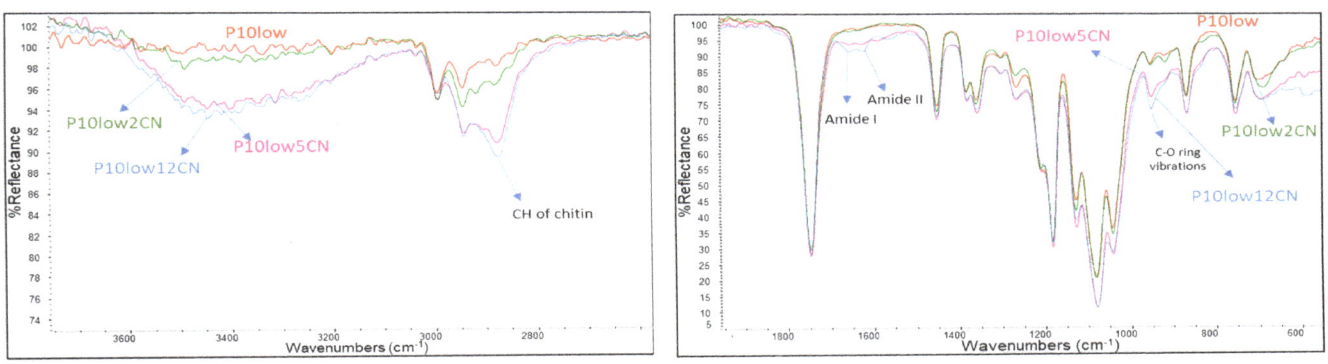

(a) (b)

Figure 9. (**a**,**b**) Infrared ATR spectra of P10low, P10low2CN, P10low5CN and P10low12CN.

2.2. Tensile Properties

Tensile tests were performed on specimens produced by injection molding and conditioned for two weeks at 50% as relative humidity. The pure PLA is brittle and shows a high value of Young's Modulus (3.5 GPa), a high value of stress at break, but a low value of elongation at break (Table 2). When the PLA is plasticized using PEG 400 (trial P10low), a strong decrease in Young's Modulus and stress at break and an increase in elongation at break (up to 180%) were observed, in agreement with an increased ductility of the material. Interestingly, the addition of CN to plasticized PLA resulted in a decrease in Young's Modulus and stress at break, whereas the elongation at break was not significantly affected by the presence of CNs, also when their content was increased in the composites. Interestingly, the addition of PEG 400 and PEG 8000 resulted in similar properties, with the latter composite showing a higher Young's Modulus and stress at break than the former.

Table 2. Tensile properties of the PLA nanocomposites: E is the Young's Modulus, σ_y is the stress at yield, σ_b is the stress at break and ε_b is the elongation at break.

Samples	E (GPa)	σ_y (MPa)	σ_b (MPa)	ε_b (%)
P	3.5 ± 0.1	60.4 ± 0.3	57 ± 1	4.1± 0.5
P2CN* [a]	2.9 ± 0.1	–	58 ± 2	2.3 ± 0.4
P10low	2.3 ± 0.3	26 ± 0.3	33 ± 2	180± 10
P10low2CN	1.8 ± 0.3	23 ± 5	32 ± 2	160 ± 10
P10high2CN	2.5 ± 0.1	45 ± 5	34 ± 2	160 ± 10
P1low2CN	3.2 ± 0.8	52 ± 6	40 ± 7	10 ± 2
P5low2CN	2.8 ± 0.8	47 ± 3	32 ± 2	11.4 ± 0.9
P10low5CN	1.8 ± 0.3	34 ± 2	23 ± 2	160 ± 10
P10low12CN	1.7 ± 0.3	23 ± 5	22 ± 6	181 ± 6

[a] CN* is a sample of Chitin nanofibrils powder obtained by simply drying the CN suspension, without using PEG.

2.3. Thermal Properties

Thermal properties were recorded on nanocomposite injection molded specimens after a few months from preparation following a methodology consisting of a first heating step, cooling, and a second heating. Regarding the first heating (Table 3), it showed a glass transition T_g with an evident peak of enthalpy relaxation due to the ordering of PLA chains during the specimen storing period. The presence of PEG determined the decrease in the glass transition temperature and the decrease of the cold crystallization temperature. The cold crystallization temperature T_{cc} did not significantly change as a function of CN content, whereas the crystallinity significantly changed (Table 3).

Table 3. DSC results related to chitin nanofibrils PLA based nanocomposites (first heating).

	T_g (°C)	T_{cc} (°C)	ΔH_{cc} (J/g)	T_m (°C)	ΔH_m (J/g)	X_c (%)
P	57.7	106.9	21.24	149.2/157.4	21.4	0.2
P2CN*	57.5	107.7	23.21	148.8/155.9	25.5	2.4
P10low	42.1	77.6	17.8	154.7	27.1	9.9
P10low2CN	40.0	74.4	18.3	152.1	26.6	7.8
P10high2CN	43.8	76.6	6.7	154.7	33.3	28.4
P1low2CN	54.9	100.9	18.6	147.9/157.9	23.7	5.5
P5low2CN	47.7	88.6	15.6	(142.3)/155.9	19.2	3.8
P10low5CN	42.8	74.1	15.5	154.2	29.0	14.5
P10low12CN	40.8	75.5	11.4	152.3	29.1	19.0

T_g = glass transition temperature; Tcc = crystallization during heating peak temperature; ΔHcc = enthalpy of crystallization during heating; Tm = melting peak temperature; ΔHm = integral of the melting peak.

Regarding the second heating (Table 4), it is evident that the cold crystallization temperature significantly decreased as a function of CN content, in agreement with a slight nucleating action of the

CNs. This effect is evident only in the second heating thanks to the lower content of crystallinity X_c developed during the controlled cooling step.

Table 4. DSC results related to chitin nanofibrils PLA based nanocomposites (second heating).

	T_g (°C)	T_{cc} (°C)	ΔH_{cc} (J/g)	T_m (°C)	ΔH_m (J/g)	X_c (%)
P	59.1	109.4	23.8	148.3/157.4	26.3	2.8
P2CN*	55.2	108.8	23.3	147.9/155.8	24.5	1.3
P10low	36.1	83.8	21.3	153.7	32.6	12.1
P10low2CN	32.7	79.8	17.1	152.1	29.2	13.0
P10high2CN	31.4	80.5	17.5	154.2	29.3	12.7
P1low2CN	55.2	104.6	22.2	147.2/156.7	30.2	8.3
P5low2CN	45.3	92.4	24.6	141.4/157.4	27.2	2.7
P10low5CN	36.3	78.8	19.1	153.6	31.1	12.8
P10low12CN	29.8	73.3	20.4	151.6	28.8	9.0

T_g = glass transition temperature; T_{cc} = crystallization during heating peak temperature; ΔHcc = enthalpy of crystallization during heating; T_m = melting peak temperature; ΔHm = integral of the melting peak.

3. Discussion

The obtained results regarding phase morphology characterization, as well as the analysis of properties, agree with the achievement of a nanoscaled dispersion of chitin nanofibrils in plasticized PLA. The SEM characterization (Figures 6 and 7) and the observation of the optical properties of the injection molded specimens (Figure 5) evidenced this achievement and showed that it is dependent on the CN content and PEG content in the composites.

Thanks to infrared ATR characterization analysis, it was possible to show the clear presence of the CN bands on the injection molded specimens surface in samples containing more than 5% by weight of CN. Conditions for dispersing CNs in the bulk that resulted in the effective presence of CNs on the surface of injection molded specimens, for potentially exploiting their functional anti-microbial properties, were thus evidenced. It is interesting to note that, in the case where injection molded products are in a hot and humid environment (e.g., in applications related to human body, like implants or surgical suture wires), the surface of the PLA can degrade by hydrolysis, leaving the chitin nanofibrils to emerge on the surface. This behavior allows the preservation over time of the functional characteristics of CNs on the surface, even in the case of slow degradation of the PLA.

The thermal properties of the nanocomposites were analyzed as a function of CN and PEG 400 content (Figure 10). The trend of the glass transition as a function of the CN content is reported in Figure 10a. The values are almost constant but higher for the first heating than for the second. This difference can be ascribed to the formation of ordered structure in the sample injection molded and stored for some months (first heating), and is also responsible of the presence of the enthalpy relaxation peak in the glass transition range. Interestingly the glass transition temperature (T_g) showed an almost linear trend both in the first and second heating as a function of the PEG content (Figure 10b). The highest slope for the second heating trend can be explained by considering that in the first heating, the enthalpy relation made the samples less sensitive to plasticizer content. The linear fitting of the two trends allowed us to determine the intercept value of the line, corresponding to the glass transition temperature of nanocomposites without PEG. The values of extrapolated T_gs for pure PLA containing 2% of CN of 56.3 and 57.7 °C for first and second heating respectively. Interestingly the values recorded for the P2CN* sample were 57,5 and 55,2 °C for the first and second heatings, respectively (Tables 3 and 4). In the first heating, due to the uncontrolled storing conditions, the difference of about 1 °C seems to be irrelevant. In the second heating, recorded after a controlled cooling step, the increase of 2.5 °C is significant, and can be ascribed to the better dispersion achieved in the sample obtained by dispersing CN by using PEG, resulting in a better interaction between CNs and PLA matrix, than in the sample P2CN*, where CNs were present in agglomerated micrometric particles (Figure 6).

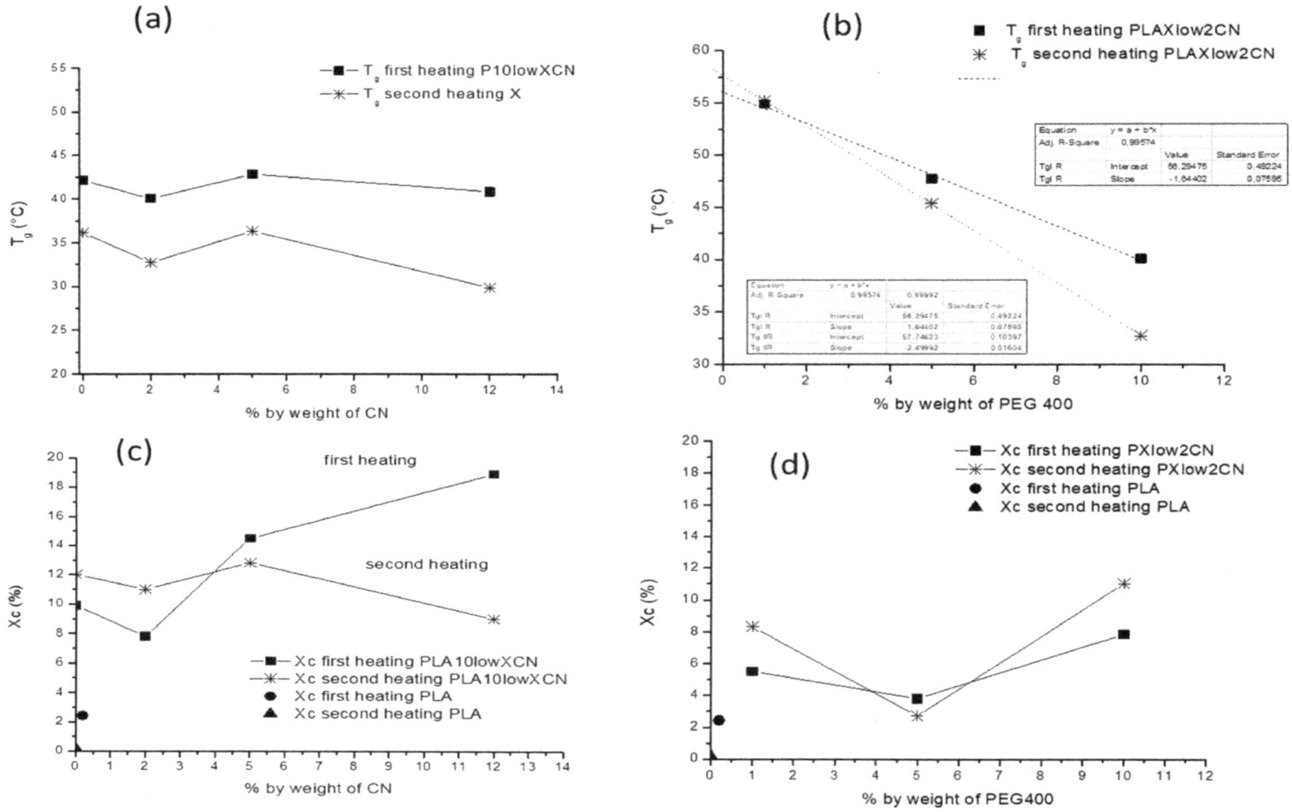

Figure 10. Analysis of thermal properties from first and second heating steps in PLA/Cn nanocomposites: (**a**) trend of Tg as a function of % by weight of CN; (**b**) trend of Tg as a function of PEG 400 content and dashed lines to extrapolate Tg at 0% by weight of PEG; (**c**) trend of crystallinity as function of % by weight of CN; (**d**) trend of crystallinity as a function of % by weight of PEG 400.

The crystallinity X_C was almost constant as a function of CN content in the second heating, showing an insignificant effect of CNs on controlled crystallization. In contrast, the crystallinity was significantly increased when the content of CN was 5% and 12% by weight. As the cooling of the injection molding process was very fast, this difference can be ascribed to the crystallinity developed during the storing of specimens. Hence, the presence of CNs in amounts higher than 2% resulted in an increase of crystallinity in the injection molding specimens during the storage that resulted in shrinkage or slow plasticizer expulsion. The surface of P10low5CN and P10low12CN became oily after some months from their preparation, whereas the P10low2CN specimens did not show any surface oiliness. This evidence may be relevant in view of the application of these nanocomposites to the injection molding sector.

The crystallinity as a function of % by weight of PEG400 for the nanocomposites at 2% by weight of CN showed a minimum value for both first and second heating at 5% by weight of PEG. The trend is like the one observed for plasticized and nucleated PLA by Fehri et al. [35]. When the concentration of the plasticizer is low (up to 5%), it hinders the crystal growth with respect to pure PLA. In contrast, when the concentration is high, the main effect of the plasticizer is to provide a higher free volume for segments motions, allowing the chain fragments to assemble more easily in crystals.

Based on the slight changes in properties observed in cases of relatively fast cooling conditions, the effect of thermal properties on tensile properties, performed a few days after the preparation of specimens by injection molding, can be considered almost negligible, in agreement with the studies carried out by Herrera et al. [31].

The stress strain curves recorded for the nanocomposites showed a trend like the one reported in Figure 10a as an example, where the stress at break resulted higher than the stress at yield. In Figure 11, a comparison between PEG 400 and PEG8000 regarding the tensile properties of the samples is proposed. The Young's Modulus (Figure 11b) decreased by adding the plasticizer alone, but the addition of CN resulted in a further decrease. Hence, the CNs, in the presence of 10% by weight of plasticizer, did not show a clear reinforcing action; this was also the case using PEG400 and PEG8000. The elongation at break (Figure 11e) increased by up to 180% by adding PEG 400 to PLA, and the presence of 2% by weight of nano-dispersed CNs did not result in a decrease in elongation at break both for PEG400 and PEG8000. The stress at break resulted in similar behavior for nanocomposites obtained with PEG400 and PEG8000, whereas the stress at yield resulted higher for PEG8000 than for PEG400. This can be ascribed to the higher mobility allowed in the system by the PEG400, with the lower molecular weight, allowing for easier sliding of macromolecules in correspondence with the beginning of the yield.

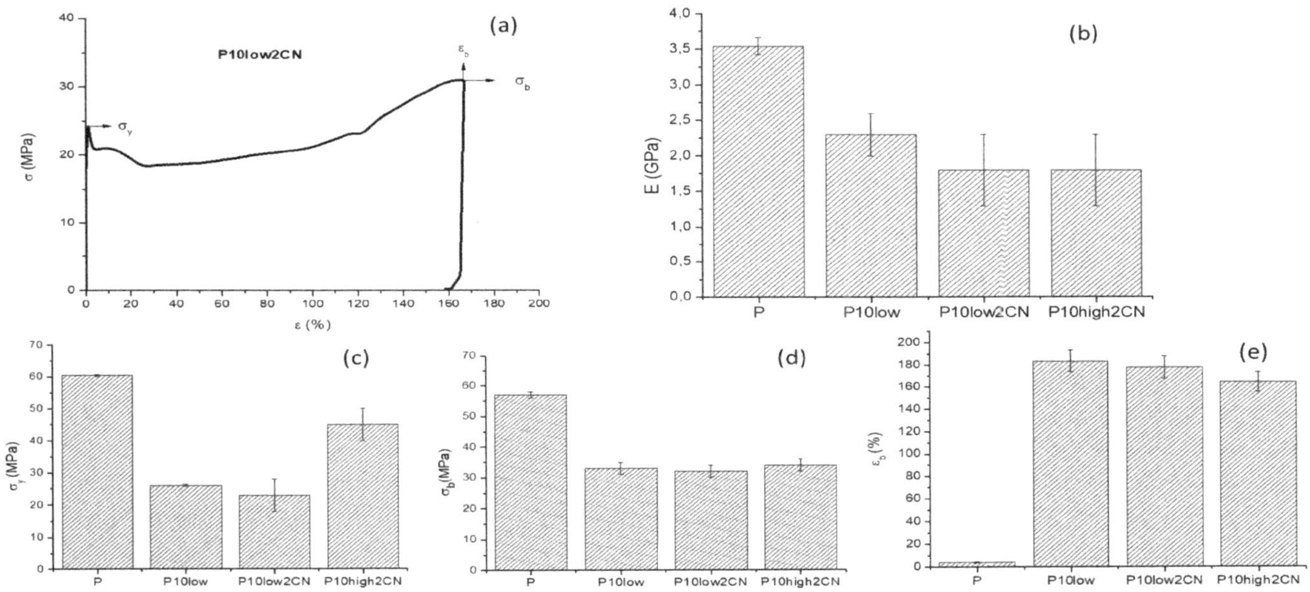

Figure 11. Tensile results to compare the use of PEG with high and low molecular weight: (**a**) example of stress vs. strain curve of the ductile composites; (**b**) Young's Modulus E; (**c**) Stress at yield σ_y; (**d**) stress at break σ_b; (**e**) elongation at break ε_b. Standard deviation is reported as error bars.

The tensile properties were also investigated by considering the trends of the different measurements as a function of chitin nanofibrils content and PEG content.

Regarding the chitin nanofibrils content investigated in the composite at 10% by weight of PEG400, the elastic Modulus slightly decreased by adding 2% by weight of CN, and remained almost constant at up to 12% by weight (Figure 12a). Stress at yield showed a maximum for the composite at 5% by weight of CN and the stress at break decreased as a function of CN content, showing the highest decrease, i.e., between 2% and 5% by weight, of CN. The nanocomposite at 5% seemed to be the most rigid; this was confirmed by the elongation at break data, showing a minimum for this composite. The variations of elongation at break are much limited in any case, which is in agreement with the good dispersion resulting from morphologic analysis.

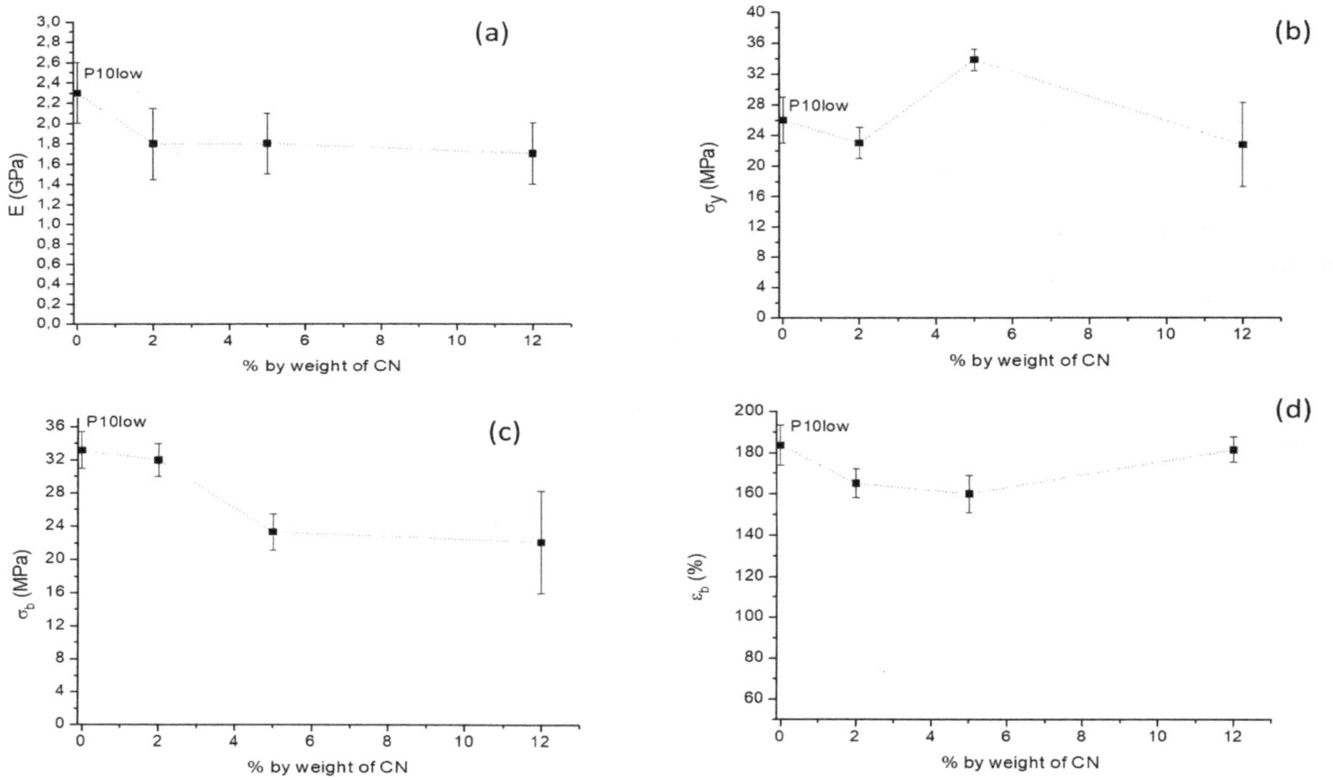

Figure 12. Tensile results as a function of CN content at constant PEG 400 concentration (10% by weight): (**a**) Young's Modulus E; (**b**) Stress at yield σ_y; (**c**) stress at break σ_b; (**d**) elongation at break ε_b.

The stress at yield significantly decreased as a function of PEG content (Figure 13b) because of the increased mobility due to plasticization contributing to decrease the energy required for the sliding of macromolecules with respect to the other. The stress at break slightly decreased as a function of PEG content because of a plasticization effect, decreasing the energy necessary for sample deformation. The elongation at break largely increased in between 5% and 10% by weight of PEG (Figure 13c). Only the nanocomposites obtained with a PEG content of 10% could reach elongation at break values above 150%. The Young's Modulus as a function of PEG content showed a decreasing linear trend (Figure 13a). The linear fitting allowed us to determine the value of the Modulus for the composite consisting of PLA and CN (Figure 13a) as the intercept of the obtained line. The obtained value by this methodology was 3.45 GPa, representing the Modulus of the PLA/CN 2% nanocomposite. It should be noted that the value obtained for the P2CN* sample presenting micrometric agglomerates of CN (Figure 6) was 2.9 GPa. The increase of 0,55 GPa can be reasonably attributed to the improved dispersion in the PLA matrix achieved thanks to the use of PEG 400. The increased interfacial surface between CN and PLA, favoring matrix-filler interactions, accounts for this improvement. It is important to note that the obtained value results were lower than the value determined for pure PLA (3.5 GPa). Hence the CNs could not reinforce the PLA matrix, either when micrometrically dispersed (P2CN*) or when nanometrically dispersed (extrapolated intercept value). These results are different from those obtained by Herrera et al. [31], who noticed an increase of Young's Modulus when only 1% of CNs was added. This effect on the mechanical properties can be attributed to the low affinity of the dispersing agent for the PLA matrix, TEC, used by these authors. This molecule, as well as other citrates, due to the presence of ester groups, showed a very high affinity for the PLA matrix [35–37], thus strongly favoring the nanodispersion of CN in the matrix and the formation of an extended CN/PLA interface. In contrast, PEG, having a high affinity for the polar groups of chitin, can support its nanodispersion, but can be more difficultly removed from CNs, remaining in between the CN surface and the PLA matrix. The PEG can thus coat the CN surface, lowering the reinforcing effect of nanofibers.

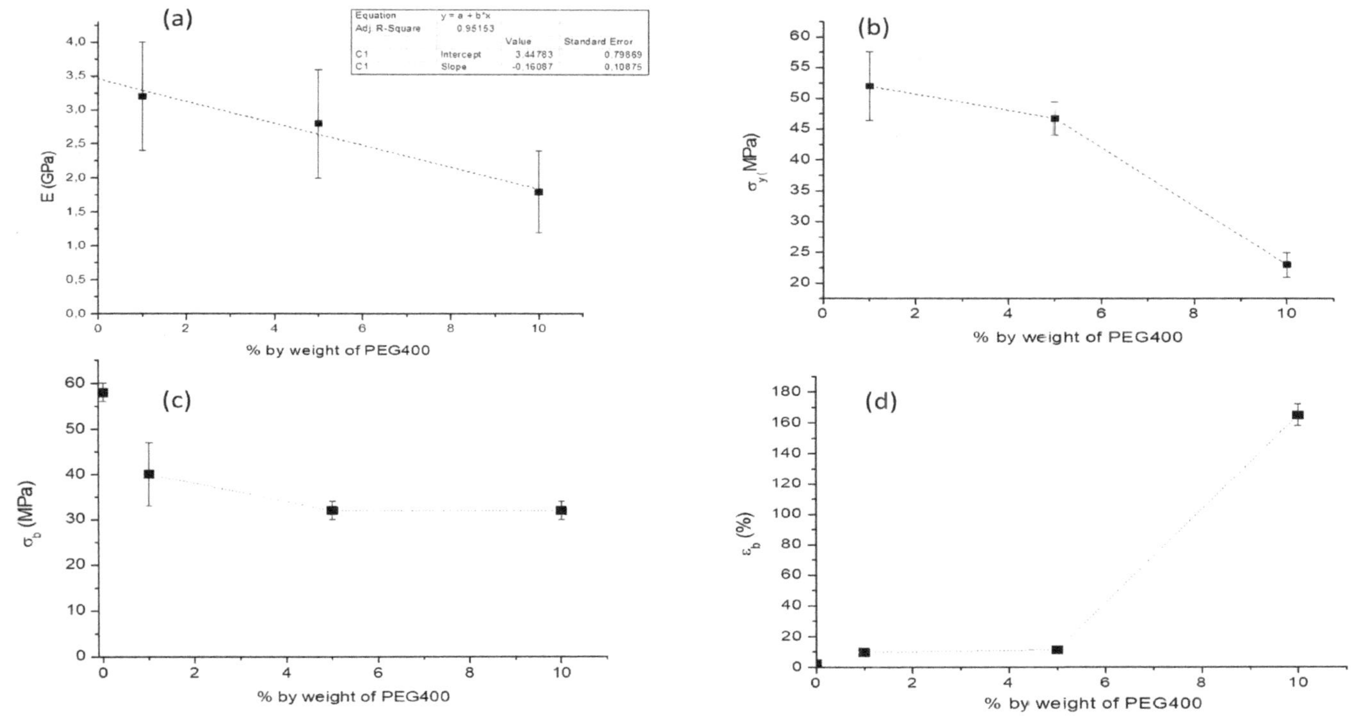

Figure 13. Tensile results as a function of PEG400 content at constant CN concentration (2% by weight): (**a**) Young's Modulus E; (**b**) Stress at yield σ_y; (**c**) stress at break σ_b; (**d**) elongation at break ε_b.

In literature, some samples prepared by the PEG method were studied, and it was noticed that some short fiber clusters are entangled with PLA matrix with a not evident phase separation between CNs and PLA. This observation revealed that PEG is a good interfacial compatibilizer for CNs and PLA. However, this reinforcing effect may be weakened due to a decrement of the CN aspect ratio. These aspects would require further research to better understand the effect of structural and morphologic parameters on the final properties.

4. Materials and Methods

4.1. Materials

Chitin nanofibrils (CN) water suspension at a concentration of 2% wt. was produced by MAVI SUD through its patented process [56], starting from chitin coming from seafood waste. For the preparation of the pre-composites, it was concentrated at 20% by weight.

Poly(ethylene glycol) (PEG), a liquid having a molecular weight of 400 (low), and PEG, a solid having a molecular weight of 1500, 4000, 6000 and 8000 (high), were purchased from Aldrich and used without any further purification.

PLA Ingeo™ 2003D, Extrusion Grade with density of 1.24 g/cm^3, a melt index of 6 g/10 min at 210 °C and 2.16 Kg, produced by NatureWorks LLC. It has a molecular weight of 170,000 g/mol and contains up to 4.1% isomeric D units. It was dried in ventilated oven at 60 °C for 16 h before extrusion trials.

4.2. Materials Preparation

PEG400 (or PEG800) were added to concentrated chitin nanofibrils suspension and stirred for two hours at room temperature. The amount was calculated considering that in the final pre-composites, the weight ratio of CN and PEG was 1:1. The obtained semiliquid emulsion was dried in a ventilated oven at 50 °C up to constant weight to obtain a solid when PEG8000, PEG6000, PEG4000 and PEG1500 were used, and a sample with liquid highly viscous consistency when PEG 400 was used.

The extrusion of the PLA 2003D (Ingeo™ Nature Works, Minnetonka, MN, USA) in the presence of PEG8000 or PEG400 was carried out after drying the material for 16 h at 60 °C in a ventilated oven, using a TwinLab II Haake™ Rheomex CTW 5 laboratory screw extruder (Vreden, Germany). After homogenizing using mortar and pestle, the materials were fed into the extruder from the hopper at the beginning of the twin screws, and were mixed into the recirculating channel of the extruder. The extrusion was carried out at 180 °C and 90 rpm for one minute. After extrusion, the molten material was transferred through a preheated cylinder into the Haake™ MiniJet II mini injection molding machine to obtain Haake type III test specimens for tensile testing. The injection molding was carried out at 180 °C, 650 bars, holding time of 15 s, mold temperature of 35 °C.

4.3. Characterizations

The analysis of the average length and width of chitin nanofibrils was carried out using ImageJ software applied on micrographs obtained by using a FEI Quanta 450 ESEM FEG field emission instrument.

The morphology of master batches and composites was studied by scanning electron microscopy (SEM) using a JEOL JSM-5600LV instrument and analyzing cryo-fractured surfaces, previously subjected to sputtering with gold.

Infrared spectra were recorded in the 550–4000 cm^{-1} range with a Nicolet 380 Thermo Corporation Fourier Transform Infrared (FTIR) Spectrometer equipped with smart Itx ATR accessory, collecting 256 scans at 4 cm^{-1} resolution

Thermogravimetric analysis was performed on 10–20 mg of sample using a Mettler-Toledo TGA/SDTA 851 instrument (Columbus, OH, USA) operating with nitrogen as the purge gas (60 mL/min) at 10 °C·min^{-1} heating rate in the 25–800 °C temperature range.

The nanocomposites samples were cut with a Reichert Ultracut E ultramicrotome into ultrafine sheet (<1 micron thickness) and collected onto double folding copper grids (50/100 mesh). The grids were closed and coated with a thin layer of carbon by a EMITECH K950 Evaporator Coater Sputter (Laughton, UK) to make them electrically conductive. The micrographs were obtained with a FEI Quanta 450 ESEM FEG in STEM mode.

Tensile tests (UNI EN ISO527) were carried out using a universal INSTRON 5500R test machine with a 1 kN load cell at a speed of 10 mm/min onto specimens conditioned for 2 weeks at 25 °C and relative humidity of 50%.

Differential scanning calorimetry analyzes (DSC) were performed on material sampled from injection molded specimens using a TA Q200 instrument with nitrogen as carrier gas and indium as a calibration standard. The samples were heated from −100 °C to 250 °C at 10 °C/min and cooled from 250 °C to −100 °C at 20 °C/min. The second heating was carried out by heating analogously from −100 °C to 200 °C at 10 °C/min. The crystallinity was calculated by the formula $Xc = [(\Delta H_m - \Delta H_{cc})/(\Delta H_0 \cdot w)] \times 100$, ΔH_m is the melting enthalpy, ΔH_{cc} is the cold crystallization enthalpy, ΔH_0 is the melting enthalpy of PLA fully crystalline (the value of 93J/g was considered [52]) and w is the weight fraction of PLA in the composite.

5. Conclusions

In the present paper, the preparation by extrusion of nanocomposites consisting of a PLA matrix and a dispersed chitin nanofibrils (CN) phase was obtained using a methodology based on the preliminary preparation of pre-composites based on CN and poly(ethylene glycol) (PEG) having a molecular weight of 8000 or 400. The presence of PEG, making it possible to keep the CNs separated, avoiding the problem of their agglomeration due to inter-macromolecular interactions.

The pre-composites were dispersed in molten PLA to obtain nanocomposites with different content of PEG and CN in a laboratory extruder. The morphologic analysis by SEM demonstrated the absence of agglomerated CNs in the sample, in agreement with the presence of a nanostructured material.

The content of PEG and CN strongly influenced the colorlessness and the transparency of the injection molded specimens. In particular, a higher PEG content and a lower CN content resulted in more transparent and colorless specimens. The presence of CNs on the surface of the injection molded specimens was detected in the nanocomposites with a content of CN higher than 5%.

The thermal properties showed few relevant changes due to the addition of CNs to the samples, being only responsible of a slight nucleating action. Interestingly, it was found that in injection molded specimens, the polymeric chains slowly reorganize into ordered structures during storage at room temperature, thus showing both enthalpy relaxation peak and increased crystallinity. This reorganization is evident above 5% by weight of CNs.

The tensile properties did not show a reinforcing effect of CNs, but the achieved good dispersion allowed us to maintain high values of elongation at break (>150%), typical of plasticized PLA, up to 12% by weight of CN. The absence of a reinforcing effect, not in agreement with literature data, was tentatively explained by considering that PEG, having a high affinity for the polar groups of chitin, can support its nano-dispersion, but can be more difficult to remove from CNs, remaining in between CNs surface and PLA matrix. The PEG can thus coat the CN surface, lowering the reinforcing effect of nanofibers. This aspect, useful for selecting the correct CN/dispersing agent system, would necessitate further elucidation.

The methodology investigated and developed here, in comparison with the other methodologies for dispersing CNs in PLA, can offer the advantage of being easily applicable also at an industrial scale, and does not modify the thermo-mechanical properties typical of plasticized PLA. Moreover, this methodology can be advantageous in the case of the study of ductile materials for injection molding or flexible materials, to be applied to plastic films, both with low contents of CNs, in order to exploit their functional properties.

Author Contributions: Conceptualization, M.-B.C. and A.L.; methodology, M.-B.C., V.G., L.P. and L.A.; formal analysis, M.-B.C. and L.P.; resources, P.C. and P.M.; data curation, V.G., L.P. and L.A.; writing—original draft preparation, M.-B.C.; writing—review and editing, L.A., P.C. and A.L.; supervision, P.M. and A.L.; project administration, M.B.C. and P.C.; funding acquisition, P.C. and A.L.

Acknowledgments: Valter Castelvetro and Sabrina Bianchi, of the Department of Chemistry and Industrial Chemistry of the University of Pisa, are thanked for supporting in TGA characterization. Irene Anguillesi is thanked for technical support. Randa Ishak is thanked for the kind assistance in electron microscopy characterization.

References

1. Kim, S.K. (Ed.) *Chitin, Chitosan, Oligosaccharides and Their Derivatives*, 1st ed.; CRC Press: Boca Raton, FL, USA, 2011; p. 241.
2. Strelcova, Z.; Kulhanek, P.; Friak, M.; Fabritius, H.O.; Petrov, M.; Neugebauer, J.; Koca, J. The structure and dynamics of chitin nanofibrils in an aqueous environment revealed by molecular dynamics simulations. *RSC Adv.* **2016**, *6*, 30710–30721. [CrossRef]
3. Fazeni, K.; Lindorfer, J. The Energy Institute at the Johannes Kepler University, 2011. Available online: http://www.energyefficiency.at/web/projekte/chibio-1-53-1-53-1-53-1-53.html (accessed on 28 February 2015).
4. Nisticò, R. Aquatic-Derived Biomaterials for a Sustainable Future: A European Opportunity. *Resources* **2017**, *6*, 65. [CrossRef]
5. Younes, I.; Rinaudo, M. Chitin and Chitosan Preparation from Marine Sources. Structure, Properties and Applications. *Mar. Drugs* **2015**, *13*, 1133–1174. [CrossRef] [PubMed]
6. Morganti, P.; Muzzarelli, C. Spray Dried Chitin Nanofibrils, Method for Production and Uses Thereof. U.S. Patent 2009/0203642, 13 August 2009.
7. Gadgey, K.K.; Bahekar, A. Studies on extraction methods of chitin from crab shell and investigation of its mechanical properties. *Int. J. Mech. Eng. Technol.* **2017**, *8*, 220–231.
8. Wu, J.; Zhang, K.; Girouard, N.; Carson Meredith, J. Facile Route to Produce Chitin Nanofibers as Precursors for Flexible and Transparent Gas Barrier Materials. *Biomacromolecules* **2014**, *15*, 4614–4620. [CrossRef] [PubMed]
9. Gao, X.; Chen, X.; Zhang, J.; Guo, W.; Jin, F.; Yan, N. Transformation of Chitin and Waste Shrimp Shells into Acetic Acid and Pyrrole. *ACS Sustain. Chem. Eng.* **2016**, *4*, 3912–3920. [CrossRef]

10. De Azeredo, H.M.C. Antimicrobial nanostructures in food packaging. *Trends Food Sci. Technol.* **2013**, *30*, 56–69. [CrossRef]
11. Tsai, G.-J.; Su, W.-H.; Chen, H.-C.; Pan, C.-L. Antimicrobial activity of shrimp chitin and chitosan from different treatments and applications of fish preservation. *Fish. Sci.* **2002**, *68*, 170–177. [CrossRef]
12. Jayakumara, R.; Prabaharan, M.; Sudheesh Kumar, P.T.; Naira, S.V.; Tamura, H. Biomaterials based on chitin and chitosan in wound dressing applications. *Biotechnol. Adv.* **2011**, *29*, 322–337. [CrossRef]
13. Garrison, T.F.; Murawski, A.; Quirino, R.L. Bio-Based Polymers with Potential for Biodegradability. *Polymers* **2016**, *8*, 262. [CrossRef]
14. Pandey, A.; Chauhan, N.P.S.; Shabafrooz, V.; Ameta, R.; Mozafari, M. Polylactic acid and polyethylene glycol as antimicrobial agents. In *Biocidal Polymers*, 1st ed.; Chauhan, N.P.S., Ed.; Smither Rapra: Shawbury, UK, 2016; pp. 131–144.
15. Rabnawaz, M.; Wyman, I.; Auras, R.; Cheng, S. A roadmap towards green packaging: The current status and future outlook for polyesters in the packaging industry. *Green Chem.* **2017**, *19*, 4737–4753. [CrossRef]
16. Chen, Y.; Geever, L.M.; Killion, J.A.; Lyons, J.G.; Higginbotham, C.L.; Devine, D.M. Review of Multifarious Applications of Poly (Lactic Acid). *Polym.-Plast. Technol. Eng.* **2016**, *55*, 1057–1075. [CrossRef]
17. Saini, P.; Arora, M.; Kumar, M.N.V.R. Poly(lactic acid) blends in biomedical applications. *Adv. Drug Deliv. Rev.* **2016**, *107*, 47–59. [CrossRef] [PubMed]
18. Yu, H.Y.; Zhang, H.; Song, M.L.; Zhou, Y.; Yao, J.; Ni, Q.Q. From Cellulose Nanospheres, Nanorods to Nanofibers: Various Aspect Ratio Induced Nucleation/Reinforcing Effects on Polylactic Acid for Robust-Barrier Food Packaging. *ACS Appl. Mater. Interfaces* **2017**, *9*, 43920–43938. [CrossRef] [PubMed]
19. Basu, A.; Nazarkovsky, M.; Ghadi, R.; Khan, W.; Domb, A.J. Poly(lactic acid)-based nanocomposites. *Polym. Adv. Technol.* **2017**, *28*, 919–930. [CrossRef]
20. Scaffaro, R.; Botta, L.; Lopresti, F.; Maio, A.; Sutera, F. Polysaccharide nanocrystals as fillers for PLA based nanocomposites. *Cellulose* **2017**, *24*, 447–478. [CrossRef]
21. Jonoobi, M.; Harun, J.; Mathew, A.P.; Oksman, K. Mechanical properties of cellulose nanofiber (CNF) reinforced polylactic acid (PLA) prepared by twin screw extrusion. *Compos. Sci. Technol.* **2010**, *70*, 1742–1747. [CrossRef]
22. Trifol Guzman, J.; Plackett, D.; Sillard, C.; Szabo, P.; Bras, J.; Daugaard, A.E. Hybrid poly(lactic acid)/nanocellulose/nanoclay composites with synergistically enhanced barrier properties and improved thermomechanical resistance. *Polym. Int.* **2016**, *65*, 988–995. [CrossRef]
23. Nagarajan, V.; Mohanty, A.K.; Misra, M. Perspective on Polylactic Acid (PLA) based Sustainable Materials for Durable Applications: Focus on Toughness and Heat Resistance. *ACS Sustain. Chem. Eng.* **2016**, *4*, 2899–2916. [CrossRef]
24. Fan, Y.M.; Saito, T.; Isogai, A. Preparation of chitin nanofibers from squid pen beta-chitin by simple mechanical treatment under acid conditions. *Biomacromolecules* **2008**, *9*, 1919–1923. [CrossRef]
25. Nakagaito, A.N.; Kanzawa, S.; Takagi, H. Polylactic Acid Reinforced with Mixed Cellulose and Chitin Nanofibers—Effect of Mixture Ratio on the Mechanical Properties of Composites. *J. Compos. Sci.* **2018**, *2*, 36. [CrossRef]
26. Araki, J. Preparation of Sterically Stabilized Chitin Nanowhisker Dispersions by Grafting of Poly(ethylene glycol) and Evaluation of Their Dispersion Stability. *Biomacromolecules* **2015**, *16*, 379–388. [CrossRef] [PubMed]
27. Araki, J.; Wada, M.; Kuga, S. Steric Stabilization of a Cellulose Microcrystal Suspension by Poly(ethylene glycol) Grafting. *Langmuir* **2001**, *17*, 21–27. [CrossRef]
28. Araki, J.; Kuga, S.; Magoshi, J. Influence of reagent addition on carbodiimide-mediated amidation for poly(ethylene glycol) grafting. *J. Appl. Polym. Sci.* **2002**, *85*, 1349–1352. [CrossRef]
29. Zhang, Q.; Wei, S.; Huang, J.; Feng, J.; Chang, P.R. Effect of Surface Acetylated-Chitin Nanocrystals on Structure and Mechanical Properties of Poly (lactic acid). *J. Appl. Polym. Sci.* **2014**, *131*, 39809. [CrossRef]
30. Rizvi, R.; Cochrane, B.; Naguib, H.; Lee, P.C. Fabrication and characterization of melt-blended polylactide-chitin composites and their foams. *J. Cell. Plast.* **2011**, *47*, 283–300. [CrossRef]
31. Herrera, N.; Singh, A.A.; Salaberria, A.M.; Labidi, J.; Mathew, A.P.; Oksman, K. Triethyl Citrate (TEC) as a Dispersing Aid in Polylactic Acid/Chitin Nanocomposites Prepared via Liquid-Assisted Extrusion. *Polymers* **2017**, *9*, 406. [CrossRef]
32. Guan, Q.; Naguib, H.E. Fabrication and Characterization of PLA/PHBV-Chitin Nanocomposites and Their Foams. *J. Polym. Environ.* **2014**, *22*, 119–130. [CrossRef]

33. Herrera, N.; Salaberria, A.M.; Mathew, A.P.; Oksman, K. Plasticized polylactic acid nanocomposite films with cellulose and chitin nanocrystals prepared using extrusion and compression molding with two cooling rates: Effects on mechanical, thermal and optical properties. *Compos. Part A* **2016**, *83*, 89–97. [CrossRef]
34. Herrera, N.; Roch, H.; Salaberria, A.M.; Pino-Orellana, M.A.; Labidi, J.; Fernandes, S.C.M.; Radic, D.; Leiva, A.; Oksman, K. Functionalized blown films of plasticized polylactic acid/chitin nanocomposite: Preparation and characterization. *Mater. Des.* **2016**, *92*, 846–852. [CrossRef]
35. Fehri, M.K.; Mugoni, C.; Cinelli, P.; Anguillesi, I.; Coltelli, M.B.; Fiori, S.; Montorsi, M.; Lazzeri, A. Composition dependence of the synergistic effect of nucleating agent and plasticizer in poly(lactic acid): A Mixture Design study. *eXPRESS Polym. Lett.* **2016**, *10*, 274–288. [CrossRef]
36. Coltelli, M.B.; Della Maggiore, I.; Bertoldo, M.; Signori, F.; Bronco, S.; Ciardelli, F. Poly(lactic acid) Properties as a Consequence of Poly(butylene adipate-co-terephthalate) Blending and Acetyl Tributyl Citrate Plasticization. *J. Appl. Polym. Sci.* **2008**, *110*, 1250–1262. [CrossRef]
37. Quero, E.; Müller, A.J.; Signori, F.; Coltelli, M.B.; Bronco, S. Isothermal Cold-Crystallization of PLA/PBAT Blends with and without the Addition of Acetyl Tributyl Citrate. *Macromol. Chem. Phys.* **2012**, *213*, 36–48. [CrossRef]
38. Scatto, M.; Salmini, E.; Castiello, S.; Coltelli, M.B.; Conzatti, L.; Stagnaro, P.; Andreotti, L.; Bronco, S. Plasticized and nanofilled poly(lactic acid)-based cast films: Effect of plasticizer and organoclay on processability and final properties. *J. Appl. Polym. Sci.* **2013**, *127*, 4947–4956. [CrossRef]
39. Coltelli, M.B.; Cinelli, P.; Anguillesi, I.; Salvadori, S.; Lazzeri, A. Structure and properties of extruded composites based on bio-polyesters and nano-chitin, Session M: Functional Textiles-from research and development to innovations and industrial uptake. In Proceedings of the Symposium E-MRS Fall Meeting, Warsaw University of Technology, Warsaw, Poland, 15–18 September 2014; p. 192.
40. Cinelli, P.; Coltelli, M.B.; Mallegni, N.; Morganti, P.; Lazzeri, A. Degradability and sustainability of nanocomposites based on polylactic acid and chitin nano fibrils. *Chem. Eng. Trans.* **2017**, *60*, 1–6.
41. Coltelli, M.B.; Gigante, V.; Panariello, L.; Aliotta, L.; Morganti, P.; Danti, S.; Cinelli, P.; Lazzeri, A. Chitin nanofibrils in renewable materials for packaging and personal care applications. *Adv. Mater. Lett.* **2019**, in press.
42. Li, J.; Gao, Y.; Zhao, J.; Sun, J.; Li, D. Homogeneous dispersion of chitin nanofibers in polylactic acid with different pretreatment methods. *Cellulose* **2017**, *24*, 1705–1715. [CrossRef]
43. Nakagaito, A.N.; Yamada, K.; Ifuku, S.; Morimoto, M.; Saimoto, H. Fabrication of Chitin Nanofiber-Reinforced Polylactic Acid Nanocomposites by an Environmentally Friendly Process. *J. Biobased Mater. Bioenergy* **2013**, *7*, 152–156. [CrossRef]
44. Shamshina, J.L.; Zavgorodnya, O.; Berton, P.; Chhotaray, P.K.; Choudhary, H.; Rogers, R.D. An Ionic Liquid Platform for Spinning Composite Chitin-Poly(lactic acid) Fibers. *ACS Sustain. Chem. Eng.* **2018**, *6*, 10241–10251. [CrossRef]
45. Jolanta Kumirska, J.; Czerwicka, M.; Kaczyński, Z.; Bychowska, A.; Brzozowski, K.; Thöming, J.; Stepnowski, P. Application of Spectroscopic Methods for Structural Analysis of Chitin and Chitosan. *Mar. Drugs* **2010**, *8*, 1567–1636. [CrossRef]
46. Pearson, F.G.; Marchessault, R.H.; Liang, C.Y. Infrared spectra of crystalline polysaccharides. V. Chitin. *J. Polym. Sci.* **1960**, *13*, 101–116. [CrossRef]
47. Brunner, E.; Ehrlich, H.; Schupp, P.; Hedrich, R.; Hunoldt, S.; Kammer, M.; Machill, S.; Paasch, S.; Bazhenov, V.V.; Kurek, D.V.; et al. Chitin-based scaffolds are an integral part of the skeleton of the marine demosponge Ianthella basta. *J. Struct. Biol.* **2009**, *168*, 539–547. [CrossRef] [PubMed]
48. Focher, B.; Naggi, A.; Torri, G.; Cosani, A.; Terbojevich, M. Structural differences between chitin polymorphs and their precipitates from solutions-evidence from CP-MAS ^{13}CNMR, FT-IR and FT-Raman spectroscopy. *Carbohydr. Polym.* **1992**, *17*, 97–102. [CrossRef]
49. Cheng, D.; Wen, Y.; Wang, L.; An, X.; Zhua, X.; Ni, Y. Adsorption of polyethylene glycol (PEG) onto cellulose nano-crystals to improve its dispersity. *Carbohydr. Polym.* **2015**, *123*, 157–163. [CrossRef] [PubMed]
50. Paul, D.R.; Robeson, L.M. Polymer nanotechnology: Nanocomposites. *Polymer* **2008**, *49*, 3187–3204. [CrossRef]
51. Biswas, S.K.; Shams, M.I.; Das, A.K.; Islam, M.N.; Nazhad, M.M. Flexible and Transparent Chitin/Acrylic Nanocomposite Films with High Mechanical Strength. *Fibers Polym.* **2015**, *16*, 774–781. [CrossRef]
52. Mikesová, J.; Hasek, J.; Tishchenko, G.; Morganti, P. Rheological study of chitosan acetate solutions containing chitin nanofibrils. *Carbohydr. Polym.* **2014**, *112*, 753–757. [CrossRef] [PubMed]

53. Xu, X.; Wang, H.; Jiang, L.; Wang, X.; Payne, S.A.; Zhu, J.Y.; Li, R. Comparison between Cellulose Nanocrystal and Cellulose Nanofibril Reinforced Poly(ethylene oxide) Nanofibers and Their Novel Shish-Kebab-Like Crystalline Structures. *Macromolecules* **2014**, *47*, 3409–3416. [CrossRef]
54. Cardinaels, R.; Moldenaers, P. Morphology development in immiscible polymer blends. In *Polymer Morphology. Principles, Characterization and Properties*; Guo, Q., Ed.; John Wiley and Sons: Hoboken, NJ, USA, 2016; Chapter 19; pp. 348–373.
55. Potschke, P.; Paul, D.R. Formation of co-continuous structures in met-mixed immiscible polymer blends. *J. Macromol. Sci. C Polym. Rev.* **2003**, *43*, 87–141. [CrossRef]
56. Muzzarelli, C.; Morganti, P. Preparation of Chitin and Derivatives Thereof for Cosmetic and Therapeutic Use. U.S. Patent 8383157 B2, 26 February 2013.

8

Characterization of *Ecklonia cava* Alginate Films Containing Cinnamon Essential Oils

Su-Kyoung Baek, Sujin Kim and Kyung Bin Song *

Department of Food Science and Technology, Chungnam National University, Daejeon 34134, Korea; sukyoungb@cnu.ac.kr (S.-K.B.); pppink32@gmail.com (S.K.)

* Correspondence: kbsong@cnu.ac.kr

Abstract: In this study, *Ecklonia cava* alginate (ECA) was used as a base material for biodegradable films. Calcium chloride ($CaCl_2$) was used as a cross-linking agent, and various concentrations (0%, 0.4%, 0.7%, and 1.0%) of cinnamon leaf oil (CLO) or cinnamon bark oil (CBO) were incorporated to prepare active films. The ECA film containing 3% $CaCl_2$ had a tensile strength (TS) of 17.82 MPa and an elongation at break (E) of 10.36%, which were higher than those of the film without $CaCl_2$. As the content of essential oils (EOs) increased, TS decreased and E increased. Addition of CLO or CBO also provided antioxidant and antimicrobial activities to the ECA films. The antioxidant activity of the ECA film with CBO was higher than that of the film containing CLO. In particular, the scavenging activities of the 2,2-diphenyl-1-picrylhydrazyl (DPPH) and 2,2′-azino-bis (3-ethylbenzothiazoline-6-sulphonic acid) (ABTS) radicals in the ECA film containing 1% CBO were 50.45% and 99.37%, respectively. In contrast, the antimicrobial activities against *Escherichia coli* O157:H7, *Salmonella* Typhimurium, *Staphylococcus aureus*, and *Listeria monocytogenes* were superior in the ECA films with CLO. These results suggest that ECA films containing CLO or CBO can be applied as new active packaging materials.

Keywords: active packaging materials; alginate films; antimicrobial agents; antioxidant activity; biodegradable films; essential oils

---

1. Introduction

In recent years, environmental pollution caused by plastic waste has worsened, and interest in eco-friendly packaging materials has increased. In particular, biodegradable films have been studied as new packaging materials in the food industry [1–3]. As biodegradable film materials, carbohydrates, proteins, and lipids are used individually or in combination.

Seaweeds can grow in abundance, and they are an excellent alternative to replace limited land resources [2]. Among them, *Ecklonia cava* is a brown alga present mainly in the East Asian region [4], and it contains approximately 26% alginate by dry weight [5]. Alginate, which is mainly extracted from brown algae, is non-toxic, relatively inexpensive, and biocompatible, and it can be considered as a base material for biodegradable films [3]. However, *Ecklonia cava* alginate (ECA) has not been studied as a biodegradable film.

Alginate is composed of guluronic acid and mannuronic acid, and their ratio affects alginate film properties. When the proportion of guluronic acid is high, the alginate film is hard and fragile. On the contrary, when the proportion of mannuronic acid is high, elastic and soft gels can be formed [6]. Ionic bonds can be formed by the interactions between divalent cations such as calcium and the carboxyl groups of guluronic acid, and the alginate polymer forms an egg-box structure surrounding the divalent cations [7,8], resulting in films with rigid structures by cross-linking. Therefore, the incorporation of calcium chloride ($CaCl_2$) in alginate films can be effective in improving the physical properties of the films.

Active packaging can protect foods during storage and extend their shelf life by the addition of active materials to packaging materials [9]. Essential oils (EOs), classified as "Generally Recognized as Safe" by the Food and Drug Administration, are added as natural active materials to produce active packaging [1]. Among EOs, cinnamon oil is extracted from various parts of cinnamon (*Cinnamomum verum*) such as leaf, bark, flower, and root, but cinnamon leaf oil (CLO) and cinnamon bark oil (CBO) are commonly used as EOs. The major compounds in CLO and CBO are eugenol and trans-cinnamaldehyde, respectively, which contribute to antioxidant and antimicrobial activities [10,11]. CLO and CBO can be added to ECA films as natural active materials to produce active packaging films.

Therefore, this study aimed to prepare alginate films extracted from *Ecklonia cava* with improved physical properties and to characterize the films containing various concentrations of CLO or CBO.

2. Results and Discussion

2.1. Attenuated Total Reflectance-Fourier Transform Infrared (ATR-FTIR) Analysis of ECA

To investigate the molecular characteristics of ECA, ATR-FTIR analysis was performed and the results are presented in Figure 1. A sharp peak at 3600–3500 cm^{-1} indicates the presence of free OH groups [12], but it was not observed in the FTIR spectrum of ECA. The broad band at 3500–3100 cm^{-1} was observed owing to the vibration of the hydroxyl group [12,13]. In the case of alginate, the stretching of protonated carboxyl groups appeared at approximately 1730 cm^{-1}. When monovalent ions (sodium) displaced the proton, peaks on the polymeric backbone of alginate appeared at approximately 1600 cm^{-1} and 1400 cm^{-1} by asymmetric and symmetric stretching vibration of the carboxylate group of sodium alginate [8]. In this study, the bands were observed at 1607.01 cm^{-1} and 1426.74 cm^{-1}. These results are in accordance with the data found in previous reports on alginate [7,8]. The band at 1027.94 cm^{-1} represented the stretching vibration of C-O, and the band at 880.16 cm^{-1} was related to the C1-H deformation vibration of β-mannuronic acid residues [7]. In general, the content of guluronic acid (G) and mannuronic acid (M) in alginate varies, depending upon the alginate species. Since G and M have characteristic bands at 1025 cm^{-1} and 1100 cm^{-1}, respectively, an approximate M/G ratio can be predicted by FTIR analysis [14], and the M/G ratio of ECA was estimated to be lower than 1 in the present study.

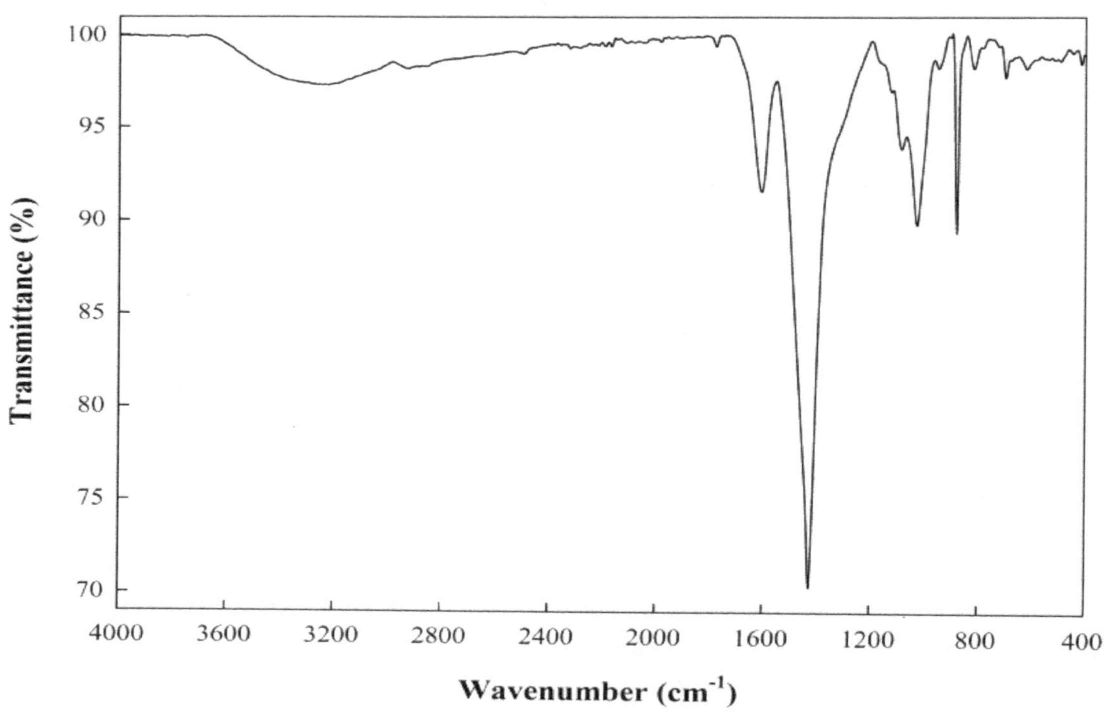

Figure 1. FTIR spectra of *Ecklonia cava* alginate (ECA).

2.2. Physical Properties of ECA Films with $CaCl_2$

The concentration of the cross-linking agent and the type and concentration of the plasticizer have been optimized based on preliminary experiments. In this study, $CaCl_2$ was incorporated as a cross-linking agent, and various concentrations of $CaCl_2$ (1%, 2%, 3%, 5%, and 7%) were added to the ECA films. The physical properties of the ECA films prepared with varying contents of $CaCl_2$ are shown in Table 1. The thickness and tensile strength (TS) of the ECA films with added $CaCl_2$ increased compared to those of films without $CaCl_2$. In particular, the TS of the ECA film containing 3% $CaCl_2$ was significantly improved from 10.49 MPa to 17.82 MPa compared to that of the ECA film without $CaCl_2$. The incorporation of $CaCl_2$ increased the TS because of the formation of more rigid polymers via interactions between the calcium ions and carboxyl groups of alginate [15]. Zactiti and Kieckbusch [16] reported that increasing calcium concentration resulted in a thicker "egg-box" structure and the formation of a stronger network. Although the addition of $CaCl_2$ improved the TS of the ECA film, it did not affect water vapor permeability (WVP). Similar results have been reported in the literature [8,17]. In the present study, the concentration of $CaCl_2$ was optimized to 3% based on the mechanical properties of the film.

Table 1. Mechanical properties of the *Ecklonia cava* alginate (ECA) films containing different amounts of $CaCl_2$.

$CaCl_2$ (%)	Thickness (μm)	Tensile Strength (MPa)	Elongation at Break (%)	Water Vapor Permeability ($\times 10^{-9}$ g /m s Pa)
0	34.60 ± 0.86^{d}	10.49 ± 1.38^{d}	9.30 ± 0.89^{bc}	1.89 ± 0.13^{a}
1	35.00 ± 0.81^{cd}	14.78 ± 2.74^{b}	9.90 ± 0.56^{ab}	1.86 ± 0.16^{a}
2	35.96 ± 1.42^{bc}	12.99 ± 1.65^{bc}	9.96 ± 0.52^{ab}	1.85 ± 0.15^{a}
3	36.68 ± 0.61^{b}	17.82 ± 1.05^{a}	10.36 ± 0.27^{a}	1.78 ± 0.02^{a}
5	37.12 ± 0.64^{b}	15.26 ± 1.80^{b}	9.69 ± 0.60^{ab}	1.83 ± 0.10^{a}
7	38.80 ± 1.20^{a}	10.79 ± 1.55^{cd}	8.85 ± 0.49^{c}	1.74 ± 0.07^{a}

Means ± S.D., $n = 5$, $^{a\text{–}d}$: any means in the same column followed by different letters are significantly ($p < 0.05$) different by Duncan's multiple range test.

2.3. Physical Properties of ECA Films Containing EOs

EOs were added to ECA films at different concentrations (0.4%, 0.7%, and 1.0%), and Table 2 shows the physical properties of ECA films incorporated with CLO or CBO. The addition of EOs increased the film thickness and elongation at break (E) values, whereas the TS decreased. In particular, the addition of CLO or CBO at 0.7% into the ECA films increased E from 10.36% to 18.25% or 18.65%, respectively, resulting in flexible ECA films. EOs can interfere with the interactions between alginate and calcium ions, thereby improving the flexibility of the films by reducing the intermolecular forces along the polymer chain. Benavides et al. [15] found that, as the content of oregano EOs increased in alginate films, TS decreased and E increased because oregano EOs could act as a plasticizer and increase flexibility and chain mobility. These findings are consistent with those in previous studies on other alginate films containing EOs [18,19].

In general, the addition of hydrophobic EOs causes a decrease in WVP in the films [1]. It has been reported that the incorporation of EOs into various films improved the WVP [20–22]. On the contrary, the WVP of ECA films containing CLO or CBO increased as the contents of EOs increased in the present study. This difference can be explained by the formation of pores via changes in the internal structure of the film matrix. Previously, it was reported that the WVP of alginate films incorporated with garlic oil increased as a result of expansion by the intermolecular interactions in the film matrix [19]. Similarly, the WVP of alginate films with other EOs was not improved [18,23]. Atarés et al. [24] also reported that the WVP of the films was not easily improved by the addition of hydrophobic materials, although the incorporation of lipids in the microstructure of the films affected the WVP.

Table 2. Physical properties of the ECA films containing essential oils (EOs).

EOs (%)		Thickness (μm)	Tensile Strength (MPa)	Elongation at Break (%)	Water Vapor Permeability ($\times10^{-9}$ g /m s Pa)
Control	0	36.68 ± 0.61^{d}	17.82 ± 1.05^{a}	10.36 ± 0.27^{c}	1.78 ± 0.02^{d}
CLO	0.4	40.12 ± 0.73^{c}	17.40 ± 0.87^{ab}	17.94 ± 2.59^{ab}	2.01 ± 0.05^{c}
	0.7	44.12 ± 0.67^{b}	15.10 ± 1.88^{cd}	18.25 ± 3.32^{ab}	2.35 ± 0.06^{b}
	1.0	49.80 ± 0.60^{a}	14.94 ± 1.22^{cd}	17.28 ± 1.40^{ab}	2.56 ± 0.05^{a}
CBO	0.4	40.40 ± 1.14^{c}	16.02 ± 0.80^{bc}	16.92 ± 1.73^{ab}	1.99 ± 0.03^{c}
	0.7	44.08 ± 0.61^{b}	15.45 ± 1.32^{c}	18.65 ± 2.06^{a}	2.30 ± 0.07^{b}
	1.0	49.68 ± 0.46^{a}	13.58 ± 0.92^{d}	15.61 ± 1.77^{b}	2.54 ± 0.09^{a}

Means ± S.D., $n = 5$, $^{a-d}$: any means in the same column followed by different letters are significantly ($p < 0.05$) different by Duncan's multiple range test. CLO: cinnamon leaf oil; CBO: cinnamon bark oil.

2.4. Optical Properties of ECA Films Containing EOs

The optical characteristics of the films are important because they directly affect the appearance and quality of food products. They influence not only customer preference but also the lipid oxidation rate, which can affect the quality of foods [20]. As shown in Table 3, the incorporation of EOs decreased the values of L * and b *, whereas the value of a * increased. The total color difference (ΔE *) in ECA films with EOs increased compared with that in films without EOs. Furthermore, the ECA films gradually became opaque with increasing EO content. The opacity values of ECA films with CLO or CBO at 1.0% were 25.12 or 22.54 A/mm, respectively, whereas ECA films without EOs had an opacity of 9.77 A/mm. Similar to our results, the addition of CBO into chitosan films decreased lightness, whereas redness and opacity increased [25].

Table 3. Optical properties of the ECA films containing EOs.

EOs (%)		L *	a *	b *	ΔE *	Opacity (A/mm)
Control	0	72.03 ± 0.27^{a}	5.16 ± 0.10^{e}	49.19 ± 0.34^{ab}	-	9.77 ± 0.31^{f}
CLO	0.4	68.52 ± 0.40^{b}	7.71 ± 0.26^{d}	49.76 ± 0.32^{a}	4.39 ± 0.48^{d}	13.43 ± 0.51^{d}
	0.7	66.40 ± 0.43^{d}	8.53 ± 0.19^{b}	45.21 ± 1.08^{d}	7.75 ± 0.35^{b}	22.64 ± 0.53^{b}
	1.0	64.73 ± 1.11^{e}	8.96 ± 0.66^{a}	44.55 ± 0.77^{d}	9.48 ± 1.20^{a}	25.12 ± 0.37^{a}
CBO	0.4	68.48 ± 0.33^{b}	7.52 ± 0.19^{d}	49.00 ± 0.97^{ab}	4.37 ± 0.36^{d}	12.12 ± 0.19^{e}
	0.7	67.12 ± 1.10^{c}	8.17 ± 0.24^{c}	48.52 ± 0.81^{b}	5.86 ± 1.10^{c}	20.85 ± 0.34^{c}
	1.0	64.01 ± 0.67^{f}	8.59 ± 0.42^{b}	46.87 ± 1.29^{c}	9.11 ± 0.79^{a}	22.54 ± 0.54^{b}

Means ± S.D., $n = 10$, $^{a-f}$: any means in the same column followed by different letters are significantly ($p < 0.05$) different by Duncan's multiple range test.

2.5. Scanning Electron Microscopy (SEM) Analysis

Cross-sectional images showing the internal structure of ECA films incorporated with CLO or CBO are presented in Figure 2. As the content of EOs increased, it was apparent that the microstructure was not compact and pores were formed. The structural change and pore formation in the film increased the WVP. Similarly, when CBO was added to gelatin films, pores were formed by aggregation caused by irregular dispersion of hydrophobic molecules as the concentration of EOs increased [26]. Zhang et al. [23] also reported that alginate films had a relatively homogeneous and smooth surface, but the addition of cinnamon EOs or soybean EOs resulted in irregular surfaces and pores. It was reported that these irregular surfaces and pores were generated by the arrangement of alginate molecules and the agglomeration of oil droplets when the films were formed.

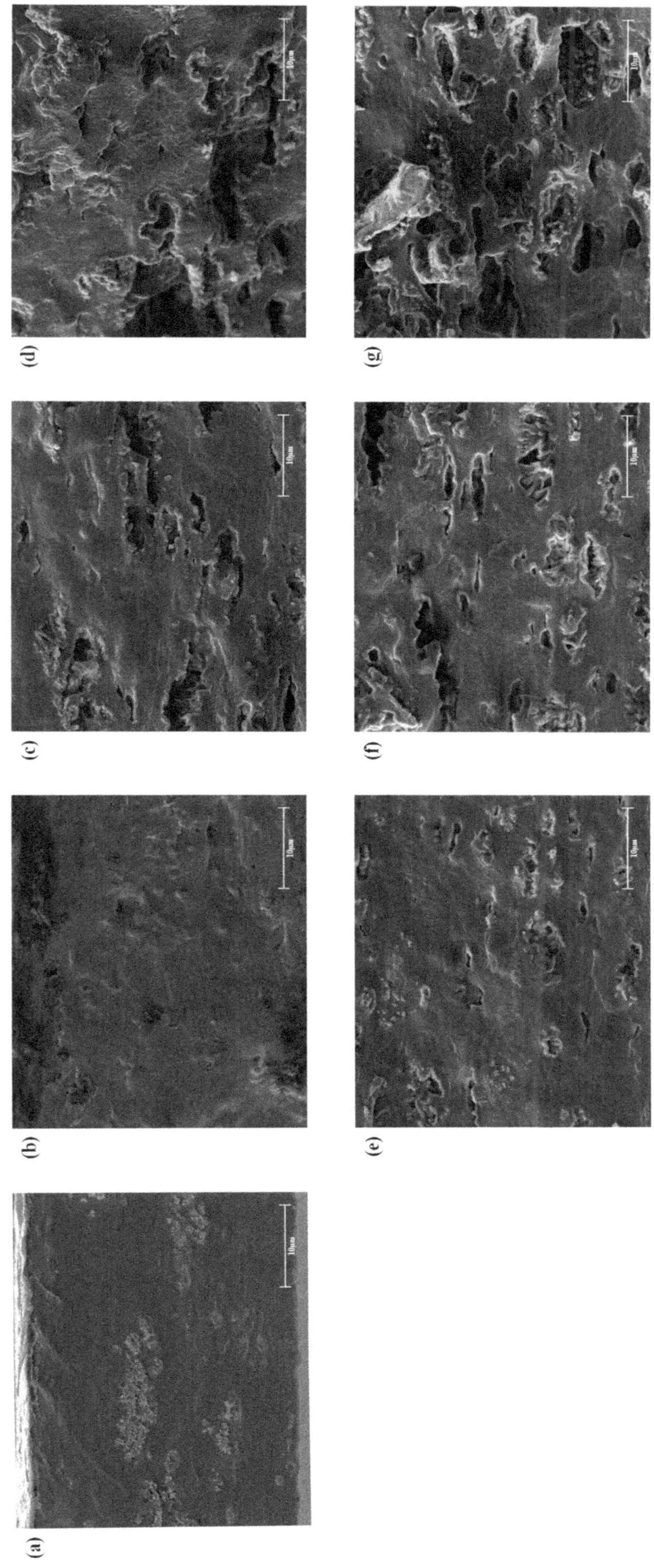

Figure 2. SEM images of the ECA films containing essential oils (EOs). (**a**) Control, (**b**) cinnamon leaf oil (CLO) 0.1%, (**c**) CLO 0.7%, (**d**) CLO 1.0%, (**e**) cinnamon bark oil (CBO) 0.4%, (**f**) CBO 0.7%, and (**g**) CBO 1.0%.

2.6. Thermal Analysis

Figure 3 shows the effect of the incorporation of CLO or CBO on the thermal stability of the ECA films using thermogravimetric analysis (TGA). The initial loss of mass appeared to be due to loss of moisture in the ECA film matrix [27]. The main reduction of mass occurred in the range of 200 °C to 300 °C, and the decrease in this region was caused by the volatilization of glycerol used as a plasticizer and the alginate polymer [28]. A loss of mass in ECA films with CLO or CBO was observed at 450 °C, probably due to the presence of highly stable substances in the EOs. At 600 °C, the residue amounts in the control, CLO 0.7%, and CBO 0.7% films were 46.93%, 40.31%, and 40.68%, respectively. The amount of residue was highest in the control, suggesting high thermal stability. The decrease in thermal stability of ECA films incorporated with EOs could be due to the structural change in the ECA film network, causing a decrease in TS by the incorporation of EOs.

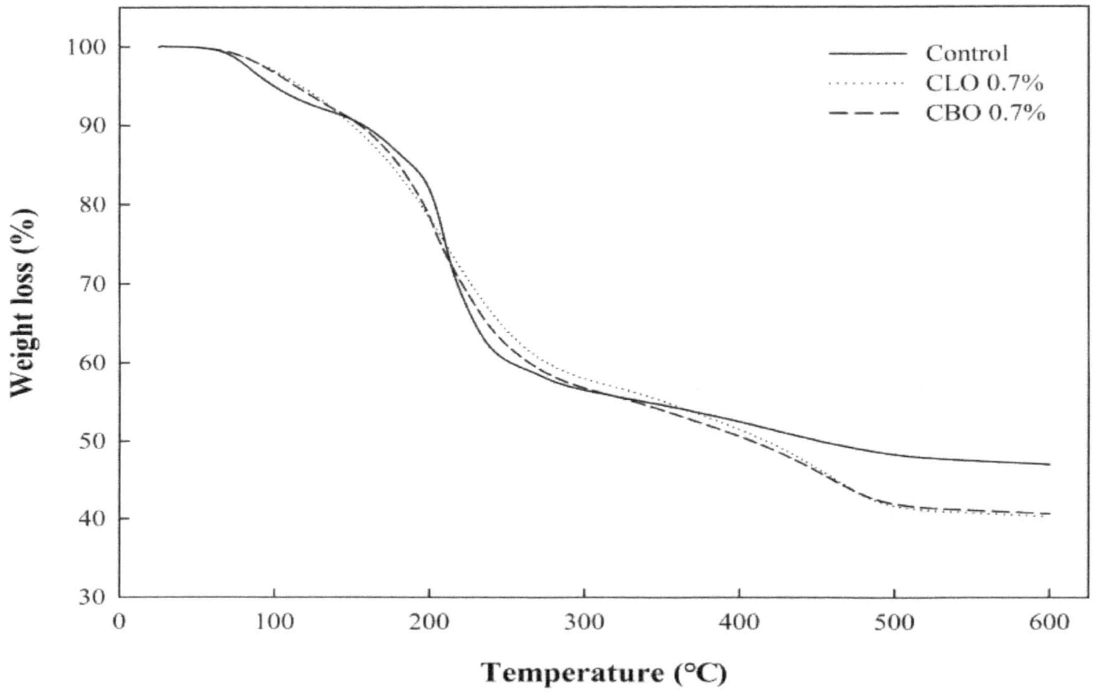

Figure 3. Thermogravimetric analysis (TGA) curves of the ECA films containing cinnamon leaf oil (CLO) and cinnamon bark oil (CBO).

The results of differential scanning calorimetry (DSC) analysis of the ECA films, the first heating scans, are presented in Figure 4. The ECA film without EOs (control) showed two T_e (endothermic peak temperature) at 86.63 (T_{e1}) and 233.65 °C (T_{e2}). This observation was similar to the results of DSC analysis of alginate having Te at 86 °C and 197 °C [29]. However, it should be noted that the T_{e2} was higher for the ECA film, probably due to the cross-linking effect of $CaCl_2$. T_{e1} reflected water loss, and the addition of CLO and CBO to the ECA film increased T_{e1} to 109.81 °C and 108.46 °C, respectively. The ECA films also showed an exothermic peak, and a second endothermic peak T_{e2} appeared at 233.65 °C in the control and at 236.66 °C and 229.14 °C for ECA films with CLO and CBO, respectively. These peaks could be attributed to recrystallization of ECA after heat induction [30]. The values of T_g (glass transition temperature) were measured in the second heating scan. In general, the T_g of alginate is around 120 °C [31]. In this study, the T_g of the control was 126.34 °C and decreased to 103.54 °C and 103.63 °C with the addition of CLO and CBO, respectively. The lower T_g was reflected by the decrease in TS due to the addition of EOs in the ECA films, thereby lowering the amount of thermal energy required to transform the ECA films from a glassy to a rubbery state. Tongnuanchan et al. [32] also observed similar results in gelatin films, where EOs interfered with the interactions of polymers and increased the mobility of the polymer by weakening the film structure and consequently lowering T_g. These results clearly support the TGA results, where the thermal stability of the ECA films with CLO or CBO did not improve compared to the control.

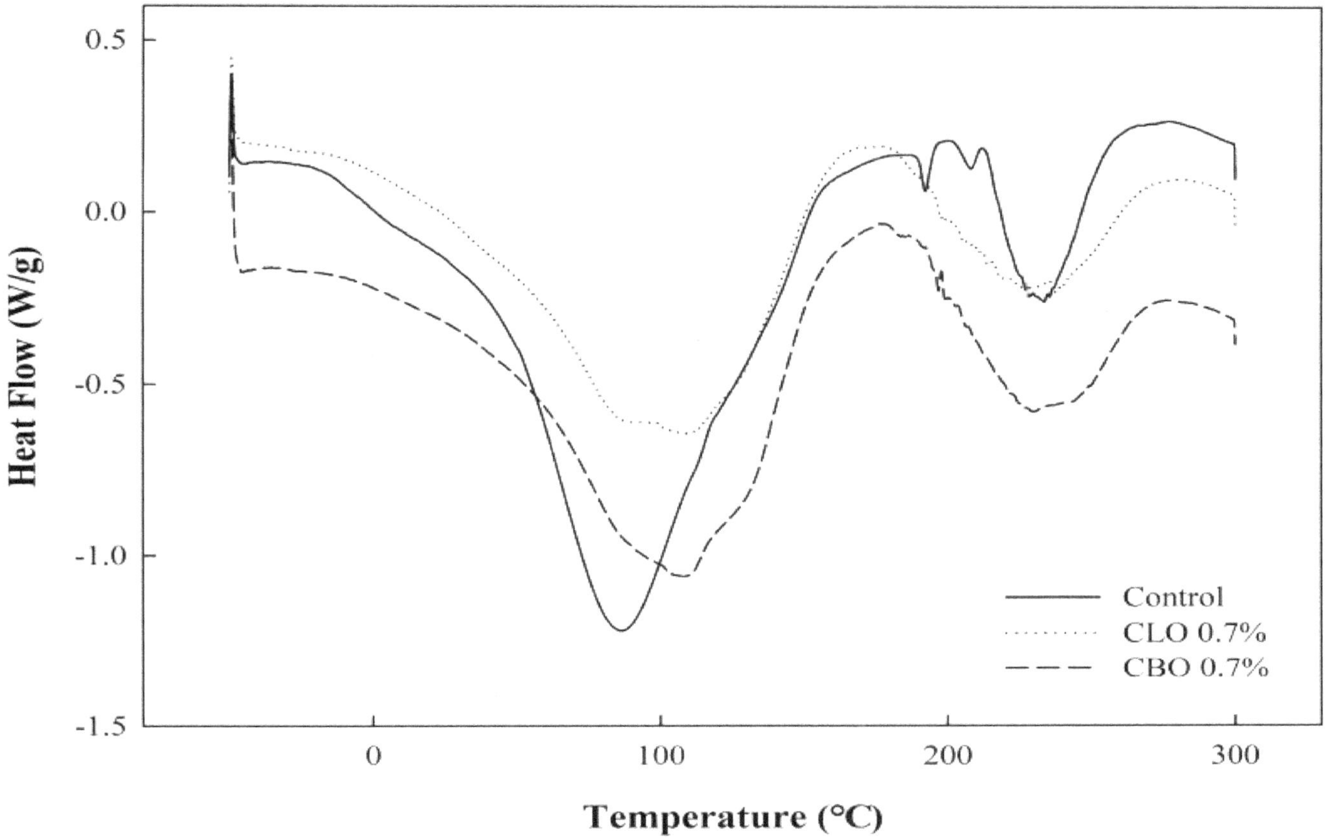

Figure 4. DSC curves of the ECA films containing CLO and CBO.

2.7. *Antioxidant Activity*

The antioxidant property of ECA films containing EOs was evaluated (Figures 5 and 6). As the content of CLO or CBO increased, the radical scavenging properties of 2,2-diphenyl-1-picrylhydrazyl (DPPH) and 2,2′-azino-bis (3-ethylbenzothiazoline-6-sulphonic acid) (ABTS) increased. The antioxidant activity of the ECA films with CBO was superior to that of the films with CLO. ECA films containing CLO at various concentrations did not exhibit significant differences in antioxidant activity, except for the film incorporated with 1.0% CLO. In contrast, ECA films with CBO exhibited significant differences in antioxidant activity as the EO content increased. The DPPH radical scavenging activities of ECA films containing CLO or CBO at 1.0% were 9.58% and 50.45%, respectively, and the ABTS radical scavenging activities were 55.05% and 99.37%, respectively. It is known that eugenol has higher radical scavenging activity than trans-cinnamaldehyde [33]. However, the effect of CBO on the radical scavenging activities of DPPH and ABTS was greater than that of CLO in the present study. Singh et al. [34] reported that the DPPH radical scavenging ability of CBO was better than that of CLO, even though eugenol has stronger DPPH radical scavenging ability than trans-cinnamaldehyde. These results suggest that antioxidant activity is affected not only by the major component of the EOs, eugenol or trans-cinnamaldehyde, but also by other constituents in the EOs. Overall, our results suggest that ECA films containing CBO are more suitable as antioxidant films than those containing CLO.

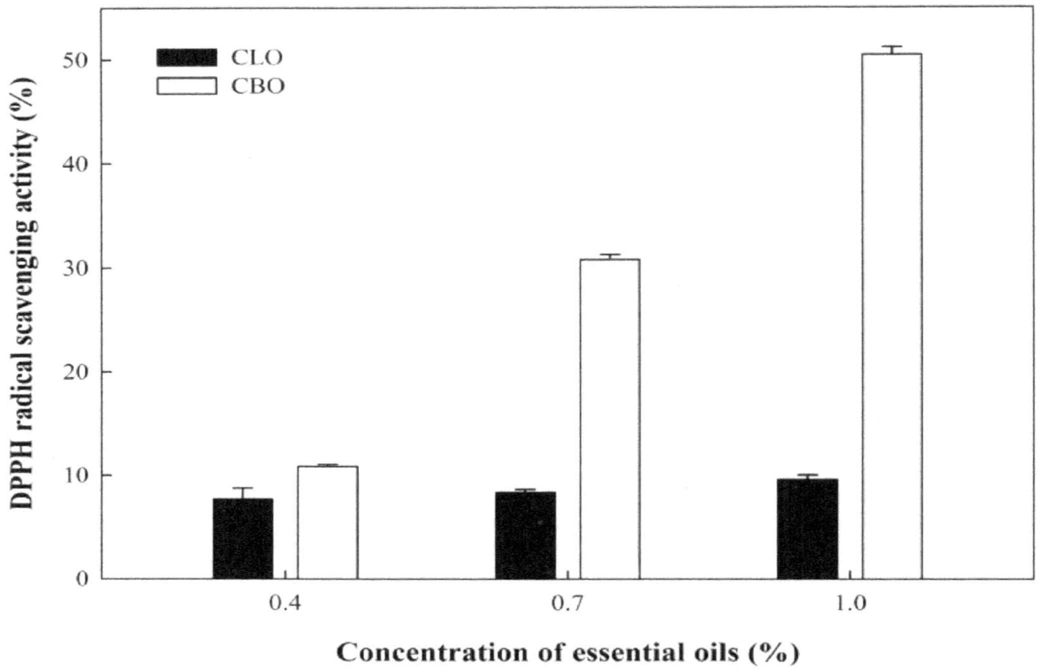

Figure 5. DPPH radical scavenging activity of the ECA films containing EOs.

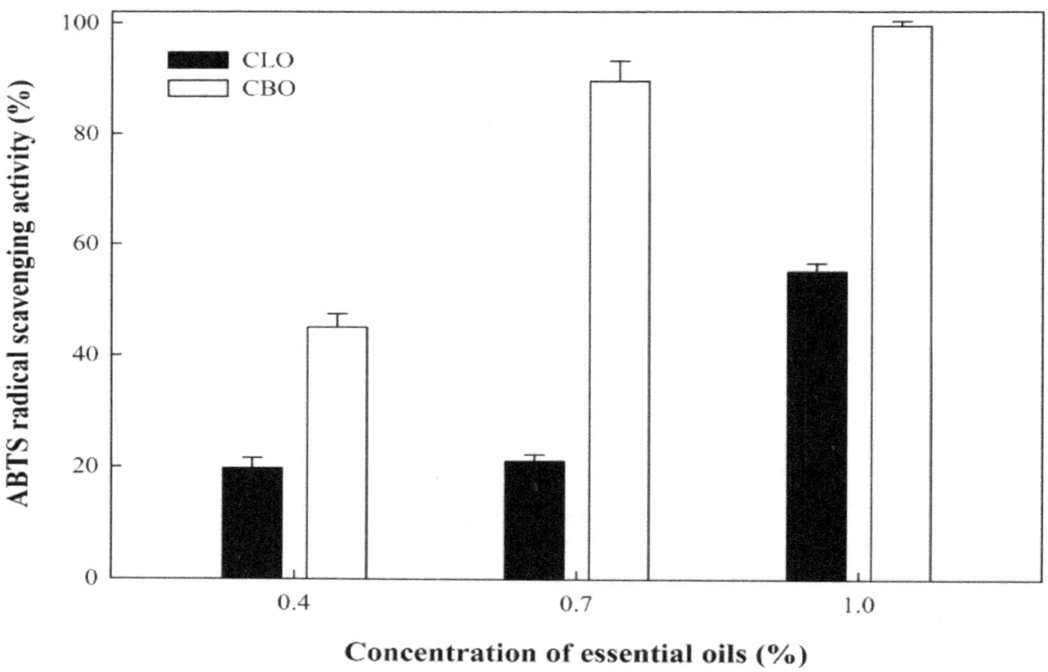

Figure 6. ABTS radical scavenging activity of the ECA films containing EOs.

2.8. Antimicrobial Activity

Table 4 shows the inhibitory effects of ECA films containing EOs against various bacteria (*Escherichia coli* O157:H7, *Salmonella* Typhimurium, *Staphylococcus aureus*, and *Listeria monocytogenes*). As expected, ECA films without CLO or CBO did not show antimicrobial activity against bacterial strains. In addition, the ECA film, containing 0.4% EO, showed no antimicrobial activity because of the low EO concentration. In contrast, ECA films containing 0.7% and 1.0% EOs exhibited antimicrobial activity, and the inhibition zones against four strains significantly increased as the amount of EOs increased. The antibacterial ability of EOs is mainly due to the presence of functional compounds, such as eugenol and trans-cinnamaldehyde, which disrupt the bacterial cell structure by penetrating cell membranes and inhibiting the production of essential enzymes in bacteria [35]. As shown in Table 3, ECA films containing CLO showed higher antimicrobial activity against four strains, when compared to the films incorporated with CBO. This is probably due to the difference in antimicrobial activity between eugenol, the main component of CLO, and trans-cinnamaldehyde, a major component of CBO [36]. Kang and Song [37] demonstrated the higher antibacterial effect of CLO (higher than that of CBO) by measuring the minimum inhibitory concentration of the EOs.

Table 4. Antimicrobial activity of the ECA films containing EOs.

EOs (%)		Inhibition Zone (mm)			
		E. coli O157:H7	*S.* Typhimurium	*S. aureus*	*L. monocytogenes*
Control	0	ND *	ND	ND	ND
CLO	0.4	ND	ND	ND	ND
	0.7	12.04 ± 0.10 [c]	11.49 ± 0.14 [c]	10.77 ± 0.07 [b]	13.17 ± 0.49 [b]
	1.0	14.50 ± 0.51 [a]	15.37 ± 0.16 [a]	14.28 ± 0.25 [a]	15.02 ± 0.17 [a]
CBO	0.4	ND	ND	ND	ND
	0.7	10.88 ± 0.08 [d]	10.77 ± 0.09 [d]	10.35 ± 0.52 [b]	12.03 ± 0.66 [c]
	1.0	13.48 ± 0.17 [b]	12.76 ± 0.22 [b]	13.93 ± 0.36 [a]	14.48 ± 0.13 [a]

Means ± S.D., $n = 5$, [a–d]: any means in the same column followed by different letters are significantly ($p < 0.05$) different by Duncan's multiple range test, * ND: not detected.

3. Materials and Methods

3.1. Materials

Ecklonia cava (Jeju-do, Korea) was processed in the laboratory, as described below. Glycerol and $CaCl_2$ were purchased from Samchun Pure Chemical Co. (Pyeongtaek-si, Gyeonggi-do, Korea). Span 80 was purchased from Yakuri Pure Chemical Co. (Osaka, Japan). CLO and CBO were obtained from Gooworl Co. (Daegu, Korea).

3.2. Alginate Extraction

Alginate extraction from *Ecklonia cava* was performed as reported previously [38,39]. Ground *Ecklonia cava* in 0.025% H_2SO_4 solution (1:50, w/v) was stirred for 3 h. After filtration, the residue was washed with distilled water and filtered again. The residue in 3% Na_2CO_3 solution (1:50, w/v) was heated at 60 °C for 2 h. Distilled water (1:25, w/v) was added and the mixture was centrifuged at 5000× g for 10 min. Methanol (95%, 1:50, v/v) was added to the supernatant and kept at 4 °C for 1 h. After centrifugation (10,000× g, 20 min), the supernatant was removed and the precipitate was lyophilized to obtain ECA.

3.3. ATR-FTIR Spectroscopy

FTIR spectrometry (Vertex 80v, Bruker, Billerica, MA, USA) was used to obtain ATR-FTIR spectra of ECA. Analysis was performed with 16 scans for each spectrum in the region between 4000 cm^{-1} and 400 cm^{-1}.

3.4. Preparation of ECA Films Containing EOs

To prepare the film-forming solution, ECA (1.5%), glycerol (0.3 g/g ECA), and $CaCl_2$ (0.03 g/g ECA) were dispersed in distilled water, followed by stirring at 70 °C for 30 min until complete dissolution was achieved. After homogenization (12,000 rpm, 3 min), various concentrations (0.4%, 0.7%, and 1.0% of total volume) of EOs and Span 80 (25% of EOs) were incorporated in the solution. Next, the mixture was homogenized at 12,000 rpm for 7 min, sonicated for 8 min, and degassed for

5 min. The filtered film solution (16 mL) was uniformly spread onto Petri dishes (90 mm diameter) and dried at 25 °C for 18 h.

3.5. Physical Properties

A micrometer (No. 7327, Mitutoyo Co., Kawasaki, Japan) was used to measure the thickness of each film. TS and E of the ECA films containing CLO or CBO were evaluated with a Testometric machine (Model 250-2.5 CT, Testometric Co., Lancashire, UK). The films were cut into a rectangular shape (2.54 cm × 10 cm) for the measurement of their physical properties such as TS and E. The initial grip distance in the instrument was 50 mm, and its cross-head speed was set at 50 mm/min. The TS of the films, which is the maximum stress to withstand before breakage and is measured as the highest point of stress–strain curve, was obtained by dividing the maximum force (N), applied when the film was broken, by the cross-sectional area of the film. In addition, the E (%) was calculated by dividing the maximum length (mm) at the moment of film rupture by the initial grip distance (mm), followed by multiplying with 100. The method used for WVP measurement was previously described by Lee and Song [40]. Before the measurement, the water content of the ECA films was controlled by conditioning in a thermos-hygrostat (25 °C, 50% relative humidity) for 24 h.

3.6. Optical Properties

To characterize the color of the ECA films containing CLO or CBO, a colorimeter (CR-400M, Minolta, Tokyo, Japan) was used. According to the CIELAB color scale, the L *, a *, and b * values of the prepared ECA films were evaluated on a white standard plate with L *, a *, and b * values of 97.37, −0.14, and 2.00, respectively. The difference in color between the ECA film without EO and those containing EO was indicated as ΔE *. The method described by Yang et al. [41] was used to obtain the opacity of the ECA films with a spectrophotometer (UV-2450, Shimadzu Co., Kyoto, Japan).

3.7. SEM Analysis

To investigate the microstructure of the ECA films, SEM analysis was conducted. Before the analysis, the ECA films containing CLO or CBO were fractured in liquid nitrogen and coated with a platinum layer. SEM images of cross sections were acquired using an ion beam scanning electron microscope (LYRA3 WMU, TESCAN, Brno, Czech Republic) under an accelerating voltage of 5 kV. All images were displayed at a magnification of 5000×.

3.8. Thermal Property Analysis

To analyze the thermal properties of the ECA films, TGA and DSC were conducted. TGA was carried out with a thermogravimetric analyzer (Mettler Toledo, Columbus, OH, USA). The films (3.8 ± 0.1 mg) were sealed in a sample pan and heated in the temperature range of 25–600 °C at a rate of 10 °C/min under nitrogen flow at 50 mL/min. DSC measurements were carried out using a DSC1 (Mettler Toledo, Schwerzenbach, Switzerland). Each sample was scanned from −50 °C to 300 °C at a rate of 10 °C/min. After the first scan was completed, the second scan was conducted in the same way.

3.9. Antioxidant Activity

To examine the antioxidant property of the ECA films, DPPH and ABTS were used. The method described by Shojaee-Aliabadi et al. [42] was used to measure the DPPH radical scavenging activities of the ECA films. The film solution (1%) for this measurement was prepared by diluting the ECA film (0.1 g) in distilled water (10 mL) and shaking in an incubator for 30 min. Next, the solution was mixed with the DPPH solution at a ratio of 1:39 and stored in the dark for 1 h, followed by measuring the absorbance at 517 nm. The ABTS radical scavenging activities of the ECA films were also determined [43]. Before the test, ABTS solution was prepared by blending 7 mM ABTS and 2.45 mM potassium persulfate (2:1). After keeping the solution in the dark for 16 h, ethanol was added to dilute the solution until it had an absorbance value of 0.7 at 734 nm. The ABTS solution was then mixed with the film solution (1%). After being stored in the dark for 10 min, the supernatant of the mixture was separated by centrifugation (3000 rpm, 5 min) and the absorbance was measured at 734 nm.

3.10. Antimicrobial Activity

The disc diffusion method was applied to confirm the antibacterial property of the ECA films. *E. coli* O157:H7 (NCTC 12079, ATCC 43889) and *S.* Typhimurium (KCTC 2421, ATCC 14028) were inoculated into tryptic soy broth. *S. aureus* (KCTC 1621, ATCC 10537) and *L. monocytogenes* (KCTC 13064, ATCC 15313) were cultured in brain heart infusion broth. Culture cocktails were prepared by combining each bacterial strain. The bacterial solutions were diluted with peptone water (0.1%) and placed on Mueller Hinton Agar (MHA) medium. The ECA film-forming solutions (80 μL) were impregnated into 8 mm paper discs and dried for 30 min. After drying, the discs were put onto the inoculated MHA medium. Thereafter, it was incubated at 37 °C for 24 h, and the size of the inhibition zone was measured with a digimatic caliper (Model 500-181-20, Mitutoyo Corp., Kawasaki, Japan).

3.11. Statistical Analysis

The SAS program (version 9.4, SAS Institute Inc., Cary, NC, USA) was used for statistical analysis of the experimental results. All values are presented as mean ± standard deviations, and $p < 0.05$ indicates statistically significant differences. All experiments were repeated at least 5 times.

4. Conclusions

New biodegradable films were prepared using ECA, which has not yet been studied, as a base material in this study. The physical properties of the ECA films were improved with the use of $CaCl_2$ as a cross-linking agent, and 3% of $CaCl_2$ was found to be optimal. CLO and CBO were incorporated to develop active packaging films and, as a result, have influenced the various properties of the ECA film. As the content of EOs increased, the TS decreased, while E increased. The incorporation of CLO or CBO into the films resulted in antioxidant activities, and the ECA films with CBO showed stronger DPPH and ABTS radical scavenging activities than did the ECA films containing CLO. Antimicrobial activities against various bacteria were confirmed, and higher antimicrobial activity was found in the ECA film containing CLO. Therefore, our studies demonstrate that the ECA films containing CLO or CBO can be applied as new active packaging materials in the food industry.

Author Contributions: S.-K.B. performed the experiments, analyzed the data, and wrote the original draft. S.-K.B. and S.K. carried out the discussion of the work results. K.B.S. supervised the study and finally revised the manuscript.

References

1. Atarés, L.; Chiralt, A. Essential oils as additives in biodegradable films and coatings for active food packaging. *Trends Food Sci. Technol.* **2016**, *48*, 51–62. [CrossRef]
2. Khalil, H.A.; Tye, Y.Y.; Saurabh, C.K.; Leh, C.P.; Lai, T.K.; Chong, E.W.N.; Syakir, M.I. Biodegradable polymer films from seaweed polysaccharides: A review on cellulose as a reinforcement material. *Express Polym. Lett.* **2017**, *11*, 244–265. [CrossRef]
3. Tavassoli-Kafrani, E.; Shekarchizadeh, H.; Masoudpour-Behabadi, M. Development of edible films and coatings from alginates and carrageenans. *Carbohydr. Polym.* **2016**, *137*, 360–374. [CrossRef] [PubMed]
4. Wijesinghe, W.A.J.P.; Jeon, Y.J. Exploiting biological activities of brown seaweed *Ecklonia cava* for potential industrial applications: A review. *Int. J. Food Sci. Nutr.* **2012**, *63*, 225–235. [CrossRef] [PubMed]
5. Park, Y.H. Seasonal variation in the chemical composition of brown algae with special reference to alginic acid. *Korean J. Fish. Aquat. Sci.* **1969**, *2*, 71–82.
6. Jost, V.; Reinelt, M. Effect of Ca^{2+} induced crosslinking on the mechanical and barrier properties of cast alginate films. *J. Appl. Polym. Sci.* **2018**, *135*, 45754. [CrossRef]

7. Fawzy, M.A.; Gomaa, M.; Hifney, A.F.; Abdel-Gawad, K.M. Optimization of alginate alkaline extraction technology from *Sargassum latifolium* and its potential antioxidant and emulsifying properties. *Carbohydr. Polym.* **2017**, *157*, 1903–1912. [CrossRef] [PubMed]
8. Sellimi, S.; Younes, I.; Ayed, H.B.; Maalej, H.; Montero, V.; Rinaudo, M.; Nasri, M. Structural, physicochemical and antioxidant properties of sodium alginate isolated from a Tunisian brown seaweed. *Int. J. Biol. Macromol.* **2015**, *72*, 1358–1367. [CrossRef] [PubMed]
9. Kuorwel, K.K.; Cran, M.J.; Orbell, J.D.; Buddhadasa, S.; Bigger, S.W. Review of mechanical properties, migration, and potential applications in active food packaging systems containing nanoclays and nanosilver. *Compr. Rev. Food Sci. F.* **2015**, *14*, 411–430. [CrossRef]
10. Ranasinghe, P.; Pigera, S.; Premakumara, G.S.; Galappaththy, P.; Constantine, G.R.; Katulanda, P. Medicinal properties of 'true'cinnamon (*Cinnamomum zeylanicum*): A systematic review. *BMC Complement Altern. Med.* **2013**, *13*, 275. [CrossRef] [PubMed]
11. Sangal, A. Role of cinnamon as beneficial antidiabetic food adjunct: A review. *Adv. Appl. Sci. Res.* **2011**, *2*, 440–450.
12. Mahmood, A.; Bano, S.; Kim, S.G.; Lee, K.H. Water–methanol separation characteristics of annealed SA/PVA complex membranes. *J. Memb. Sci.* **2012**, *415*, 360–367. [CrossRef]
13. Falkeborg, M.; Paitaid, P.; Shu, A.N.; Pérez, B.; Guo, Z. Dodecenyl succinylated alginate as a novel material for encapsulation and hyperactivation of lipases. *Carbohydr. Polym.* **2015**, *133*, 194–202. [CrossRef] [PubMed]
14. Pereira, L.; Sousa, A.; Coelho, H.; Amado, A.M.; Ribeiro-Claro, P.J. Use of FTIR, FT-Raman and 13 C-NMR spectroscopy for identification of some seaweed phycocolloids. *Biomol. Eng.* **2003**, *20*, 223–228. [CrossRef]
15. Benavides, S.; Villalobos-Carvajal, R.; Reyes, J.E. Physical, mechanical and antibacterial properties of alginate film: Effect of the crosslinking degree and oregano essential oil concentration. *J. Food Eng.* **2012**, *110*, 232–239. [CrossRef]
16. Zactiti, E.M.; Kieckbusch, T.G. Potassium sorbate permeability in biodegradable alginate films: Effect of the antimicrobial agent concentration and crosslinking degree. *J. Food Eng.* **2006**, *77*, 462–467. [CrossRef]
17. Zactiti, E.M.; Kieckbusch, T.G. Release of potassium sorbate from active films of sodium alginate crosslinked with calcium chloride. *Packag. Technol. Sci.* **2009**, *22*, 349–358. [CrossRef]
18. Han, Y.; Yu, M.; Wang, L. Physical and antimicrobial properties of sodium alginate/carboxymethyl cellulose films incorporated with cinnamon essential oil. *Food Packag. Shelf Life* **2018**, *15*, 35–42. [CrossRef]
19. Pranoto, Y.; Salokhe, V.M.; Rakshit, S.K. Physical and antibacterial properties of alginate-based edible film incorporated with garlic oil. *Food Res. Int.* **2005**, *38*, 267–272. [CrossRef]
20. Abdollahi, M.; Rezaei, M.; Farzi, G. Improvement of active chitosan film properties with rosemary essential oil for food packaging. *Int. J. Food Sci. Technol.* **2012**, *47*, 847–853. [CrossRef]
21. Shojaee-Aliabadi, S.; Hosseini, H.; Mohammadifar, M.A.; Mohammadi, A.; Ghasemlou, M.; Hosseini, S.M.; Khaksar, R. Characterization of κ-carrageenan films incorporated plant essential oils with improved antimicrobial activity. *Carbohydr. Polym.* **2014**, *101*, 582–591. [CrossRef] [PubMed]
22. Tongnuanchan, P.; Benjakul, S.; Prodpran, T. Properties and antioxidant activity of fish skin gelatin film incorporated with citrus essential oils. *Food Chem.* **2012**, *134*, 1571–1579. [CrossRef] [PubMed]
23. Zhang, Y.; Ma, Q.; Critzer, F.; Davidson, P.M.; Zhong, Q. Physical and antibacterial properties of alginate films containing cinnamon bark oil and soybean oil. *LWT Food Sci. Technol.* **2015**, *64*, 423–430. [CrossRef]
24. Atarés, L.; De-Jesús, C.; Talens, P.; Chiralt, A. Characterization of SPI-based edible films incorporated with cinnamon or ginger essential oils. *J. Food Eng.* **2010**, *99*, 384–391. [CrossRef]
25. Peng, Y.; Li, Y. Combined effects of two kinds of essential oils on physical, mechanical and structural properties of chitosan films. *Food Hydrocoll.* **2014**, *36*, 287–293. [CrossRef]
26. Wu, J.; Sun, X.; Guo, X.; Ge, S.; Zhang, Q. Physicochemical properties, antimicrobial activity and oil release of fish gelatin films incorporated with cinnamon essential oil. *Aquac. Fish.* **2017**, *2*, 185–192. [CrossRef]
27. Yadav, M.; Rhee, K.Y.; Park, S.J. Synthesis and characterization of graphene oxide/carboxymethylcellulose/alginate composite blend films. *Carbohydr. Polym.* **2014**, *110*, 18–25. [CrossRef] [PubMed]
28. Shankar, S.; Wang, L.F.; Rhim, J.W. Preparations and characterization of alginate/silver composite films: Effect of types of silver particles. *Carbohydr. Polym.* **2016**, *146*, 208–216. [CrossRef] [PubMed]
29. Soares, J.P.; Santos, J.E.; Chierice, G.O.; Cavalheiro, E.T.G. Thermal behavior of alginic acid and its sodium salt. *Eclética Química* **2004**, *29*, 57–64. [CrossRef]

30. Pongjanyakul, T.; Priprem, A.; Puttipipatkhachorn, S. Investigation of novel alginate−magnesium aluminum silicate microcomposite films for modified-release tablets. *J. Control Release.* **2005**, *107*, 343–356. [CrossRef] [PubMed]
31. Naidu, B.V.K.; Sairam, M.; Raju, K.V.; Aminabhavi, T.M. Thermal, viscoelastic, solution and membrane properties of sodium alginate/hydroxyethylcellulose blends. *Carbohydr. Polym.* **2005**, *61*, 52–60.
32. Tongnuanchan, P.; Benjakul, S.; Prodpran, T. Structural, morphological and thermal behavior characterizations of fish gelatin film incorporated with basil and citronella essential oils as affected by surfactants. *Food Hydrocoll.* **2014**, *41*, 33–43. [CrossRef]
33. Sharma, U.K.; Sharma, A.K.; Pandey, A.K. Medicinal attributes of major phenylpropanoids present in cinnamon. *BMC Complement Altern. Med.* **2016**, *16*, 156. [CrossRef] [PubMed]
34. Singh, G.; Maurya, S.; Catalan, C.A. A comparison of chemical, antioxidant and antimicrobial studies of cinnamon leaf and bark volatile oils, oleoresins and their constituents. *Food Chem. Toxicol.* **2007**, *45*, 1650–1661. [CrossRef] [PubMed]
35. Sanla-Ead, N.; Jangchud, A.; Chonhenchob, V.; Suppakul, P. Antimicrobial Activity of cinnamaldehyde and eugenol and their activity after incorporation into cellulose-based packaging films. *Packag. Technol. Sci.* **2012**, *25*, 7–17. [CrossRef]
36. Nazzaro, F.; Fratianni, F.; De Martino, L.; Coppola, R.; De Feo, V. Effect of essential oils on pathogenic bacteria. *Pharmaceuticals* **2013**, *6*, 1451–1474. [CrossRef] [PubMed]
37. Kang, J.H.; Song, K.B. Inhibitory effect of plant essential oil nanoemulsions against *Listeria monocytogenes*, *Escherichia coli* O157: H7, and *Salmonella* Typhimurium on red mustard leaves. *Innov. Food Sci. Emerg. Technol.* **2018**, *45*, 447–454. [CrossRef]
38. Cho, M.; Yoon, S.J.; Kim, Y.B. The nutritional composition and antioxidant activity from *Undariopsis peterseniana*. *Ocean Polar Res.* **2013**, *35*, 273–280. [CrossRef]
39. You, B.J.; Jeong, I.H.; Lee, K.H. Effect extraction conditions on bile acids binding capacity in vitro of alginate extracted from sea tangle (*Laminaria* spp.). *Korean J. Fish. Aquat. Sci.* **1997**, *30*, 31–38.
40. Lee, K.Y.; Song, K.B. Preparation and characterization of an olive flounder (*Paralichthys olivaceus*) skin gelatin and polylactic acid bilayer film. *J. Food Sci.* **2017**, *82*, 706–710. [CrossRef] [PubMed]
41. Yang, S.Y.; Lee, K.Y.; Beak, S.E.; Kim, H.; Song, K.B. Antimicrobial activity of gelatin films based on duck feet containing cinnamon leaf oil and their applications in packaging of cherry tomatoes. *Food Sci. Biotechnol.* **2017**, *26*, 1429–1435. [CrossRef] [PubMed]
42. Shojaee-Aliabadi, S.; Hosseini, H.; Mohammadifar, M.A.; Mohammadi, A.; Ghasemlou, M.; Ojagh, S.M.; Khaksar, R. Characterization of antioxidant-antimicrobial κ-carrageenan films containing *Satureja hortensis* essential oil. *Int. J. Biol. Macromol.* **2013**, *52*, 116–124. [CrossRef] [PubMed]
43. Bitencourt, C.M.; Fávaro-Trindade, C.S.; Sobral, P.J.A.; Carvalho, R.A. Gelatin-based films additivated with curcuma ethanol extract: Antioxidant activity and physical properties of films. *Food Hydrocoll.* **2014**, *40*, 145–152. [CrossRef]

9

Chitin Nanofibrils and Nanolignin as Functional Agents in Skin Regeneration

Serena Danti [1,2,*], Luisa Trombi [2], Alessandra Fusco [2,3], Bahareh Azimi [2], Andrea Lazzeri [1,2], Pierfrancesco Morganti [3], Maria-Beatrice Coltelli [1,2] and Giovanna Donnarumma [2,3]

1 Department of Civil and Industrial Engineering, University of Pisa, 56122 Pisa, Italy; andrea.lazzeri@unipi.it (A.L.), maria.beatrice.coltelli@unipi.it (M.-B.C.)

2 Consorzio Interuniversitario Nazionale per la Scienza e Tecnologia dei Materiali (INSTM), 50121 Florence, Italy; l.trombi@yahoo.it (L.T.), alessandra.fusco@unicampania.it (A.F.), b.azimi@ing.unipi.it (B.A.), giovanna.donnarumma@unicampania.it (G.D.)

3 Department of Experimental Medicine, University of Campania "Luigi Vanvitelli", 80138 Naples, Italy; pierfrancesco.morganti@mavicosmetics.it

* Correspondence: serena.danti@unipi.it

Abstract: Chitin and lignin, by-products of fishery and plant biomass, can be converted to innovative high value bio- and eco-compatible materials. On the nanoscale, high antibacterial, anti-inflammatory, cicatrizing and anti-aging activity is obtained by controlling their crystalline structure and purity. Moreover, electropositive chitin nanofibrlis (CN) can be combined with electronegative nanolignin (NL) leading to microcapsule-like systems suitable for entrapping both hydrophilic and lipophilic molecules. The aim of this study was to provide morphological, physico-chemical, thermogravimetric and biological characterization of CN, NL, and CN-NL complexes, which were also loaded with glycyrrhetinic acid (GA) as a model of a bioactive molecule. CN-NL and CN-NL/GA were thermally stable up to 114 °C and 127 °C, respectively. The compounds were administered to in vitro cultures of human keratinocytes (HaCaT cells) and human mesenchymal stromal cells (hMSCs) for potential use in skin contact applications. Cell viability, cytokine expression and effects on hMSC multipotency were studied. For each component, CN, NL, CN-NL and CN-NL/GA, non-toxic concentrations towards HaCaT cells were identified. In the keratinocyte model, the proinflammatory cytokines IL-1α, IL-1 β, IL-6, IL-8 and TNF-α that resulted were downregulated, whereas the antimicrobial peptide human β defensin-2 was upregulated by CN-LN. The hMSCs were viable, and the use of these complexes did not modify the osteo-differentiation capability of these cells. The obtained findings demonstrate that these biocomponents are cytocompatible, show anti-inflammatory activity and may serve for the delivery of biomolecules for skin care and regeneration.

Keywords: chitin; lignin; anti-inflammatory; immunomodulation; keratinocytes; mesenchymal stem cells; cosmetics; glycyrrhetinic acid; skin regeneration; bio-based; nanomaterials

1. Introduction

In recent years, the development of natural biopolymer-based products suitable for cosmetic, skin care and regeneration have received considerable attention [1]. The key specifications of these products are biocompatibility and biodegradation, both in the environment and in the human body. The constituent polymers and their decomposition products are thus expected to be safe and without side effects. Chitin and lignin, by-products of fishery and plant biomass, can be reused and converted to novel high value materials for biomedical and cosmetic applications, which are bio- and eco-compatible [2,3]. On the nanoscale, chitin and lignin crystalline structure and purity can be controlled, resulting in several interesting properties, such as antibacterial [4], anti-inflammatory [5], cicatrizing and anti-aging effectiveness [6,7]. Chitin nanofibrils (CN) represent the purest crystal form of chitin and show positive surface charges. As such, CN are able to combine with electronegative compounds entrapping different ingredients for skin-friendly applications, such as innovative cosmetics for aged [6,8], as well as problematic and sensitive skin [9]. Because of these properties, CN have been used in combination with biopolymers [3,10], such as with poly (lactic acid) (PLA) [11–15], by several researchers. Morganti et al. [6] developed CN-Hyaluronan block copolymeric nanoparticles, entrapping several active ingredients in order to evaluate their efficiency and safety as biologically active rejuvenation treatments able to accelerate skin regeneration. They concluded that these biodegradable polymer particles have some advantages over other colloidal carrier systems thanks to their higher stability and versatility in tailoring the ingredient load and its release rate. In another attempt [7], the same authors investigated CN-hyaluronan block-copolymeric nanoparticles for their ability to load lutein, as an active anti-wrinkle agent, and demonstrated their effectiveness as anti-aging cosmetic emulsion compounds. Combining electropositive CN and electronegative nanolignin (NL) also gives rise to microcapsules able to entrap both hydrophilic and lipophilic molecules such as vitamins, microelements, anti-inflammatory drugs, antioxidants, anti-ageing substances, immunomodulating agents and enzymes [3,16].

In view of the abovementioned first studies, since cosmetic product consumers are becoming more and more aware of the environmental impact of products and increasingly trust the efficacy of green ingredients, the interest in such biopolymeric nanomaterials is expected to grow in the coming years. Therefore, it is significantly important to assess the specific properties of CN and NL and CN-NL complexes, including their material characterization, in order to improve or enable efficient processing technologies and biological characterization, including usable concentrations and the disclosure of their advantages in skin contact products as well as potential side effects, if any. To this purpose, advanced in vitro platforms can be efficiently exploited as ethical models for material screening. The HaCaT cell line represents an established model of human keratinocytes (i.e., epidermis) which can give preliminary information about new materials with respect to their inflammatory reactions through the expression of several cytokines, and of antimicrobial response through the expression of antimicrobial peptides of the innate immunity, such as defensins.

The aim of this study was to test in vitro the efficacy and safety of CN, NL and CN-NL complexes, obtained in powder by spray drying, with human keratinocytes (HaCaT cells) and human mesenchymal stromal cells (hMSCs). CN-LN was also investigated for the delivery of glycyrrhetinic acid (GA), as a biomolecule with anti-inflammatory activity deriving from licorice plants (Glycyrrhiza). GA (also known as enoxolone or glycyrrhetic acid) is a pentacyclic triterpenoid obtained from the hydrolysis of glycyrrhizic acid. GA and its salts and esters are cosmetic ingredients which act as flavoring or skin-conditioning agents and have excellent antimicrobial, anti-inflammatory, antioxidant, anti-ulceration, antiviral and analgesic properties [17,18]. In this context, hMSCs represent a model of immature cells present in the connective tissue (e.g., dermis), which may be affected by the addition of nanocomponents, as they can penetrate across skin layers. The hMSC multipotency (i.e., capability of multiple lineage differentiation proper of stem cells) must be preserved for a correct tissue turn over.

To our knowledge, this is the first study investigating the efficacy of CN-NL complexes as carriers of GA. The successful fabrication of innovative complexes entrapping antibacterial and anti-inflammatory ingredients would enable the introduction of novel products and their industrial upscale for potential use in dermatologic and cosmetic applications.

2. Results

2.1. Morphological Characterization of CN, NL and CN-NL Complexes

CN was spray-dried in the presence of 2% (*w*/*w*) of poly(ethylene glycol) (PEG) to obtain CN samples and consisted of scraps of micrometric dimensions (Figure 1A). The use of PEG can be advantageous to avoid CN aggregation when the solution is concentrated [15]. In fact, this aggregation, by giving rise to particles of micrometric size, would result in a lower effectiveness of CNs because of the reduction of their active surface. The spray-dried CN-NL complex in powder, pretreated with PEG, showed a specific morphology, as it consisted of almost round particles of micrometric dimensions (Figure 1B). The supplied lignin in powder, NL, appeared as nanostructured micrometric aggregates (Figure 1C,D).

Figure 1. FE-SEM micrographs of the powders: (**A**) Spray dried CN; (**B**) Spray dried CN-NL complex; (**C**) Morphology of NL powder taken from the pristine NL sample; (**D**) Zoomed-in magnification of NL (16,000×) in back scattered modality to observe the NL nanostructure.

For better investigating the morphology of CN, pure CN suspensions at 2% (*w*/*w*) were diluted 1:1000 in distilled water and the suspension was deposited on a glass window on which the sample was dried. CN appeared as "whiskers" having a nanometric thickness and a micrometric length (Figure 2A). The spray dried complex CN-NL in powder (Figure 1B) was suspended in water (at 80 ppm) and analyzed after deposition on a glass window followed by drying. This sample showed a completely different morphology, consisting in micrometric disks having a round or ellipsoidal shape (Figure 2B). At higher magnification, the presence of a nanostructured system was highlighted (Figure 2C,D), in which the presence of both CN and NL particles could be observed.

The spray-dried CN-NL complex was easily suspended in water to obtain flat micrometric nanostructured agglomerates that can deposit onto a surface, and be used to modify its properties thanks to CN and NL functionalities.

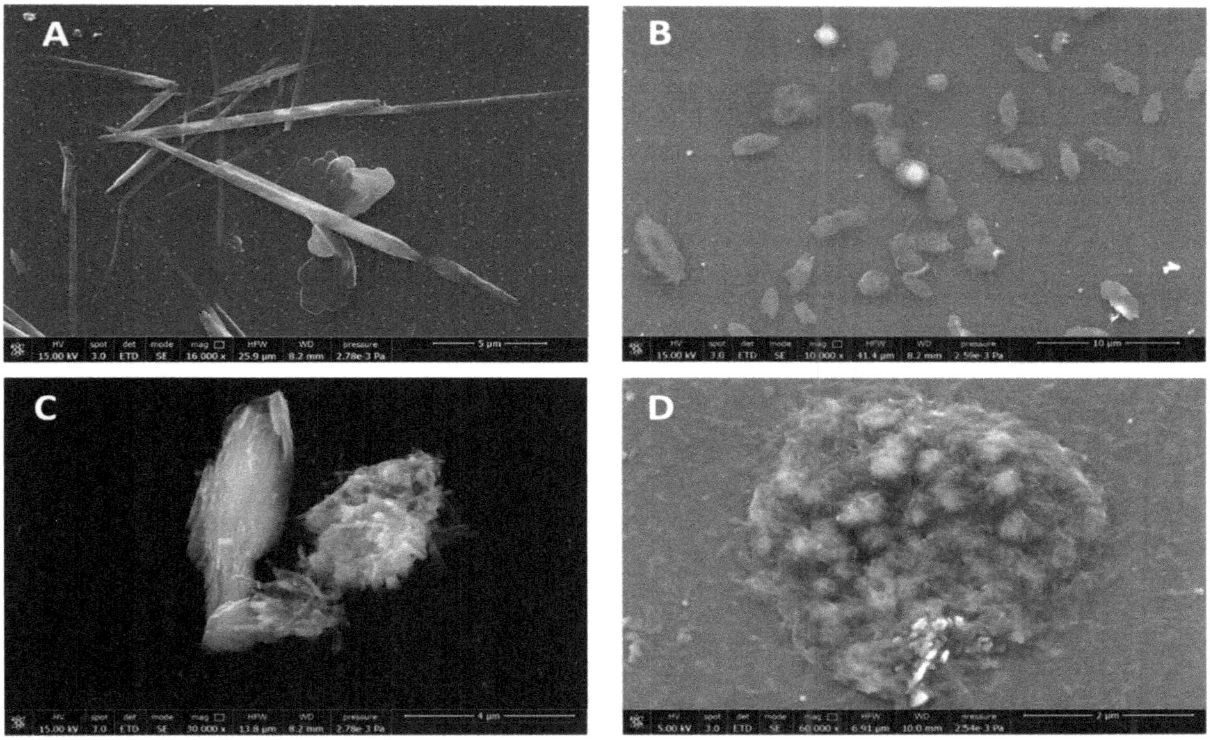

Figure 2. FE-SEM micrographs of (**A**) pure CN from deposition from diluted water suspension; (**B**) CN-NL complex; (**C**) and (**D**) CN-NL complex at higher magnification.

2.2. Chemical Structure and Thermal Stability of CN, NL, GA and CN-NL/GA Complexes

The starting materials and complexes CN-NL and CN-NL/GA were characterized by infrared spectroscopy. Spray-dried CN powder showed a spectrum (Figure 3A) with characteristic amide I and Amide II bands at 1619 and 1552 cm^{-1}, respectively. The Amide I band is split into two components at 1656 cm^{-1} and 1619 cm^{-1}. This is typical of α-chitin [19]. The most intense bands at 1010 cm^{-1} and 1070 cm^{-1} are typical of C-O stretching and agree with the polysaccharidic nature of this biopolymer. The band at 3439 cm^{-1} is attributable to O-H stretching, the bands at 3256 cm^{-1} and 3102 cm^{-1} to N-H stretching of amine and amide groups, respectively. The band at 2874 cm^{-1} indicates C-H stretching.

NL shows a spectrum (Figure 3B) characterized by the presence of strong band s at 1030 cm^{-1}, 1120 cm^{-1} and 1222 cm^{-1}. The band at 1222 cm^{-1} can be associated with C–C plus C–O plus C=O stretching.

The band at 1030 cm^{-1} can be attributed to aromatic C–H deformation that appears as a complex vibration associated with the C–O, C–C stretching and C–OH bending in polysaccharides. Carbohydrate originating vibrations are associated to the band at 1120 cm^{-1} [20]. In the carbonyl/carboxyl region, medium bands are found at 1705–1720 cm^{-1}, attributable to unconjugated carbonyl/carboxyl stretching. The bands at 1601 cm^{-1} and 1511 cm^{-1} can be associated with aromatic skeleton vibrations.

The GA spectrum (Figure 3C) is characterized by a band at 3430 cm^{-1} typical of OH stretching and a band at 2945 cm^{-1} attributable to CH stretching; both the bands at 1700 cm^{-1} and 1660 cm^{-1} are assigned to the C=O stretching (not conjugated and conjugated with the double bond, respectively) and the band at 1460 cm^{-1} to CH bending; the band at 1025 cm^{-1} can be attributed to C-O stretching and the band at 990 cm^{-1} can be attributed to the rocking of methyl groups [21].

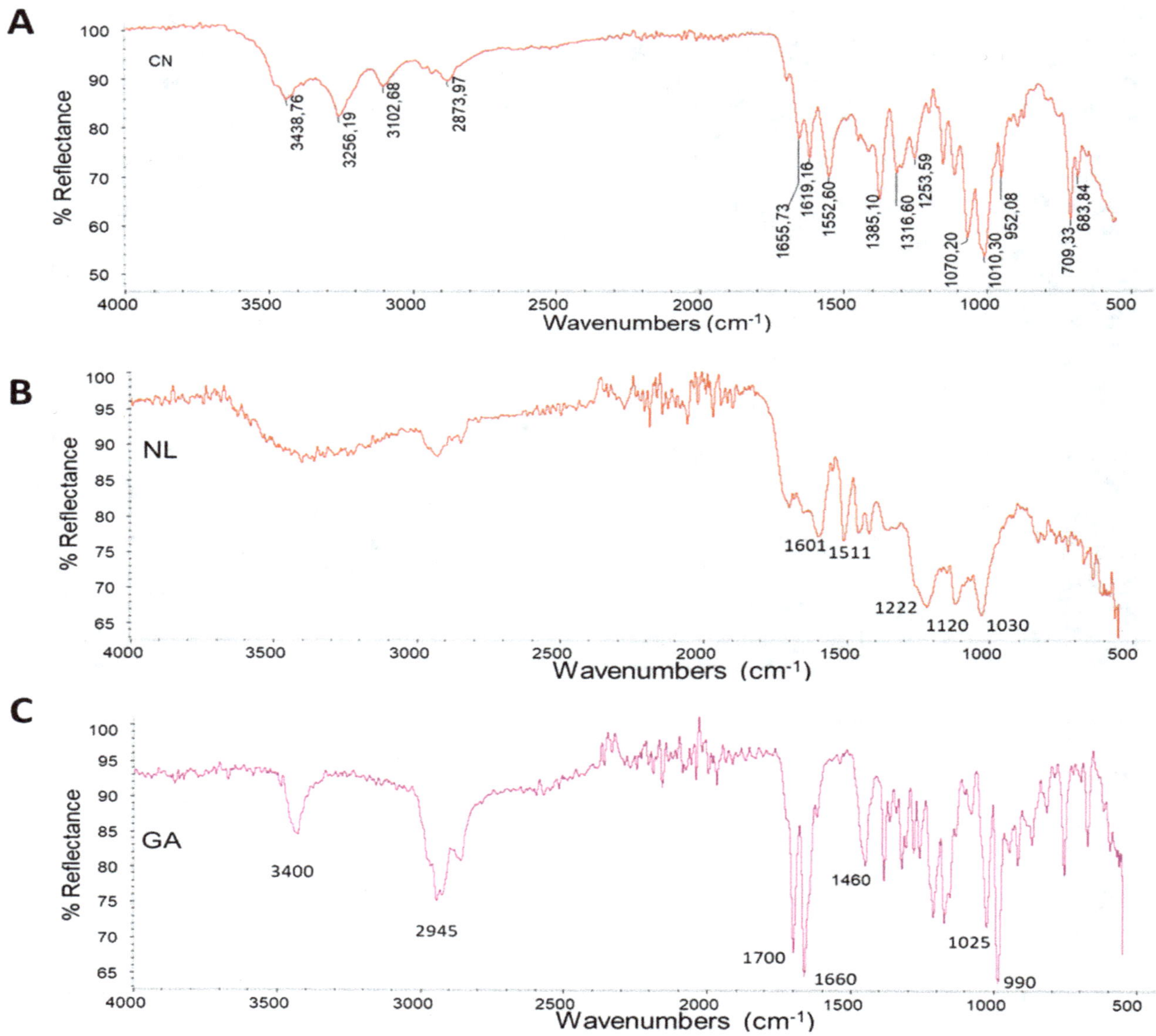

Figure 3. Infrared ATR spectra of (**A**): CN; (**B**) NL; (**C**) GA.

The complex CN-NL and the same complex entrapping GA were characterized by ATR spectroscopy (Figure 4). The obtained spectra were quite similar and they showed the bands already observed in the previous spectra of CN and NL, but slightly shifted because of the interactions occurring between CN and NL. The presence of GA can be revealed despite of its low concentration in the sample (0.2% w/w). As it can be observed in Figure 4B where the spectrum of GA, the spectrum of CN-NL and the spectrum of CN-NL/GA are compared in the region 2400–4000 cm^{-1}, the higher intensity of the CN-NL/GA with respect to CN-NL at 3400 cm^{-1} is in agreement with the presence of GA in the CN-NL/GA complex. A similar comparison is proposed in Figure 4C for the region 1100-2000 cm^{-1}. The increase of intensity of the band at 1650 cm^{-1} and 1470 cm^{-1} clearly observed in the spectrum of CN-NL/GA complex, can be attributed to the presence of GA in the powder. Interestingly the band at 1700 cm^{-1} of the carboxylic group of GA is not much evident in Figure 4C. In agreement with the work of Cheng et al. [22] this can be attributed to the interactions or reactions of the carboxylic groups with the functional group of CN and NL.

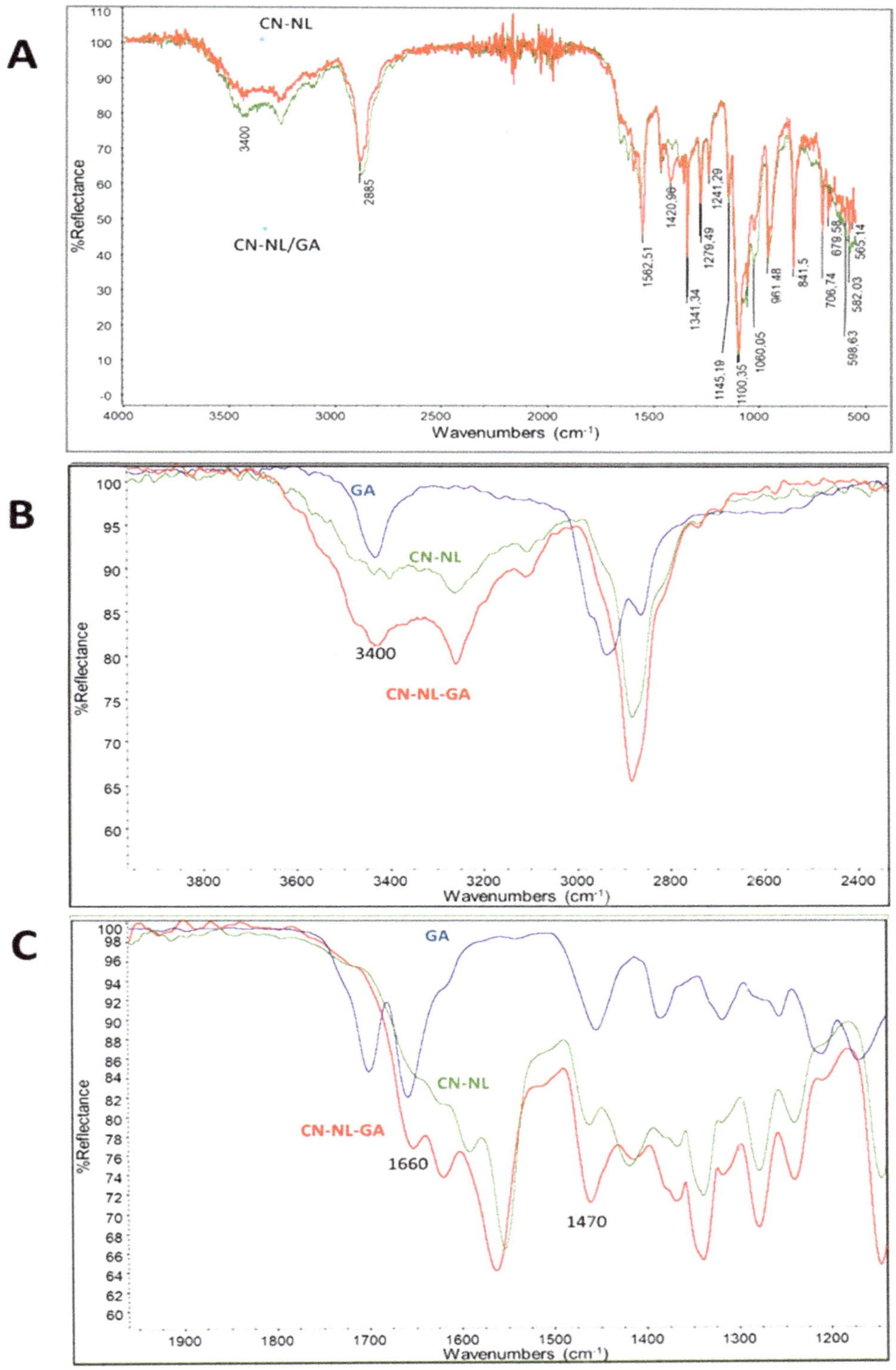

Figure 4. Infrared ATR spectra of (**A**) CN-NL and CN-NL/GA powders; (**B**) CN-NL, CN-LN/GA and GA in the region 2400-4000 cm^{-1}; (**C**) CN-NL, CN-LN/GA and GA in the region 1100-2000 cm^{-1}.

The thermogravimetric analysis (TGA) in nitrogen flow of pure components and of the two complexes was investigated and it was observed that GA, having an onset temperature at 387 °C, has a higher thermal stability than CN and NL. This different behavior is attributable to its action as a radical scavenger [23]. GA decomposition occurred in a single step and resulted in an almost complete decomposition. As observed in Figure 5A and in Table 1, the final residue is only 1.97% by weight.

The behavior of CN and NL was different, as several mass loss steps were observed (Figure 5A) and the final residue was 38.9% for NL and 17.4% for CN. These results evidenced that NL and CN formed a high amount of carbonaceous residue during thermal decomposition. Interestingly, NL showed a not negligible mass loss at 40 °C, probably because of partial decomposition, resulting in volatile fragments.

The thermogravimetric trend of the complex did not include this mass loss. Hence, NL appeared more stable thanks to the formation of the complex with CN. However, the thermal stability of CN was reduced due to the presence of NL (the onset 1 and onset 2 temperatures were both decreased comparing those of CN-NL with CN) (Table 1). By the addition of GA, the stability of the complex was improved. This could be reasonably due to the radical scavenger activity of GA. Although the thermogravimetric trends of CN-NL and CN-NL/GA were similar, the mass loss recorded for CN-NL/GA was clearly lower than the one of CN-NL. This evidence corroborated the improved stability of CN-NL/GA thanks to the presence of GA.

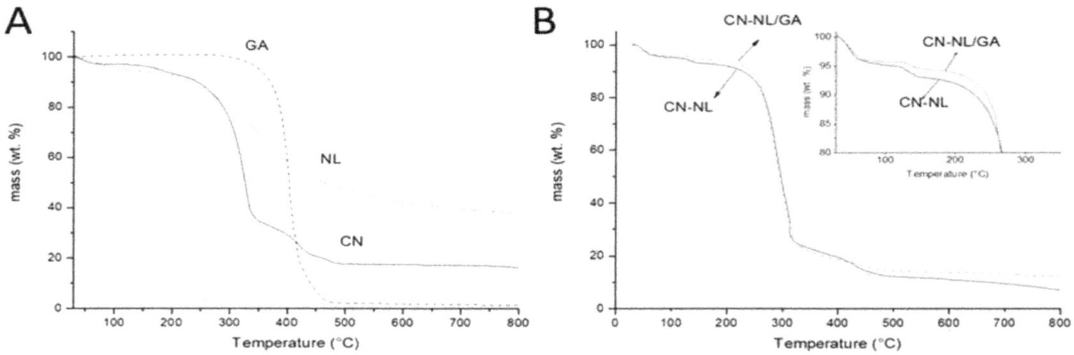

Figure 5. Thermogravimetric trends related to: (**A**) GA, NL and CN; (**B**) CN-NL and CN-NL/GA.

Table 1. Results of TGA analysis.

	Onset 1 (°C)	Mass Loss 1 (%)	Onset 2 (°C)	Mass Loss 2 (%)	Onset 3 (°C)	Mass Loss 3 (%)	Residue (%)
GA	386.95	−98.92	-	-	-	-	1.97
NL	40	−5.26	262	−55.63	-	-	38.93
CN	159	−5.35	300	−57. 54	387	−17.85	17.42
CN-NL	114	−2.60	266	−69.25	406	−10.39	12.37
CN-NL/GA	127	−1.92	251	−68.53	472	−11.63	13.84

CN, CN-NL and CN-NL/GA represent the spray-dried samples containing 2% (*w*/*w*) of PEG.

2.3. *HaCaT Cell Viability*

HaCaT cells were seeded at a density of $1 \cdot 10^3$/well in 96-well culture plates. After 24 hours, the cells were treated with CN, CN-NL and CN-NL/GA at the following concentrations: 10 μg/mL, 5 μg/mL, μg/mL, 1 μg/mL, 0.5 μg/mL, 0.2 μg/mL, 0.1 μg/mL, 75 ng/mL, 50 ng/mL, 25 ng/mL.

After 24 h, the cells were incubated with MTT at 0.5 mg/mL at 37 °C for four 4 h and, subsequently, with DMSO at room temperature for 15 min. The spectrophotometric absorbance of the samples was determined by using an Ultra Multifunctional Microplate Reader (Biorad) at 570–655 nm [21]. Results, expressed as percentage of viable treated cell to respect with untreated control, are shown in Figure 6. In vitro studies performed on HaCaT cells enabled the establishment of the safe concentration of these green nano- and micro-compounds to be used. In particular, CN and CN-NL, did not decrease cell viability at 10 μg/mL and 0.2 μg/mL, respectively. Furthermore, CN-NL/GA complexes did not decrease cell viability at 0.5 μg/mL.

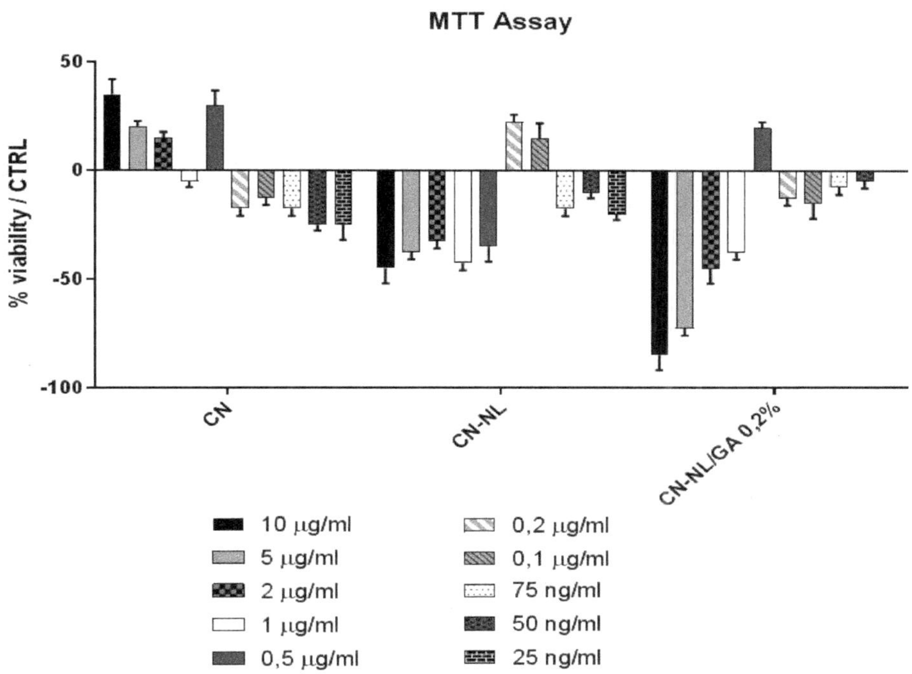

Figure 6. Results of MTT assay for different concentrations of CN, CN-NL, CN-NL/GA (0.2%), using HaCaT cells. The results were normalized by the viability of untreated cells as control.

2.4. HMSC Morphology, Viability and Osteodifferentiative Potential

Basing on the findings reported in Figure 6, hMSCs were treated with the above reported concentrations employed for HaCaT cells and the cell culture results are shown in Figures 7 and 8.

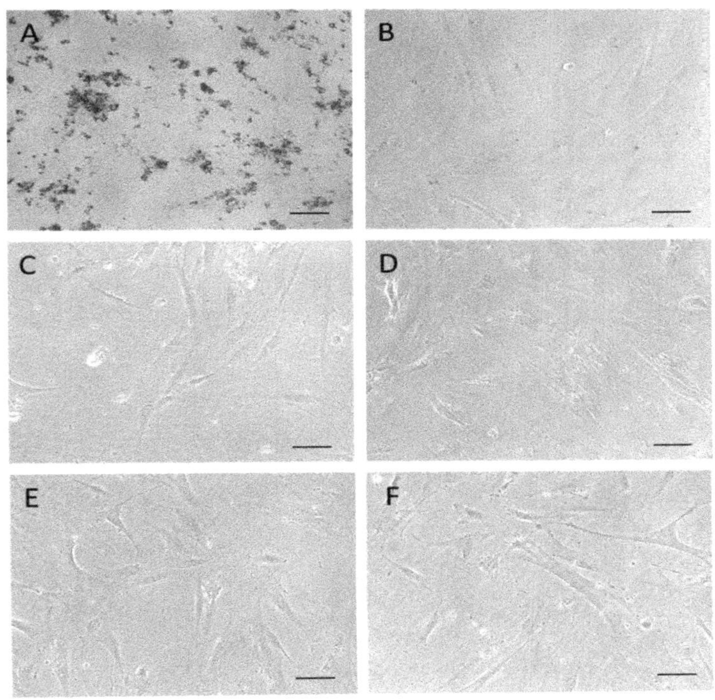

Figure 7. Micrographs at the inverted optical microscope of hMSCs in culture after 24 h of treatment with: (**A**) CN at 10 µg/mL; (**B**) NL at 10 µg/mL; (**C**) CN-NL at 0.2 µg/mL; (**D**) CN-NL/GA at 0.5 µg/mL; (**E**) GA at 0.5 µg/mL; (**F**) control. Micrographs are representative of each compound administrated at the intermediate concentration in the tested ranges. Original magnification 100×, scale bar = 100 µm.

Lower and higher concentrations of CN-NL (0.1 μg/mL and 0.5 μg/mL) and both CN-NL/GA and GA alone (0.2 μg/mL and 1 μg/mL) were tested with hMSCs. CN and NL were assayed at the maximum concentration. Treatments with CN, LN, CN-LN, CN-LN/GA complexes and GA did not affect the morphology of hMSCs at all the tested concentrations (Figure 7). CN treated samples showed micrometric aggregations that did not alter cell adhesion and growth (Figure 7A). Beside using pure CN, the other samples appeared clean and the spindle-like shape of hMSCs was clearly visible. AlamarBlue®test was performed on undifferentiated hMSCs in growth conditions in order to assess the cytocompatibility of the CN-NL complex and GA separately, and the combined CN-NL/GA complex. The obtained results are shown in Figure 8.

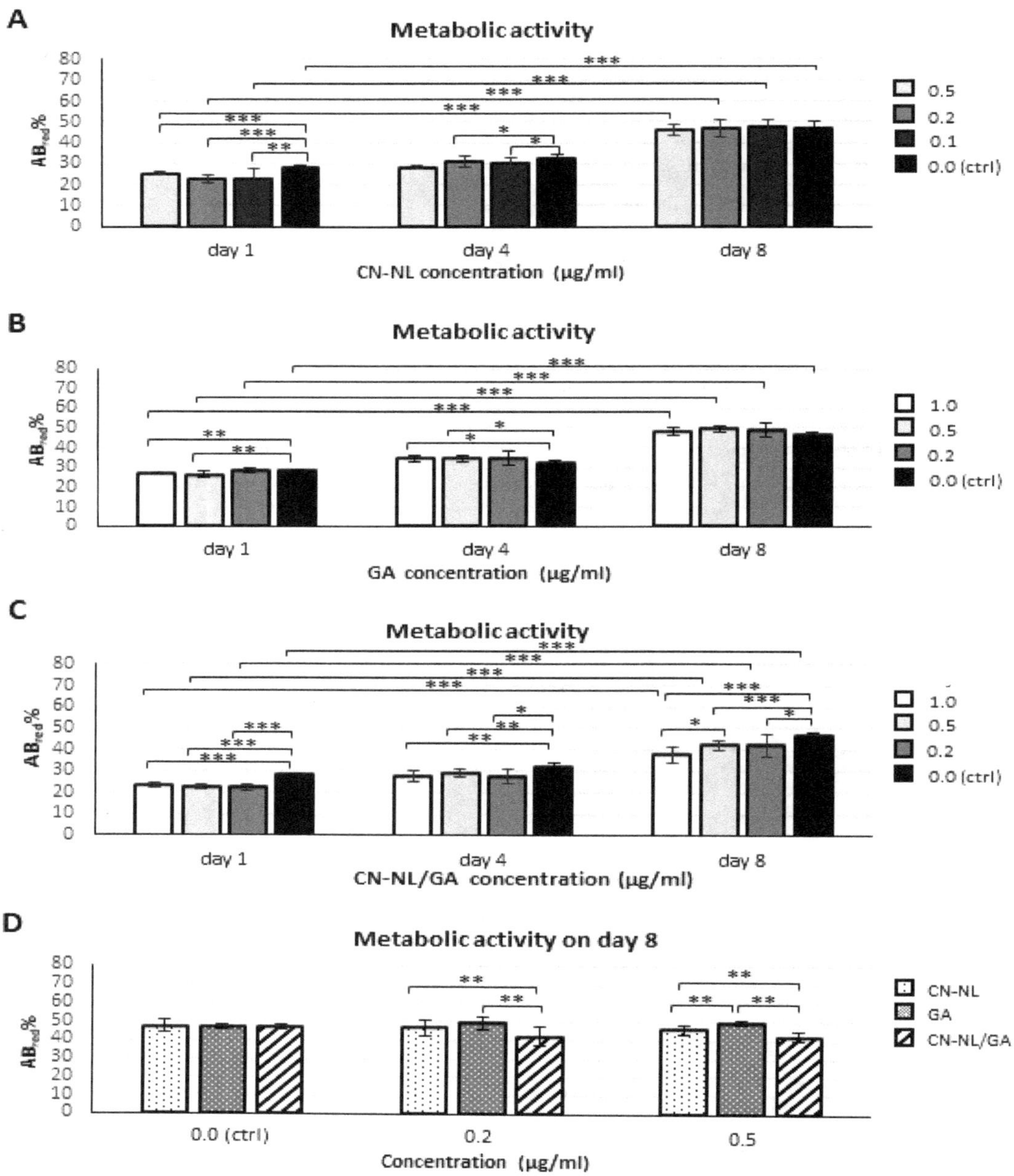

Figure 8. Bar graphs showing metabolic activity, as obtained by alamarBlue®test, performed on undifferentiated hMSCS using CN-NL, GA and CN-LN/GA complexes: (**A**–**C**) at different time points (1, 4 and 8 days in culture) and selected concentrations for each compound; (**D**) on day 8, comparing the three compounds within homogeneous concentration groups (* $p < 0.01$, ** $p < 0.001$, *** $p < 0.0001$).

In all the samples, the metabolic activity significantly increased over time ($p < 0.0001$) and was not significantly different from that of control samples (p = n.s), but using CN-NL/GA. In these samples, at the endpoint, the metabolic activity statistically decreased with respect to controls by increasing CN-NL/GA concentration ($p < 0.01$ at 0.2 μg/mL; $p < 0.0001$ at 0.5 μg/mL and 1.0 μg/mL) (Figure 8C). The highest GA concentrations (0.5 μg/mL and 1.0 μg/mL) increased hMSC metabolic activity with statistical significance only at day 4 ($p < 0.01$) (Figure 8B). On day 8, GA resulted the most performing treatment over CN-NL and CN-NL/GA both at 0.2 μg/mL and 0.5 μg/mL; instead, CN-NL/GA reduced hMSC metabolic activity ($p < 0.001$) (Figure 8D). To assess if the multipotency of hMSCs was maintained by the administration of these green compounds, the hMSCs were differentiated towards the osteogenic lineage in presence of all the additives and compared to undifferentiated hMSCs as a control. Mineral matrix deposition was investigated as a proof of osteo-differentiation using von Kossa staining (Figure 9).

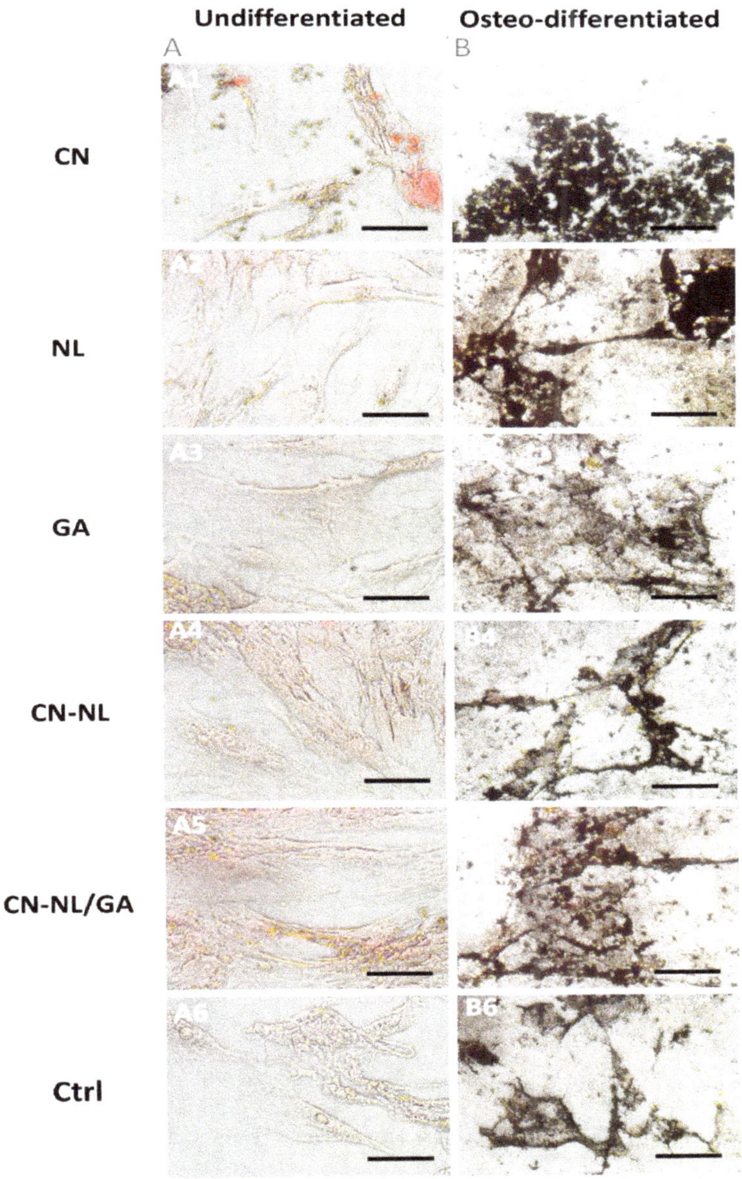

Figure 9. Light micrographs of von Kossa staining performed on (**A**) undifferentiated (control) and (**B**) osteodifferentiated hMSCs after incubation with all the nanocomponents (CN, NL, GA, CN-NL, CN-NL/GA) for 15 days at the highest concentrations in the tested ranges. Original magnification 200×; scale bar = 100 μm.

The obtained findings highlighted that both treated and untreated hMSCs, cultured in osteodifferentiation medium, produced mineral granules, visible in black, with no appreciable differences according to morphology and extension (Figure 9).

2.5. Anti-Inflammatory and Immune Responses of HaCaT Cells

Our data indicated that CN, CN-NL and GA markedly down-regulated the expression of pro-inflammatory cytokines interleukin-1 alpha (IL-1α), IL-1β, IL-6, IL-8 and tumor necrosis factor alpha (TNF-α) at mRNA level, especially after 24 h of treatment. The observed panel is in line with modulation of inflammatory activity exerted by all these components on human keratinocytes (Figure 10). The CN-NL/GA complex was the most effective in modulating the expression of the largest number of pro-inflammatory cytokines: IL-1α, IL-6, and IL-8. CN-NL was able to up-regulate the expression of an antimicrobial peptide, the human β-defensin (HBD-2), which is involved in innate immune response (Figure 10). Differently, transforming growth factor beta (TGF-β) was not modulated (data not shown).

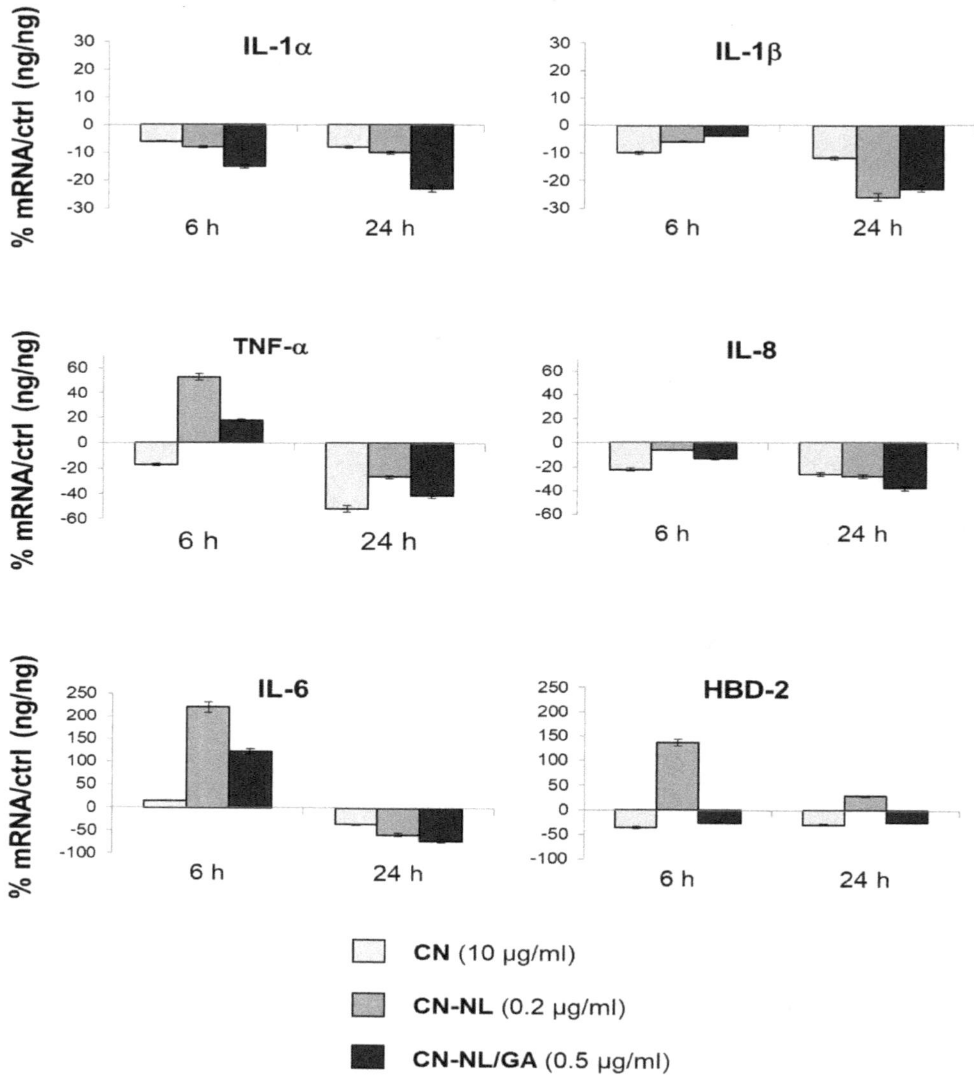

Figure 10. Bar graphs showing the results of RT-PCR performed on HaCaT cells exposed to the biopolymer nanomaterials al 6 h and 24 h for different cytokines involved in inflammatory response and HBD-2 and an antimicrobial peptide. The results are normalized by the expression in untreated cells as control.

These results are very encouraging for the use of these natural products for cosmetic applications and tissue repair, by acting on the modulation of cytokines and antimicrobial peptide HBD-2. On the other hand, these composites did not modify the proliferative and differentiative potential of hMSCs, thus demonstrating a good biocompatibility and cell affinity.

3. Discussion

The spray-dried CN-NL complex in powder was found to be water-suspendable and applicable on a surface in the form of micrometric flat nanostructured ellipsoids. Regulating the concentration of the water suspension can be fundamental to treat and confer to a surface some functional properties, such as anti-microbial and antioxidant properties. The resultant NL thermal stability was lower than that of CN. However, the CN-NL complex displayed an improved thermal stability with respect to pure NL. In the presence of GA, despite of its low amount in the powder (2% by weight), the thermal stability was further improved, in agreement with the radical scavenging properties of this substance. On the whole the complex CN-NL in powder, also in its version loaded with GA, can be considered a functionalizing agent much useful for functionalizing various substrates.

Interaction with skin cells is fundamental to assess the functional properties and demonstrate safety in skin contact applications, as they may include cosmetics, personal care as well as skin regeneration. It is a fact that nanosized objects can migrate into tissues [7], potentially leading to unknown reactions. The nanomaterials investigated in this study are biodegradable and therefore their interaction with the human body is temporary, so they do not pose harmful risks due to organ accumulation. The initial interactions with skin cells was investigated to understand their safety and disclose possible inflammatory reactions. In our study, we used HaCaT cells, as a model of human keratinocytes (i.e., epithelial cells), to select the optimal concentrations to be used for each of these biocompounds, which were 10 μg/mL for CN and NL alone, 0.2 μg/mL for CN-NL and 0.5 μg/mL for CN-NL/GA complexes. The higher concentration usable for CN and NL in comparison with those of the complexes may depend on the larger size (micrometric) of the CN-NL.

HMSCs were used as a model of mesenchymal (stem) cells present in the innermost layers of the skin, such as dermis and hypodermis (where they originate fibroblasts, among other cell types), as well as connective tissue in general [24], in order to assess the cytocompatibility and safety. Starting from the optimal concentrations chosen for HaCaT cells, each component was diluted at higher and lower concentrations to investigate the sensitivity of hMSCs to the same component. In the selected concentration ranges, the hMSCs were viable, with no appreciable differences in their morphology and osteo-differentiation capability with respect to untreated samples [25], thus demonstrating good biocompatibility and cell affinity of these green components, even at two times higher concentrations than the optimal ones selected using HaCaT cells. The wide range of cytocompatibility with hMSCs ensures a safe use of these bionanomaterials in the selected ranges. The significant reduction in hMSC metabolic activity following the administration of CN-NL/GA samples highlighted at the endpoint (day 8) with respect to GA and CN-LN administrated singularly can be a consequence of the synergistic activity exerted by the two components in the complex, and in particular by CN-NL, which alone largely decreased the metabolic activity of HaCaT cells at 1 μg/mL concentration. GA is known to affect some cellular functions in hepatocytes [17], and in our study showed a positive effect on hMSC metabolism. It is thus possible that the size and the neutral charge of the complex can affect cell metabolic activity by promoting other cellular functions.

HaCaT cells were also used to understand the reaction of the epidermis layer by investigating an array of cytokines involved in inflammation and immune response. HBD-2 is an inducible antimicrobial peptide with a molecular mass of 4–6 kDa acting as an endogenous antibiotic in the defense of host against Gram-positive and Gram-negative bacteria, fungi and the envelope of some viruses, and is involved in the innate immune response because its release is induced by pro-inflammatory cytokines, endogenous stimuli, infections or wounds [26]. Cytokines are multi-functional biological molecules that play a key role in the innate immune response, and that are involved in important biological activities such as hematopoiesis, infectious diseases, tumorigenesis, homeostasis and tissue repair, growth and cell development [27]. In particular, IL-1 promotes local inflammation and coagulation, increases the expression of adhesion molecules, and causes the release of chemokines and recruitment of leukocytes to the site of inflammation. In addition, it is an endogenous pyrogen, stimulates the synthesis of acute phase proteins and induces cachexia [28]. TNF-α is essential mediator in inflammation [29]. Its release in the site of the inflammation involves a localized vascular endothelial activation and vasodilation with increased vascular permeability. TNF-α also acts on platelet adhesiveness, favoring the formation of thrombus and occlusion of blood vessels and, therefore, reducing infection but also the tissue necrosis. IL-6 is a pleiotropic molecule that stimulates hepatocytes to synthetize many plasma proteins, such as fibrinogen, which ultimately contribute to the inflammatory acute phase response. Finally, IL-8 is a chemokine with many functions, including the attraction and activation of polymorphonuclear leukocytes, chemotaxis of basophils and a role in angiogenesis [30]. Remarkably, CN-NL and CN-NL/GA upregulated the mRNA expression of TNF-α and IL-6 at 6 h, which resulted subsequently downregulated after 24 h. This behavior can be explained by giving the nanocomposites a role in the wound repair process. In fact, wound healing is characterized a series of phases, and in particular in a first phase a number of overlapping events occur, including the production of pro-inflammatory cytokines. TNF-α represents the primary cytokine for pro-inflammatory responses; the direct effect of its release is the upregulation of IL-6 [31]. The obtained results showed that all the selected components were able to modulate the inflammatory response by decreasing the expression of proinflammatory cytokines. This fact was particularly evident in the CN-NL/GA complex, which downregulated the mRNA expression of the largest number of pro-inflammatory cytokines tested: IL-1α, IL-6, and IL-8. The unloaded complex, CN-NL alone, was able to up-regulate the expression of HBD-2. The reason why CN-NL/GA does not maintain the same functional activity towards HBD-2 may depend on specific cellular interactions with the GA-loaded complex. Indeed, GA shows different antibacterial mechanisms that are exerted directly towards bacteria and are not cell-mediated [32]. The obtained findings showed that CN, NL and specifically CN-NL complexes are very promising for skin contact applications, such as for coating biomedical, personal care and cosmetic products. Further investigation using primary human keratinocytes will provide a comprehensive validation of these compounds.

4. Materials and Methods

4.1. Materials

CN, NL, and GA were supplied by Mavi Sud, Aprilia (LT), (Milan, Italy). CN, CN-NL and CN-NL/GA were prepared in powder by using a Buchi Mini B-190 spray drier (Flawil, Switzerland) [33] by adding 2% with respect to CN-NL of PEG8000 from Sigma-Aldrich (Milan, Italy). Complexes CN-NL and CN-NL/GA were thus prepared in accordance with previous works [34,35]. The ratio between CN and NL is 2:1 by weight. The content of GA in CN-NL/GA is 0.2% by weight. HMSCs were supplied from Merck Millipore S.A.S., (Burlington, MA, USA). HaCaT, Dulbecco's Minimal Essential Medium (DMEM), L-glutamine, penicillin, streptomycin and fetal calf serum were purchased from Invitrogen, (Carlsbad, CA, USA). Fetal Bovine Serum and AlamarBlue®were purchased from Thermo Fisher Scientific, (Waltham, MA, USA). The HMSC Differentiation Osteogenic Bulletkit was purchased from Lonza, (Basel, Switzerland). Dulbecco's phosphate-buffered saline (DPBS), silver nitrate, pyrogallol, sodium thiosulphate, nuclear fast red, dimethyl sulfoxide (DMSO), $MgCl_2$ and MTT were purchased from Sigma-Aldrich (Milan, Italy). Aluminum sulphate was purchased from Carlo Erba (Milan, Italy).

4.2. Morphological Characterization of CN, NL and CN-NL Complexes

The morphology of the materials samples was investigated by field emission scanning electron microscopy (FESEM) using a FEI FEG-Quanta 450 instrument (Field Electron and Ion Company, Hillsboro, OR, USA). The samples were sputtered with Gold (Gold Edwards SP150B, England) before analysis. Inverted optical microscope (Nikon Ti, Nikon Instruments, Amsterdam, The Netherlands) was used to evaluate the morphology of hMSCs on CN, NL and CN-NL and CN-NL/GA complexes.

4.3. *Chemical Structure and Thermal Stability Charachterisation of CN, NL and CN-NL Complexes*

The powders as provided by MAVI were characterized by infrared spectroscopy using a Nicolet T380 Thermo Scientific instrument equipped with a Smart ITX ATR accessory with diamond plate.

Thermogravimetric tests onto the powders were performed on 4–10 mg of sample using a Mettler-Toledo Thermogravimetric Analysis/Scanning Differential Thermal Analysis (TGA/SDTA) 851 instrument operating with nitrogen as the purge gas (60 mL/min) at 10 °C/min heating rate in the 25–800 °C temperature range.

4.4. *In vitro Culture of hMSCs and HaCaT*

CN, NL, CN-NL, 0.2%GA and CN-NL/0.2%GA were solubilized in DPBS at 0.1 mg/mL and then diluted in the culture media to be tested with the cells. HMSCs isolated from the bone marrow were cultured in 24 well plate at 20,000 cells/well in DMEM Low Glucose with 2 mM L-glutamine, 100 IU/mL penicillin, 100 mg/mL streptomycin and 10% heat-inactivated FBS, with the above-mentioned components at different concentrations for 1 week to select the best concentration to be used. Immortalized human keratinocytes (HaCaT cell line) were cultured in DMEM supplemented with 1% Pen-Strep, 1% glutamine and 10% fetal calf serum at 37 °C in air and 5% CO_2. Subsequently, cells were dispensed into 6-well culture plates and left to grow until 80% of confluence.

4.5. *MTT Assay*

To establish the optimal non-toxic concentrations to be used in subsequent treatments, HaCaT cells were seeded at a density of 1×10^3/well in 96-well culture plates (n = 2). After 24 h, the cells were treated with CN, CN-NL and CN-NL/GA at different concentrations (from 10 µg/mL to 25 ng/mL) for 24 h and then incubated with MTT (0.5 mg/mL) at 37 °C for 4 h and, subsequently, with DMSO at room temperature for 5 min. The spectrophotometric absorbance of the samples was determined by using Ultra Multifunctional Microplate Reader (Bio-Rad, Hercules, California, USA) at 570–655 nm [36]. Results were given as average in Table 2.

Table 2. Concentrations of compounds tested with hMSCs.

CN-NL	0.5 µg/mL	0.2 µg/mL	0.1 µg/mL
GA	1 µg/mL	0.5 µg/mL	0.2 µg/mL
CN-NL/GA	1 µg/mL	0.5 µg/mL	0.2 µg/mL

4.6. *Evaluation of hMSCs Viability and Differentiation Potential*

AlamarBlue®test was performed at days 1, 4 and 8 to monitor hMSC viability in 24 well plates (n = 6) following the manufacturer's instructions. HMSCs were cultured for one week with the following concentrations of the components (Table 2):

Subsequently, hMSCs were osteoinduced using the differentiation Bulletkit osteogenic medium for one week in order to test hMSC regeneration potential. The samples plates (n = 3) were fixed with 1% *w/v* neutral buffered formalin. Mineralized matrix production was detected by von Kossa staining: cells were incubated with 1% *w/v* silver nitrate exposed to light for 15 min, 0.5% *w/v* Pyrogallol for 2 minutes and 5% *w/v* sodium thiosulphate for 2 min. All the solutions were in distilled water. The counterstaining was performed incubating cells with 0.1% *w/v* nuclear fast red diluted in a distilled water solution containing 5% *w/v* aluminum sulphate, for 5 min and washing in tap water for 5 min in order to reveal the reaction. The staining was observed with a Nikon Eclipse Ci microscope (Nikon Instruments, Amsterdam, The Netherlands) equipped with a digital camera by three independent observers for a qualitative analysis.

4.7. Anti-Inflammatory and Immune Responses Evaluation of HaCaT Cells

HaCaT cells, seeded in 6-well culture plates (n = 3) until 80% of confluence, were treated with CN, CN-NL and CN-NL/GA at selected concentrations (i.e., 10 μg/mL, 0.2 μg/mL and 0.5 μg/mL, respectively) for 6 h and 24 h. At the end of the experiment, total ribonucleic acid (RNA) was isolated and one microgram of this were reverse-transcribed into complementary deoxyribonucleic acid (cDNA) using random hexamer primers (Promega, Italy) at 42 °C for 45 min, according to the manufacturer's instructions. The anti-inflammatory and immune responses of HaCaT cells were evaluated by assaying the expression of pro-inflammatory cytokines IL-1α, IL-1β, IL-6, IL-8 and TNF-α, anti-inflammatory cytokine TGF-β, and antimicrobial peptide HBD-2 by Real time PCR with the LC Fast Start DNA Master SYBR Green kit from Roche Applied Science (Euroclone S.p.A., Pero, Italy) using 2 μl cDNA, corresponding to 10 ng of total RNA in a 20 Mm μl final volume, 3 mM Magnesium Chloride ($MgCl_2$) and 0.5 μM of sense primer and antisense primers (Table 3).

Table 3. Primer sequences and RT-PCR conditions.

Gene	Primers Sequence	Conditions	Product Size (bp)
IL-1α	5′-CATGTCAAATTTCACTGCTTCATCC -3′ 5′-GTCTCTGAATCAGAAATCCTTCTATC -3′	5 s at 95 °C, 8 s at 55 °C, 17 s at 72 °C for 45 cycles	421
TNF-α	5′-CAGAGGGAAGAGTTCCCCAG -3′ 5′-CCTTGGTCTGGTAGGAGACG -3′	5 s at 95°C, 6 s at 57°C, 13 s at 72°C for 40 cycles	324
IL-6	5′-ATGAACTCCTTCTCCACAAGCGC-3′ 5′-GAAGAGCCCTCAGGCTGGACTG-3′	5 s at 95°C, 13 s at 56°C, 25 s at 72°C for 40 cycles	628
IL-8	5-ATGACTTCCAAGCTGGCCGTG -3′ 5-TGAATTCTCAGCCCTCTTCAAAAACTTCTC-3′	5 s at 94°C, 6 s at 55°C, 12 s at 72°C for 40 cycles	297
TGF-β	5′-CCGACTACTACGCCAAGGAGGTCAC-3′ 5′-AGGCCGGTTCATGCCATGAATGGTG-3′	5 s at 94°C, 9 s at 60°C, 18 s at 72°C for 40 cycles	439
IL-1β	5′-GCATCCAGCTACGAATCTCC-3′ 5′-CCACATTCAGCACAGGACTC-3′	5 s at 95°C, 14 s at 58°C, 28 s at 72°C for 40 cycles	708

At the end of each run, the melting curve profiles were achieved by cooling the sample to 65° C for 15 s and then heating it slowly at 0.20 °C/s up to 95 °C with continuous measurement of fluorescence to confirm the amplification of specific transcripts. Cycle-to-cycle fluorescence emission readings were monitored and analyzed using LightCycler®software (Roche Diagnostics GmbH). Melting curves were generated after each run to confirm the amplification of specific transcripts. We used the b-actin coding gene, one of the most commonly used housekeeping genes, as an internal control gene. All reactions were carried out in triplicate, and the relative expression of a specific mRNA was determined by calculating the fold change relative to the b-actin control. The fold change of the tested gene mRNA was obtained with LightCycler®software by using the amplification efficiency of each primer, as calculated by the dilution curve. The specificity of the amplification products was verified by subjecting the amplification products to electrophoresis on 1.5% agarose gel and visualization by ethidium bromide staining [37].

4.8. Statistical Analysis

Statistical analyses were carried out by SPSS (SPSS v.16.0; IBM). All data were analyzed using a one-way analysis of variance (ANOVA) and post hoc test (Duncan) for multiple comparisons. Probability (p) values < 0.05 were considered as statistically significant differences.

5. Conclusions

CN-NL and CN-NL containing GA, obtained in powder by a spray drier technology can be optimal agents to form nanostructured and functional surfaces through a simple deposition by water suspension. The thermal stability of the complexes CN-NL and CN-NL/GA were investigated, and it was found that they can be considered thermally stable up to 114 °C and 127 °C, respectively. The performed characterization can give indications for a suitable application of CN-NL complexes as coating for material surfaces. All the investigated biopolymeric components were cytocompatible with HaCaT cells and hMSCs and optimal concentrations were selected for each of them. These findings demonstrate that CN-NL/GA complexes are able to downregulate a panel of anti-inflammatory cytokines in human keratinocytes and do not modify the proliferative and osteo-differentiative capacity of hMSC. Thus, these materials are very promising for skin contact applications, such as in biomedical, personal care and cosmetic products.

Author Contributions: Conceptualization, P.M., M.-B.C., G.D. and S.D.; methodology, L.T., A.F. and G.D.; formal analysis, B.A.; validation, L.T., G.D. and M.-B.C.; investigation, L.T., A.F.; resources, M.-B.C., A.L., P.M.; writing—original draft preparation, S.D.; writing—review and editing, M.B.-C., G.D., P.M.; visualization, S.D.; supervision, S.D.; project administration, M.-B.C. and S.D.; funding acquisition, M.-B.C., A.L, P.M.

Acknowledgments: The authors gratefully acknowledge Delfo D'Alessandro, University of Pisa, for his precious technical contribution to histology and Prof. Stefano Berrettini, University of Pisa, for providing access to equipment and facilities. Sabrina Bianchi, of the Department of Chemistry and Industrial Chemistry of the University of Pisa, and Randa Ishak, of the Department of Civil and Industrial Engineering of the University of Pisa, are kindly acknowledged for the support in thermogravimetry and FESEM investigations respectively.

Abbreviations

CN	Chitin nanofibrlis
NL	Nanolignin
GA	Glycyrrhetinic acid
HaCaT	Human keratinocytes
HMSCs	Human mesenchymal stromal cells
IL	Interleukin
TNF-α	Tumor necrosis factor alpha
PEG	Poly(ethylene glycol)
HBD-2	Human beta-defensin 2
TGF-β	Transforming growth factor beta
DMEM	Dulbecco's Minimal Essential Medium
DPBS	Dulbecco's phosphate-buffered saline
RT-PCR	Reverse transcription polymerase chain reaction
DMSO	Dimethyl sulfoxide
RNA	Ribonucleic acid
cDNA	Complementary deoxyribonucleic acid
MgCl2	Magnesium chloride
FESEM	Field Emission Scanning Electron Microscopy
ATR	Attenuated Total Reflection
TGA/SDTA	Thermogravimetric Analysis/Scanning Differential Thermal Analysis

References

1. Babu, R.P.; O'connor, K.; Seeram, R. Current progress on bio-based polymers and their future trends. *Prog. Biomate.* **2013**, *2*, 8–24. [CrossRef] [PubMed]
2. Morganti, P.; Danti, S.; Coltelli, M.B. Chitin and lignin to produce biocompatible tissues. *Res. Clin. Dermatol.* **2018**, *1*, 5–11.
3. Morganti, P.; Febo, P.; Cardillo, M.; Donnarumma, G.; Baroni, A. Chitin nanofibril and nanolignin: Natural polymers of biomedical interest. *J. Clin. Cosmet. Dermatol.* **2017**, 1. [CrossRef]
4. Beisl, S.; Friedl, A.; Miltner, A. Lignin from micro-to nanosize: Applications. *Int. J. Mol. Sci.* **2017**, *18*, 2367. [CrossRef] [PubMed]
5. Morganti, P.; Del Ciotto, P.; Stoller, M.; Chianese, A. Antibacterial and anti-inflammatory green nanocomposites. *Chem. Eng. Trans.* **2016**, *47*, 61–66. [CrossRef]
6. Morganti, P.; Svolacchia, F.; Del Ciotto, P.; Carezzi, F. New insights on anti-aging activity of chitin nanofibril-hyaluronan block copolymers entrapping active ingredients: *In vitro* and *in vivo* study. *J. Appl. Cosmetol.* **2013**, *31*, 1–29.
7. Morganti, P.; Palombo, M.; Tishchenko, G.; Yudin, V.E.; Guarneri, F.; Cardillo, M.; Del Ciotto, P.; Carezzi, F.; Morganti, G.; Fabrizi, G. Chitin-hyaluronan nanoparticles: A multifunctional carrier to deliver anti-aging active ingredients through the skin. *Cosmet* **2014**, *1*, 140–158. [CrossRef]

8. Morganti, P.; Carezzi, F.; Del Ciotto, P.; Tishchenco, G.; Chianese, A.; Yudin, V.E. A green multifunctional polymer from discarded material: Chitin Nanofibrils. *Br. J. Appl. Sci. Technol.* **2014**, *4*, 4175–4190. [CrossRef]
9. Morganti, P.; Tishchenko, G.; Palombo, M.; Kelnar, I.; Brozova, L.; Spirkova, M.; Pavlova, H.; Kobera, L.; Carezzi, F. Chitin nanofibrils for biomimetic products: nanoparticles and nanocomposite chitosan films in health care. In *Marine Biomaterials: Isolation, Characterization and Application*, 1st ed.; Kim, S., Ed.; CRC Press: Boca Raton, FL, USA, 2013; pp. 681–715. ISBN 9781138076389.
10. Morganti, P.; Coltelli, M.B.; Danti, S. Biobased Tissues for innovative Cosmetic products: Polybioskin as an EU Research project. *Glob. J. Nano.* **2018**, *3*, 555620. [CrossRef]
11. Rizvi, R.; Cochrane, B.; Naguib, H.; Lee, P.C. Fabrication and characterization of melt-blended polylactide-chitin composites and their foams. *J. Cell. Plast.* **2011**, *47*, 283–300. [CrossRef]
12. Herrera, N.; Singh, A.A.; Salaberria, A.M.; Labidi, J.; Mathew, A.P.; Oksman, K. Triethyl Citrate (TEC) as a Dispersing Aid in Polylactic Acid/Chitin Nanocomposites Prepared via Liquid-Assisted Extrusion. *Polymers* **2017**, *9*, 406. [CrossRef] [PubMed]
13. Guan, Q.; Naguib, H.E. Fabrication and Characterization of PLA/PHBV-Chitin Nanocomposites and Their Foams. *J. Polym. Environ.* **2014**, *22*, 119–130. [CrossRef]
14. Herrera, N.; Roch, H.; Salaberria, A.M.; Pino-Orellana, M.A.; Labidi, J.; Fernandes, S.C.M.; Radic, D.; Leiva, A.; Oksman, K. Functionalized blown films of plasticized polylactic acid/chitin nanocomposite: Preparation and characterization. *Mater. Des.* **2016**, *92*, 846–852. [CrossRef]
15. Coltelli, M.B.; Cinelli, P.; Gigante, V.; Aliotta, L.; Morganti, P.; Panariello, L.; Lazzeri, A. Chitin Nanofibrils in Poly(Lactic Acid) (PLA) Nanocomposites: Dispersion and Thermo-Mechanical Properties. *Int. J. Mol. Sci.* **2019**, *20*, 504. [CrossRef]
16. Morganti, P. Composition and material comprising chitin nanofibrils, lignin and a co-polymer and their uses. Patent WO 2016/042474 Al, 2015.
17. Andersen, F.A. Final report on the safety assessment of Glycyrrhetinic Acid, Potassium Glycyrrhetinate, Disodium Succinoyl Glycyrrhetinate, Glyceryl Glycyrrhetinate, Glycyrrhetinyl Stearate, Stearyl Glycyrrhetinate, Glycyrrhizic Acid, Ammonium Glycyrrhizate, Dipotassium Glycyrrhizate, Disodium Glycyrrhizate, Trisodium Glycyrrhizate, Methyl Glycyrrhizate, and Potassium Glycyrrhizinate. *Int. J. Toxicol.* **2007**, *26*, 79–112. [CrossRef]
18. Hussain, H.; Green, I.R.; Shamraiz, U.; Saleem, M.; Badshah, A.; Abbas, G.; Rehman, N.U.; Irshad, M. Therapeutic potential of glycyrrhetinic acids: a patent review (2010-2017). *Expert. Opin. Ther. Pat.* **2018**, *28*, 383–398. [CrossRef] [PubMed]
19. Kumirska, J.; Czerwicka, M.; Kaczyński, Z.; Bychowska, A.; Brzozowski, K.; Thöming, J.; Stepnowski, P. Application of Spectroscopic Methods for Structural Analysis of Chitin and Chitosan. *Mar. Drugs* **2010**, *8*, 1567–1636. [CrossRef]
20. Boeriu, C.G.; Bravo, D.; Gosselink, R.J.A.; Van Dam, J.E.G. Characterisation of structure-dependent functional properties of lignin with infrared spectroscopy. *Ind. Crops Prod.* **2004**, *20*, 205–218. [CrossRef]
21. Zu, Y.; Meng, L.; Zhao, X.; Ge, Y.; Yu, X.; Zhang, Y.; Deng, Y. Preparation of 10-hydroxycamptothecin- loaded glycyrrhizic acid-conjugated bovine serum albumin nanoparticles for hepatocellular carcinoma-targeted drug delivery. *Int. J. Nanomed.* **2013**, *8*, 1207–1222. [CrossRef]
22. Cheng, M.; Gao, X.; Wang, Y.; Chen, H.; He, B.; Xu, H.; Li, Y.; Han, J.; Zhang, Z. Synthesis of Glycyrrhetinic Acid-Modified Chitosan, 5-Fluorouracil Nanoparticles and Its Inhibition of Liver Cancer Characteristics in Vitro and in Vivo. *Mar. Drugs* **2013**, *11*, 3517–3536. [CrossRef]
23. Ablis, M.; Leininger-Muller, B.; Wong, C.D.; Siest, G.; Loppinet, V.; Visvikis, S. Synthesis and *in Vitro* Antioxidant Activity of Glycyrrhetinic Acid Derivatives Tested with the Cytochrome P450/NADPH System. *Chem. Pharm. Bull.* **2004**, *52*, 1436–1439. [CrossRef]
24. Nilforoushzadeh, M.A.; Ahmadi Ashtiani, H.R.; Jaffary, F.; Jahangiri, F.; Nikkhah, N.; Mahmoudbeyk, M.; Fard, M.; Ansaria, Z.; Zare, S. Dermal Fibroblast Cells: Biology and Function in Skin Regeneration. *J. Skin Stem Cell* **2017**, *4*, e69080. [CrossRef]
25. Danti, S.; D'Acunto, M.; Trombi, L.; Berrettini, S.; Pietrabissa, A. A micro/nanoscale surface mechanical study on morpho-functional changes occurred to multilineage-differentiated human mesenchymal stem cells. *Macromol. Biosci.* **2007**, *7*, 589–598. [CrossRef] [PubMed]
26. Donnarumma, G.; Paoletti, I.; Fusco, A.; Perfetto, B.; Buommino, E.; de Gregorio, V.; Baroni, A. β-Defensins: Work in Progress. *Adv. Exp. Med. Biol.* **2016**, *901*, 59–76. [CrossRef] [PubMed]

27. Mantovani, A.; Sica, A.; Sozzani, S.; Allavena, P.; Vecchi, A.; Locati, M. The chemokine system in diverse forms of macrophage activation and polarization. *Trends Immunol.* **2004**, *25*, 677–686. [CrossRef] [PubMed]
28. Dinarello, C.A. Interleukin-1. *Cytokine Growth Factor Rev.* **1997**, *8*, 253–265. [CrossRef]
29. Esposito, E.; Cuzzocrea, S. TNF-alpha as a therapeutic target in inflammatory diseases, ischemia-reperfusion injury and trauma. *Curr. Med. Chem.* **2009**, *16*, 3152–3167. [CrossRef] [PubMed]
30. Koch, A.E.; Polverini, P.J.; Kunkel, S.L.; Harlow, L.A.; DiPietro, L.A.; Elner, V.M.; Elner, S.G.; Strieter, R.M. Interleukin-8 as a macrophage-derived mediator of angiogenesis. *Science* **1992**, *258*, 1798–1801. [CrossRef]
31. Morganti, P.; Fusco, A.; Paoletti, I.; Perfetto, B.; Del Ciotto, P.; Palombo, M.; Chianese, A.; Baroni, A.; Donnarumma, G. Anti-inflammatory, immunomodulatory, and tissue repair activity on human keratinocytes by green innovative nanocomposites. *Materials* **2017**, *10*, 843. [CrossRef] [PubMed]
32. Oyama, K.; Kawada-Matsuo, M.; Oogai, Y.; Hayashi, T.; Nakamura, N.; Komatsuzawa, H. Antibacterial effects of glycyrrhetinic acid and its derivatives on staphylococcus aureus. *PLoS ONE* **2016**, *11*, e0165831. [CrossRef]
33. Morganti, P.; Muzzarelli, C. Spray-dried chitin nanofibrils, method for production and uses thereof. Patent WO2007060628A1, 23 November 2005.
34. Morganti, P. Method of preparation of chitin and active principles complexes and the so obtained complexes. Patent WO2012143875A1, 19 April 2011.
35. Brooker, A.D.M.; Vaccaro, M.; Scialla, S.; Walker, S.J.; Morganti, P.; Carezzi, F.; Benjelloun-Mlayah, B.; Crestini, C.; Lange, H.; Bartzoka, E. A consumer goods product comprising chitin nanofibrils, lignin and a polymer or co-polymer. Patent EP2995321B1, 15 April 2014.
36. Mosmann, T. Rapid colorimetric assay for cellular growth and survival: application to proliferation and cytotoxicity assays. *J. Immunol. Methods* **1983**, *1*, 55–63. [CrossRef]
37. Fusco, A.; Coretti, L.; Savio, V.; Buommino, E.; Lembo, F.; Donnarumma, G. Biofilm formation and immunomodulatory activity of Proteus mirabilis clinically isolated strains. *Int. J. Mol. Sci.* **2017**, *18*, 414. [CrossRef] [PubMed]

10

Biocompatibility Investigation of Hybrid Organometallic Polymers for Sub-Micron 3D Printing via Laser Two-Photon Polymerisation

Evaldas Balčiūnas [1], Nadežda Dreižė [1], Monika Grubliauskaitė [1], Silvija Urnikytė [1], Egidijus Šimoliūnas [1], Virginija Bukelskienė [1], Mindaugas Valius [1], Sara J. Baldock [2,3], John G. Hardy [2,3,*] and Daiva Baltriukienė [1,*]

1 Institute of Biochemistry, Life Sciences Centre, Vilnius University, 10257 Vilnius, Lithuania; ev.balciunas@gmail.com (E.B.); nadezda.dreize@gmail.com (N.D.); monika.grub@gmail.com (M.G.); urniksi@gmail.com (S.U.); egidijus.simoliunas@gmail.com (E.Š.); virginija.bukelskiene@bchi.vu.lt (V.B.); mindaugas.valius@bchi.vu.lt (M.V.)

2 Department of Chemistry, Lancaster University, Lancaster LA1 4YB, UK; s.baldock@lancaster.ac.uk

3 Materials Science Institute, Lancaster University, Lancaster LA1 4YB, UK

* Correspondence: j.g.hardy@lancaster.ac.uk (J.G.H.); daiva.baltriukiene@bchi.vu.lt (D.B.)

Abstract: Hybrid organometallic polymers are a class of functional materials which can be used to produce structures with sub-micron features via laser two-photon polymerisation. Previous studies demonstrated the relative biocompatibility of Al and Zr containing hybrid organometallic polymers in vitro. However, a deeper understanding of their effects on intracellular processes is needed if a tissue engineering strategy based on these materials is to be envisioned. Herein, primary rat myogenic cells were cultured on spin-coated Al and Zr containing polymer surfaces to investigate how each material affects the viability, adhesion strength, adhesion-associated protein expression, rate of cellular metabolism and collagen secretion. We found that the investigated surfaces supported cellular growth to full confluency. A subsequent MTT assay showed that glass and Zr surfaces led to higher rates of metabolism than did the Al surfaces. A viability assay revealed that all surfaces supported comparable levels of cell viability. Cellular adhesion strength assessment showed an insignificantly stronger relative adhesion after 4 h of culture than after 24 h. The largest amount of collagen was secreted by cells grown on the Al-containing surface. In conclusion, the materials were found to be biocompatible in vitro and have potential for bioengineering applications.

Keywords: bioactive surfaces; biomaterials; hybrid organometallic polymers; laser two-photon polymerisation; tissue engineering

---

1. Introduction

The concept of growing human replacement parts in the lab has been around for several decades [1,2]. Researchers have used different approaches for the engineering of artificial tissues—from allogeneic or xenogeneic tissue decellularisation [3] to approaches based on additive manufacturing [4], the latter of which offers a route to the generation of an optimal scaffold for a specific tissue type and patient which has significant potential for economic/health/societal impacts.

Laser two-photon polymerisation (LTPP) is a 3D fabrication technique capable of producing materials with fine details in their structures at sub-micron resolutions [5,6]. Various materials can be used to make 3D structures using this technique including derivatives of natural polymers (e.g., hyaluronic acid [7], gelatin [8]) and synthetic ones (e.g., derivative of polyethylene glycol [9] and SU-8 [10]).

The LTPP technique is highly versatile and allows several materials to be used in the same sample [11,12]. Compared to other fabrication techniques like stereo-lithography [13], fused deposition modelling [14,15] and selective laser sintering [16], LTPP is the only technique that allows for resolutions below the diffraction limit of light to be fabricated [17]. In addition, LTPP allows for fine-tuning of structural motifs as opposed to randomised porous structures obtained by other means, like template-casting [18] or particulate-leaching [19]. A group of interesting materials for LTPP are hybrid organometallic polymers [20]. To date, materials based on Al [21], Ge [22], Ti [23], V [24] and Zr [25] have been shown to be structurable using LTPP systems. An in vitro biocompatibility screening study showed that Al and Ti hybrids supported a comparable number of cells to glass, while the Zr-based hybrid exceeded the biocompatibility of all the other surfaces [21]. An in vivo study of our Zr-based hybrids showed them to be relatively biocompatible when implanted in rabbit muscle and that they did not cause inflammation or foreign body reaction as demonstrated by hematoxylin and eosin staining [26].

Cell-extracellular matrix interactions are among the most important processes to attenuate in attempting to recreate structurally and functionally viable tissue constructs analogous to natural tissues (where the resident cells adhere to the ECM and take part in remodelling it over time). For many applications, it is important to have materials that support a native-like cellular response and integration. To understand the influence of the metals contained in the organometallic hybrids on cellular behaviour, we have investigated the process of adhesion and adhesion-associated kinase expression, as well as collagen secretion of primary rat myogenic cells grown on these surfaces. We believe that this work will prove useful for tissue engineering researchers focusing on artificial scaffold-based tissue remodelling techniques in that the materials investigated in this work are highly biocompatible, precisely structurable at sub-micron resolutions and simple to prepare.

2. Materials and Methods

2.1. Material Synthesis

The hybrid organometallic polymers were synthesised according to published protocols [21,25] Briefly, the Al-based material was prepared by dissolution of aluminium isopropoxide (AIP, ≥98%, Merck, Kenilworth, NJ, USA) in toluene (ACS, ISO, Reag. Ph Eur, Merck, Kenilworth, NJ, USA). In parallel, 3-(trimethoxysilyl)propyl methacrylate (MAPTMS, 98%, Sigma-Aldrich, St. Louis, MO, USA) was hydrolysed using HCl (0.1 M, Applichem, Darmstadt, Germany). Methacrylic acid (MAA, 99%, Sigma-Aldrich, USA) was then added to the solution of aluminium isopropoxide in toluene at a 1:1 molar ratio and subsequently, hydrolysed MAPTMS was added to the mixture at a 1:1:4 AIP:MAA:MAPTMS molar ratio. Finally, 1% of photoinitiator (4,4′-bis(diethylamino)benzophenone, Sigma-Aldrich, St. Louis, MO, USA) was added to the weight of AIP, MAA and 3-(trihydroxysilyl)propyl methacrylate (the product of MAPTMS hydrolysis) and stirred, while shielding from ambient light to prevent undesired crosslinking.

The Zr-based material was prepared in an analogous manner, with molar ratios of zirconium (IV) propoxide, MAA and MAPTHS being 1:1:4 with 1% of photoinitiator by weight (excluding solvents).

2.2. 2D Sample Preparation

To secure polymer bonding to glass, MAPTMS-treated glass slides were prepared according to a protocol adopted from Kapyla et al. [27]. Briefly, circular 12 mm diameter borosilicate glass slides (Thermo Fisher Scientific, Waltham, MA, USA) were washed in ethanol, then immersed in a mixture of MAPTMS/ethanol/acetic acid/water overnight and finally washed in ethanol in an ultrasonic bath (EMAG, Mörfelden-Walldorf, Germany).

The hybrid materials were then spin-coated @ 3000 rpm for 30 s per sample and left at room temperature in the dark overnight for the solvents to evaporate. The samples were then polymerised using a UV lamp (UV-C, G15W T8, Sylvania, London, UK) for at least 5 h at around 20 cm distance, corresponding to about 1.8 mW/cm^2. The samples were sterilised under UV for at least one hour on each side, subsequently washed in sterile PBS to remove residual initiator and low molecular weight components. They were then immersed in sterile growth medium prior to cell culture.

2.3. 3D Sample Fabrication via Laser Two-Photon Polymerisation

For laser two-photon polymerisation, a droplet of a hybrid material was placed on a glass slide and then left in a fume hood to evaporate overnight while covered from ambient light. After 24 h such samples were used for structure fabrication on a commercially available Nanoscribe® Photonic Professional GT system (Nanoscribe GmbH, Eggenstein-Leopoldshafen, Germany) that is based at the Department of Chemistry at Lancaster University. A 63X 1.4 NA oil immersion lens was used to focus the laser beam into the polymer. Stereolithography (STL) flies were downloaded from the internet [28] and sliced using proprietary Nanoscribe software (DeScribe 2.3.3). Hatching and slicing distances in X, Y and Z coordinates were set to 100 nm. The laser scanning speed was set to 10 mm/s and the laser power was in the range of 20% to 100% (corresponding to 10–50 mW). After polymerization, the Al based hybrid material was developed in toluene for at least 15 min, followed by air-drying. The structures were then sputter-coated at a thickness of 10 nm using a Q150 RS coater (Quorum Technologies, Lewes, UK) and subsequently observed using a SEM (JSM 7800F, JEOL, Tokyo, Japan) operating at 10–15 kV at the Department of Chemistry at Lancaster University.

2.4. Collagen Adsorption and Secretion Assays

A protocol for assessing the concentration of collagen was adapted from Tullberg-Reinert and Jundt [29] with slight modifications. The protocol used is based on in situ staining of cell monolayers using Sirius Red. This dye has strong interactions with collagen types I and III, but weak interaction with collagen type IV.

First, a solution of collagen type I (1 mg/mL in 0.1 N acetic acid, Sigma-Aldrich, St. Louis, MO, USA) was added to the sample and incubated for 1 h at 37 °C. Then, the collagen solution was gently replaced with fixating Bouin's solution (15 mL saturated aqueous picric acid (Sigma-Aldrich, St. Louis, MO, USA) with 5 mL 35% formaldehyde (Sigma-Aldrich, St. Louis, MO, USA) and 1 mL glacial acetic acid (Sigma-Aldrich, St. Louis, MO, USA), incubated for 1 h and then washed using PBS. The samples were then transferred to new wells with a solution of Sirius Red (1 mg/mL in picric acid) for 1 h under mild rocking conditions (30 rpm). Afterwards, the samples were washed with 0.01 N hydrochloric acid to remove unbound dye. The dye was then dissolved in 0.2–0.3 mL of 0.1 N sodium hydroxide by shaking at room temperature for 30 min. Non-coated samples were stained and measured in the same way and their signals were subtracted from collagen-coated sample signals prior to analysis.

The optical density of the solution was measured using a Varioskan Flash microplate reader (Thermo Fisher Scientific, Waltham, MA, USA) at 550 nm. For reference, a calibration curve was prepared by using known concentrations of collagen type I in order to link them to the absorbance of Sirius Red at 550 nm (Figure S1 in Supplementary Materials). A total of 3 independent experiments were carried out and the results are presented as averages +/- standard errors.

Collagen synthesis was assessed in an analogous way to adsorption. The cells were seeded on the test surfaces at a density of 20,000 cells/mL/sample in 24 well tissue culture plates. After 1, 7 and 14 days of incubation, the samples were fixated using Bouin's solution, then dyed using a solution of Sirius Red (Sigma-Aldrich, St. Louis, MO, USA) and subsequently washed with a solution of HCl (Sigma-Aldrich, St. Louis, MO, USA). Finally, the dye was dissolved in a solution of NaOH (Merck, Kenilworth, NJ, USA) and measured spectrophotometrically against pure NaOH, and the amount of collagen was determined from the calibration curve. 3 independent experiments were performed with 3 repetitions per material within each experiment.

2.5. Cell Isolation, Culture and Characterisation

A Wistar rat was euthanised and a small piece of skeletal muscle tissue was removed. The experiments were approved by License of Animal Research Ethics Committee (Lithuania) No. G2-39, 03/08/2016. Large blood vessels were separated, then skeletal muscle tissue was minced and incubated in a solution of EDTA-trypsin (Gibco, Thermo Fisher Scientific, Waltham, MA, USA) with collagenase and hyaluronidase (0.5% and 0.3%, respectively, Sigma-Aldrich, St. Louis, MO, USA) for 30 min at 37 °C under mild shaking conditions. The resulting cell suspension was centrifuged, mixed with growth medium and seeded to tissue culture plates. The growth medium was replaced every 3–4 days and after 5 passages the cells were cloned by serial dilution. A highly proliferative colony forming unit was selected as a cell source, multiplied in vitro and banked in liquid nitrogen for further use.

The cells were cultured in Iscove's modified Dulbecco's medium (IMDM), supplemented with 10% Foetal bovine serum (Gibco, Thermo Fisher Scientific, Waltham, MA, USA) and Penicillin-Streptomycin (100 U/mL and 100 µg/mL, Gibco, Thermo Fisher Scientific, Waltham, MA, USA). They were subcultured every 3–4 days, by detaching with EDTA-trypsin and resuspending in fresh medium. The cells were grown in an incubator (Thermo Fisher Scientific, Waltham, MA, USA) at 37 °C with 5% CO_2.

For characterization, cells were grown in 30 mm diameter Petri dishes with glass slides on the bottom, fixed with 4% paraformaldehyde for 15 min. Then the cells were washed with PBS and incubated with 0.2% Triton X-100 in PBS for 15 min to permeabilise the membranes. After blocking in 1% BSA in PBS, the cells were incubated with primary antibodies against CD34, C45 (both from Abcam, Cambridge, UK), Myf5 and c-kit (both from Thermo Fisher Scientific, Waltham, MA, USA) overnight according to manufacturer's instructions at 4 °C. The samples were then rinsed with 1% BSA in PBS and incubated with Cy3-conjugated secondary antibodies (Merck, Kenilworth, NJ, USA). Analysis was performed using an Olympus IX71 (Olympus, Tokyo, Japan) fluorescence microscope.

2.6. Analysis of Cell Viability

AO/EB staining distinguishes cells into four categories: live (bright green, intact nucleus), early apoptotic (bright green, but fragmented nucleus), late apoptotic (orange, fragmented nucleus) and necrotic (bright orange, intact nucleus). Cells were cultured on the hybrid polymer and glass surfaces for 24, 48, 72 and 96 h. Then, the growth medium with any unattached cells was collected and the monolayer of cells were treated with EDTA-trypsin and then collected as well. All of them were subsequently stained by acridine orange and ethidium bromide (Sigma-Aldrich, St. Louis, MO, USA) as described in [30]. The cells were analysed using a BD FACSCanto™ II (BD Biosciences, San Jose, CA, USA) flow cytometer, registering 10,000 events per sample and time point. Green fluorescence of acridine orange was detected using FITC channel and red fluoresce of ethidium bromide was detected using PE channel. A total of 3 independent experiments were performed with 3 repetitions per material within each experiment. The cells were split into four categories based on their fluorescence profile.

2.7. MTT Assay

MTT assay is based on cellular mitochondrial reductase activity. The absorption measured using this technique is directly proportional to the amount of active enzyme, which gives an idea on the number of viable cells and their metabolic activity.

The cells were seeded at a density of 20,000 cells/mL/sample in 24 well tissue culture plates on the hybrid material and glass surfaces and cultured for 24, 48, 72 and 96 h at 37 °C. The samples were then gently transferred to new tissue culture plates containing a solution of MTT (0.2 mg/mL in PBS) and incubated for 2 h at 37 °C. The solution was subsequently removed and the formazan crystals dissolved in ethanol (96%, Vilniaus degtinė, Vilnius, Lithuania). The absorption was measured using a Varioskan Flash microplate reader (Thermo Fisher Scientific, Waltham, MA, USA) at 570 nm with ethanol as a reference. A total of 3 independent experiments were performed with 3 repetitions per material within each experiment.

2.8. Adhesion Strength

The cells were seeded at a density of 100 000 cells/mL/sample in 24 well tissue culture plates and cultured for either 4 or 24 h at 37 °C with 5% CO_2. At those time points, samples were transferred to new tissue culture plates, and half of the plates were subjected to shaking at 500 rpm for 5 min using a tissue culture plate shaker (Thermomixer Comfort, Eppendorf, Hamburg, Germany). The reference samples (unshaken) were incubated at 37 °C. The number of cells remaining adhered to the surface was measured by replacing the growth medium with 0.1% crystal violet solution in 20% ethanol for 30 min. Then cells were washed with tap water. Before the measurement of absorption, dye was solubilised with 0,1% acetic acid solution in 50% ethanol. Optical density proportional to cell number was measured using a Varioskan Flash microplate reader (Thermo Fisher Scientific, Waltham, MA, USA) at 570 nm and comparing it to analogously dyed, unshaken cell monolayers. A total of 3 independent experiments were performed with 3 repeats per material within each experiment.

2.9. Signalling Protein Expression and Phosphorylation

The cells were seeded on the samples at a density of 100,000 cells/mL/sample and cultured for either 4 or 24 h. At these timepoints, the samples were transferred to new tissue culture plates and gently washed with PBS. The PBS was then replaced with a lysis buffer consisting of 8 M urea, 2 M thiourea and 50 mM DTT. Cells were collected from 5 samples for each timepoint by pipetting 5–10 times. The lysates were then centrifuged for 10 min at 20,000 × G at RT. The supernatants of each vial were then transferred to new vials and frozen at −20 °C until further use. A total of 3 independent experiments was carried out.

Protein concentrations were normalised by running an SDS-PAGE gel, staining with Coomassie brilliant blue, taking images using a transilluminator (UVP, Upland, CA, USA) using analysing by ImageJ software. After diluting the highest concentration samples using lysis buffer, the concentration-equalised samples were subjected to gel electrophoresis again at 200 V for 45 min using a BioRad (Hercules, CA, USA) electrophoresis apparatus. The proteins were then transferred to a PVDF membrane (Carl Roth, Karlsruhe, Germany) using a Biometra Fastblot transfer device (Biometra GmbH, Göttingen, Germany) at 25 V and 300 mA. The membranes were blocked using 1% BSA (Sigma-Aldrich, St. Louis, MO, USA) in TBS with 0.1% Tween 20 (Sigma-Aldrich, St. Louis, MO, USA).

The membranes were subsequently treated with primary antibodies against p-Akt (Ser473 and Thr308, Cell Signalling technology, Danvers, MA, USA), Akt (Molecular Probes, Eugene, OR, USA), FAK (BD Biosciences, San Jose, CA, USA) and fluorescently-labelled anti-α-tubulin (Sigma-Aldrich, St. Louis, MO, USA) overnight according to manufacturer's instructions at 4 °C.

Next day, the membranes were washed three times using wash buffer and incubated with secondary antibodies. FAK and Akt were treated with HRP-conjugated goat anti-mouse (Invitrogen, Carlsbad, CA, USA) and HRP-conjugated anti-rabbit (Invitrogen, Carlsbad, CA, USA) according to manufacturer's instructions for 1 h. Finally, the membranes were washed again three times using wash buffer. FAK and Akt were detected using by treating with ECL reagent, which upon catalysis by HRP yields chemiluminescence that was detected using a transilluminator (UVP, Upland, CA, USA).

P-Akt (Ser473) and p-Akt (Thr308) were treated with secondary goat-anti-rabbit antibodies conjugated to a fluorescent infrared dye (IRDye 800CW, LI-COR, Lincoln, NE, USA) and detected using an infrared imaging system (Odyssey, LI-COR, Lincoln, NE, USA).

Membrane image analysis was performed using ImageJ software (National Institutes of Health, USA).

2.10. Statistical Analysis

All experiments were repeated at least 3 times independently with 3 repetitions within each experiment. The results are presented as averages +/− standard deviations (n = 0–5) or errors

($n = 6+$). Statistical significance was assessed using one-way or two-way ANOVA and Tukey's HSD post-hoc tests using RStudio (RStudio Inc., Boston, MA, USA) and plotted using the ggplot2 package. Statistical significance was considered to be achieved with $p < 0.05$.

3. Results and Discussion

3.1. Laser Two-Photon Polymerisation

As mentioned in the introduction and methods, the possibility of structuring the Al-based hybrid material has already been described previously [21]. However, we wanted to further demonstrate our ability to fabricate complex 3D shapes and thus have fabricated a micro-structure reminiscent of Zerg (StarCraft, Blizzard Entertainment©, Irvine, CA, USA)—see Figure 1. High volume suspended features were reproducibly fabricated, like the claws of the Zerg hydralisk. Literature shows that by precisely controlling the 3D cell microenvironments, one can achieve complex cellular responses, like stem cell homing towards artificially designed niches [31]. Another important group of factors in designing artificial cell niches are associated to the material chemistry and properties [32]. Integration of several materials in a single structure will hopefully one day help to guide stem cell differentiation towards different lineages on the same sample.

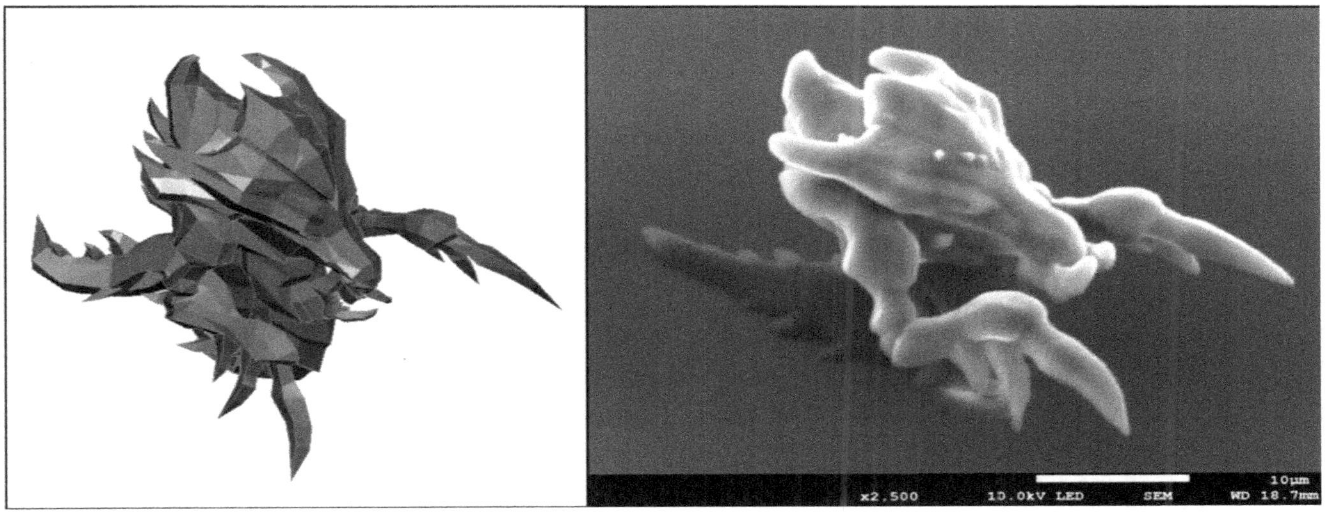

Figure 1. Right, A 3D micro-structure reminiscent of Zerg fabricated using a commercially available Nanoscribe Photonic Professional GT system. The scale bar corresponds to 10 microns. Left, A CAD model of the structure.

Multiple materials can be structured with laser two-photon polymerisation and the Al-based hybrid organometallic polymer presented in this work expands the selection of possible cellular niche materials even further. To date, a number of reports have been made of multiple-material structure fabrication using laser two-photon polymerization—including some of our own, in which PDMS, hybrid organometallic polymer based on Zr, PEG-DA and commercially-available OrmoComp have been integrated together [12]. Lamont et al. have developed a technique to exchange the material that is being used to fabricate structures on-the-go, allowing up to five different materials to be incorporated within the same structure with increased fabrication speed and potentially, multiple functionalities for applications like drug delivery, advanced optics, meta-materials and microrobotics [33]. Additionally, two or even more disparate fabrication techniques can be employed within the same structure as demonstrated with fused filament fabrication 3D-printed structure modification using laser ablation [34].

The addition of yet another 3D-printable material with new properties to the engineer's menu will allow for a wider variety of new structures and applications to be envisaged, particularly in fields that require high precision micro-fabrication, like tissue engineering scaffolds and drug delivery devices.

3.2. Protein Adsorption Assay

Extracellular matrix protein adhesion to the different surfaces may affect the subsequent adhesion of cells to the surfaces. Collagen type I was chosen as a model system for this experiment—its adsorption to simple spin-coated samples was assessed due to collagen type I abundance in the extracellular matrix. Collagen supports cellular adhesion via RGD sequences, which are specifically recognised by cellular adhesion proteins integrins. These interactions facilitate strong integration between a tissue engineered scaffold and the surrounding tissues, so it is important to have a material that supports collagen adsorption to its surface well for applications where cell adhesion is beneficial.

Collagen itself has been used as a tissue engineering building block in various studies, for example that of Ber et al., where osteoblast growth was guided on 2D surfaces of collagen [35] or work by Ramanathan et al., who have investigated 3D hybrid collagen matrixes as antibacterial dermal substitutes [36]. Collagen and its derivatives are a widely investigated group of materials for soft tissue engineering, while for hard tissue, like bone, collagen coating is a highly desirable approach that can be used in conjunction with other materials. Multiple reviews have focused on the use of collagen as a tissue engineering material either in a pure or composite form [37–39]. Consequently, a possibility of coating hybrid polymer structures with collagen seems like a good approach for improving cellular integration and therefore chosen in this study.

In our experiments, the choice of protein quantification technique was limited by residual photo-initiators within the materials which rendered them fluorescent. Thus, the signal of fluorescence-based techniques for protein visualisation would have been hindered by sample autofluorescence. To overcome this issue, a colorimetric assay was employed.

Collagen adsorption was measured on both the hybrid polymers and reference glass slides by immersion in a solution of collagen, then dying the bound collagen and measuring the optical absorption of the dye. We found that even though the largest amount of collagen was found on the Zr surface and the smallest amount on Al surface, the results in each group were highly disperse, and no statistically significant differences were determined with $p = 0.35$ between Al and Zr, $p = 0.72$ between Glass and Zr and $p = 0.81$ between Glass and Al. The results are presented in Figure 2.

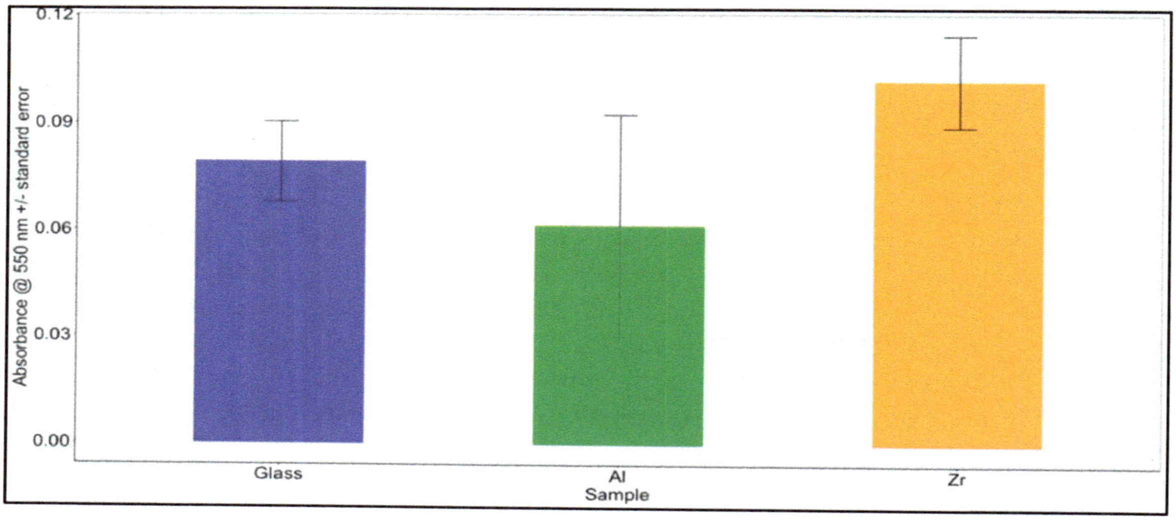

Figure 2. Collagen I adsorption to Al and Zr based hybrid organometallic polymers and glass. The results are presented as average Sirius Red absorbances + standard errors for three independent experiments with n = 3 per each experiment (a total of n = 9 per material).

A calibration curve between the total amount of collagen I and Sirius Red absorption was prepared and is presented in Supplementary Materials Figure S1. A highly linear relation was found in the range of 0–100 μg with an R^2 value of 0.994. The amounts of collagen adsorbed to the samples were calculated according to the linear model and were found to be 5.59 μg for Glass, 4.40 μg for Al and 6.73 μg for Zr on average per sample.

In our previous work [21], we investigated the contact angles of these materials and found them to be 39° for Glass, 72° for Al-based hybrid materials and 71° for Zr-based hybrid materials. The collagen adsorption data does not seem to correlate with the surface contact angles. Even though there is a significant difference between the contact angle of glass to Al or Zr surfaces, the difference in collagen I adsorption could not be distinguished due to a high deviation from the mean. A paper by Ying et al. [40] indicates that around a two-fold increase in the adsorption of collagen I to the glass surface is expected with an increase of the surface contact angle from 40° to 70°. An analogous result would be expected on other surfaces, since protein adsorption is governed by the same forces as surface contact angle—electrostatic and hydrophobic interactions.

3.3. Cell Viability

Having established the structurability of the materials as well as their collagen adsorption capacity, we investigated how their surfaces influence cellular viability. In this particular study, a primary rat muscle cell line was used to model a situation that would be as close as possible to a clinical one. Using primary muscle-derived stem cells is a safe choice in designing a possible future tissue engineering strategy due to the abundance of donor tissue sites, relatively mild donor site morbidity and high proliferative capacity as well as multipotency of the cells. A great example by Nieponice et al. demonstrates how rat muscle-derived stem cells could be used together with elastomeric scaffold-building materials in constructing vascular grafts [41]. The study successfully showed that the cells tended to home to the lumen of the vascular grafts and started to express α-actin, calponin as well as secrete collagen. Endothelial differentiation was supported via the presence of von Willebrand factor. Our isolated rat muscle-derived progenitor cells were positive for CD34, Myf5 and c-kit and negative for CD45 (Figure S2).

This experiment aimed to find the ratios between the numbers of viable, early apoptotic, necrotic and late apoptotic cells at different timepoints and on different surfaces. Flow cytometry data shows that no statistically significant differences between the numbers of cells in each category can be observed, since the majority of cells at each time point was viable (Figure 3). Additional data is presented in the Supplementary Materials, providing light microscopy images of cell culture at different time points as well as statistical analysis (Figure S3 and Figure S4 in the Supplementary Materials).

The results are in accord with our previous work on these materials with NIH/3T3 cells, which showed that the cells remain viable in culture for extended periods of time [21]. Data obtained in this experiment open the way for the future applications of both Al and Zr based hybrid materials. Long-term viability support of the cells is the first step towards engineering adequate implants for patients.

3.4. Cell Proliferation and Metabolism

As a subsequent step after viability assessment, the rate of cellular metabolism was measured using MTT assay. This test is a direct indicator of mitochondrial activity, since the dye is being digested by mitochondrial reductase to yield a colorimetric signal. This can be used as an indicator for overall cellular metabolism. It is important to know if cellular metabolism rates are different among the various surfaces, since significant deviations in cellular metabolism rates could potentially yield new properties of the cell population on that particular surface. Increased rates of metabolism could be associated with malignant formations and dysregulation of the cell cycle, while down-regulated rates of metabolism could indicate low cellular viability or transfer into a more quiescent state via differentiation or other mechanisms [42].

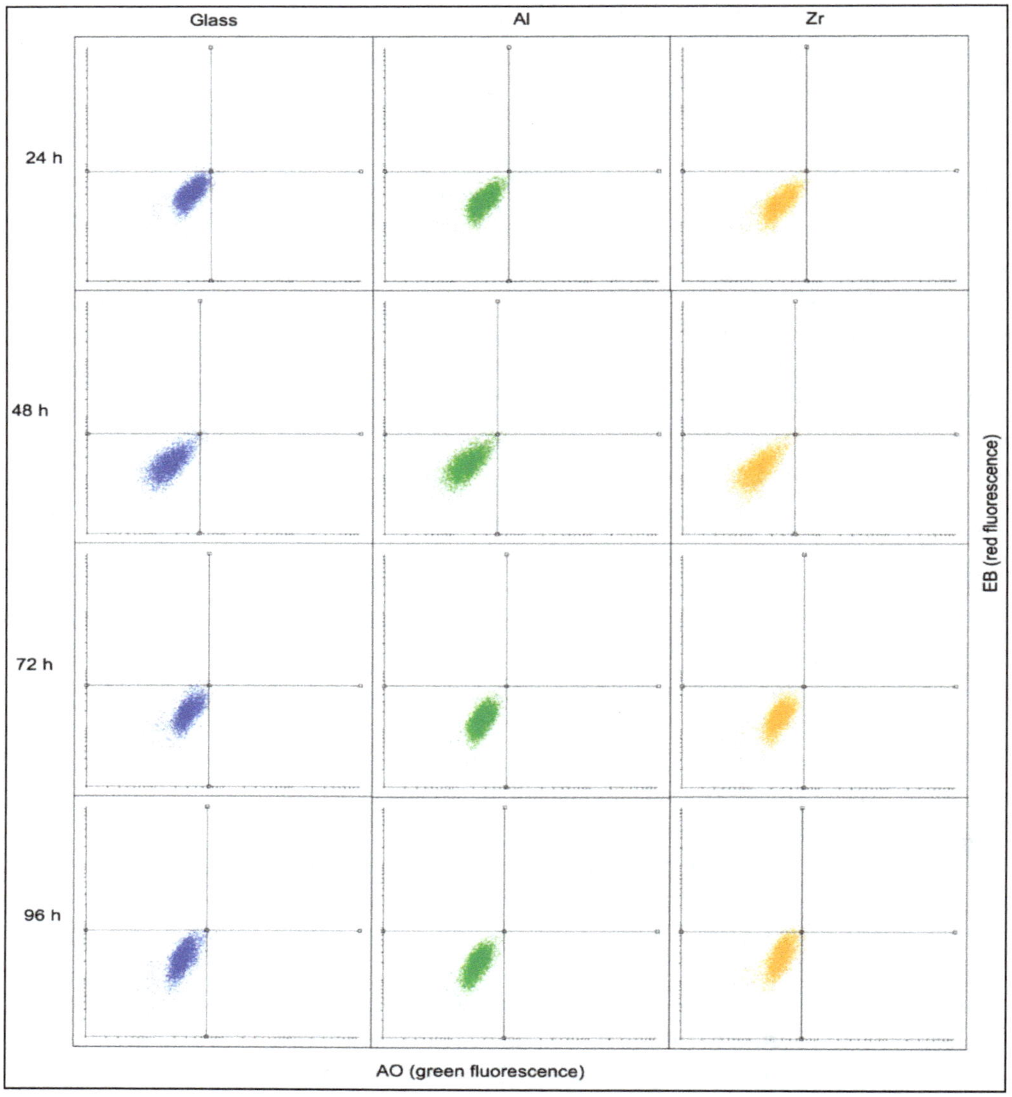

Figure 3. Flow cytometry data on cellular viability on Glass, Al and Zr surfaces after 24, 48, 72 and 96 h. The four quadrants in each case show the following: lower left—viable cells, upper left—necrotic cells, upper right—late apoptotic cells, lower right—early apoptotic cells. The lower left quadrant corresponds to 95% of all signals in all cases.

Light microscopy images show cells grown on both the hybrid polymer and glass surfaces to have formed confluent monolayers after 48 to 72 h (Supplementary Materials, Figure S3). No significant differences were observed neither in daily culture check-ups nor in the images taken of those cells. The cells had healthy spindle-shaped morphologies and stopped dividing after having reached confluency.

The assessment of total metabolic activity revealed that the metabolic rate of cells grown on glass was significantly higher than on the hybrid organometallic polymers after 96 h of culture ($p < 0.001$) and significantly higher than on the Al-based hybrid starting from 48 h ($p < 0.05$). These results are presented in Figure 4. Our previous study [21] investigated the proliferation rate of NIH/3T3 fibroblasts on analogous surfaces after 120 h and showed the Zr surface to support the highest rate, while no statistically significant differences were observed between Glass and Al.

The results obtained in this experiment show promise in using both the Al and Zr hybrid polymers in tissue engineering applications. Both materials supported cell growth, even though the total rate of metabolism was lower than on reference glass surfaces.

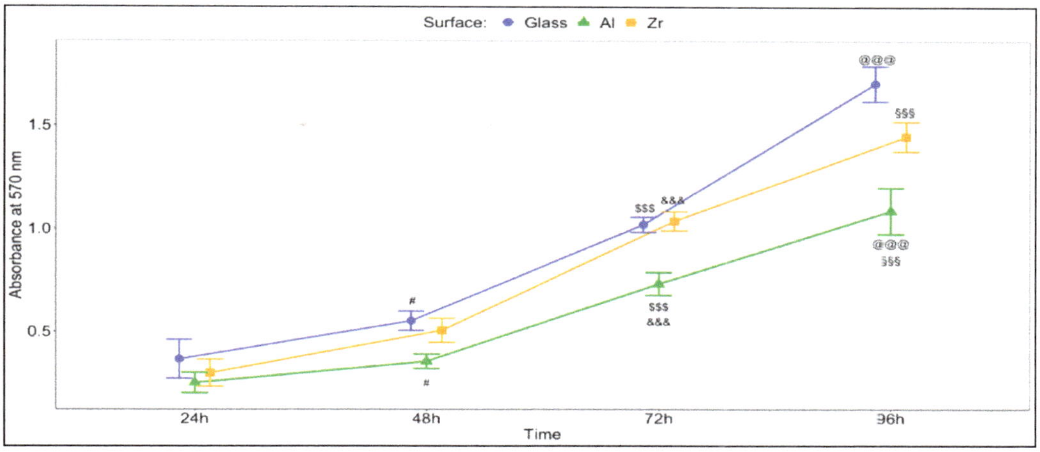

Figure 4. Cell proliferation as measured by MTT. The results are presented as average +/- standard error. A total of 9 measurements were performed per material per time point (3 independent experiments with 3 repetitions in each). #—significant difference between glass and Al with $p < 0.05$; $$$—significant difference between glass and Al with $p < 0.001$; &&&—significant difference between Al and Zr with $p < 0.001$; @@@—significant difference between glass and Al with $p < 0.001$; §§§—significant difference between Al and Zr with $p < 0.001$.

3.5. Cell Adhesion Strength

Cellular viability, proliferation and adhesion are highly intertwined processes. Knowing that the viability does not change with surface and that the rate of cellular metabolism grown on the Al hybrid is lower, we sought to understand if the reduced metabolic activity was associated with weaker adhesion. The cells were seeded on the samples and then shaken vigorously at two timepoints. The number of remaining cells was counted and calculated against the number of cells on reference surfaces without shaking. The results are presented in Figure 5. There was a clear trend that showed a decrease in the number of adhered cells after 24 h, suggesting that the initial stage of attachment after 4 h was stronger. Even though there were substantial differences between the number of adhered cells after 24 h, the differences were not statistically significant. However, in spite of a lack of statistical significance, the number of cells that remained attached to the Al surface was the lowest and thus in accord with the proliferation data, that showed the lowest rate of MTT metabolism on the Al- based hybrid. Overall, the Al surface seems to be slightly less supportive of cell adhesion, though insignificantly. The reason for that could potentially be the lowest adsorption of collagen to the Al surface, as presented in Figure 2. To further investigate this difference, an experiment was devised to assess the expression of adhesion-associated proteins.

3.6. FAK/Akt Expression and Activation

When integrin-mediated cell adhesion takes place, a cascade of biochemical reactions is activated that delivers a signal about cellular attachment to the cell nucleus, which in turn activates the transcription and translation of certain proteins that are required for these processes. The cascade is activated when cell surface integrins dimerise after binding to a specific adhesion sequence in the ECM [43]. To initiate intracellular signalling integrin dimer induces conformational changes in focal adhesion kinase (FAK), which is one of the first proteins to join the newly forming focal adhesion. Subsequently, this leads to the recruitment of multiple other adhesion-associated proteins to the focal adhesion site. Upon formation of a focal adhesions, multiple pathways can be activated, one of which is the Akt pathway. This pathway is responsible for transferring a survival and proliferation signal to the nucleus via the mTOR pathway [44]. Akt can be activated by either phosphorylation on Ser473 or Thr308—this depends on the phosphorylating protein. In the case of integrin-mediated adhesion,

PI3K is being activated and thus, only Ser473 is phosphorylated [45]. Phosphorylation on the Thr308 occurs from mTORC2, which is associated with cellular metabolism and cytoskeletal reorganisation.

To assess the expression and activation of adhesion-associated FAK pathway proteins, a Western blot analysis was performed on cells grown on the investigated Glass, Al and Zr surfaces. The results are presented in Figure 6 with membrane images presented in Supplementary Materials Figure S5. Even though no statistically significant results were observed on the investigated surfaces, a tendency for reduced phosphorylation of Akt kinase can be seen in most of the cases when comparing 4 and 24 h. This is to be expected as the adhesion-associated Akt kinase's role is mostly pronounced during the initial cell-to-surface and cell-cell interactions that take place upon cell seeding.

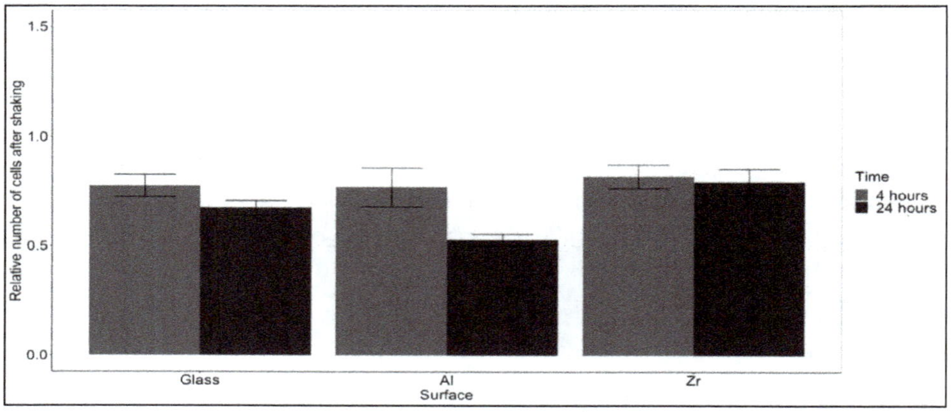

Figure 5. Evaluation of cell adhesion strength on the tested surfaces. The results are presented as a ratio between the average number of cells after shaking to that of cells before shaking after 4 and 24 h of culture + standard errors. A total of 9 measurements per material were performed (3 independent experiments with 3 repetitions each). No statistically significant differences were observed between the different time points on each surface.

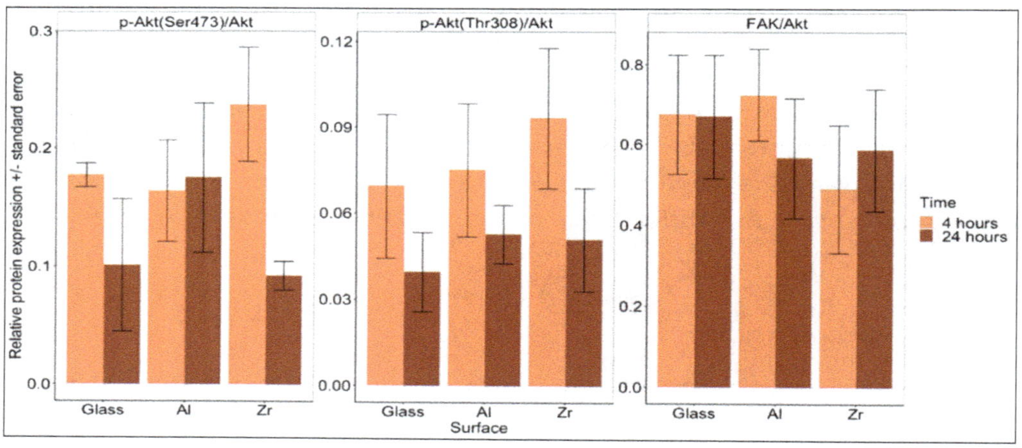

Figure 6. Western blot data, showing the expression of FAK and the expression and activation of Akt kinase after 4 and 24 h of culture on glass, Al and Zr surfaces. The amount of p-Akt (Ser 473), p-Akt (Thr308), Akt and FAK were observed. The results are presented as relative expression +/- standard error. Three independent experiments were performed. No statistically significant differences were observed.

The ratio between FAK and Akt kinases was mostly unchanged between time points and only slightly and insignificantly lower on Zr surfaces. This is again to be expected as the total amount of FAK and Akt kinases should remain unchanged.

On the glass surface, the cells tended to attach stronger during the initial 4 h—this is in accord with a higher level of Akt kinase activation in this time point. After 24 h the drop in the number of adhered cells and p-Akt (both Ser473 and Thr308) is clearly visible, though insignificant.

On the Al surface, the situation is similar in terms of p-Akt (Thr308), where the amount of phosphorylated Akt drops after 24 h in accord with a decreased number of cells. However, this was not observed in the p-Akt (Ser473) test. The level of FAK decreased as well, suggesting that the number of focal adhesions had been reduced as well.

The highest drop in terms of level of phosphorylation was observed on the Zr surface. From 4 to 24 h, the level of p-Akt (Ser473) dropped with a $p = 0.25$ and the level of p-Akt (Thr308) dropped with a $p = 0.67$. A slight, but insignificant increase was observed in the level of total FAK. Even though the changes in phosphorylation of Akt on Zr surfaces was the highest, the change in observed cell detachment was lowest in this case.

Even though some tendencies in cell attachment and adhesion-associated protein expression can be seen, these are not statistically significant.

Strong cellular adhesion is an essential prerequisite for successful tissue engineering techniques as discussed in a review by Lee et al. [46]. Whenever adhesion-associated signalling is insufficient, anchorage-dependent cells undergo anoikis, a form of programmed cell death that occurs in anchorage-dependent cells when they detach from the surrounding ECM.

3.7. Collagen Synthesis

Finally, we wanted to investigate whether the surface had any effect on the synthesis of collagen type I. Collagen is one of the main components of the extracellular matrix. Healthy tissues constantly undergo remodelling of their protein structures with the creation and digestion of new proteins by parenchymal cells, like fibroblasts. In tissue engineering, it is essential to have cells be able to model their environments by secreting ECM proteins, like collagen I.

Even though the adsorption was weakest to the Al-based hybrid and so was the cellular adhesion and proliferation, the synthesis of collagen in cells grown on the Al hybrid was the highest. The results are presented in Figure 7. The difference is already visible after one week, but it becomes significant after 2 weeks. This is to be expected as the process of extracellular matrix remodelling takes a substantial amount of time.

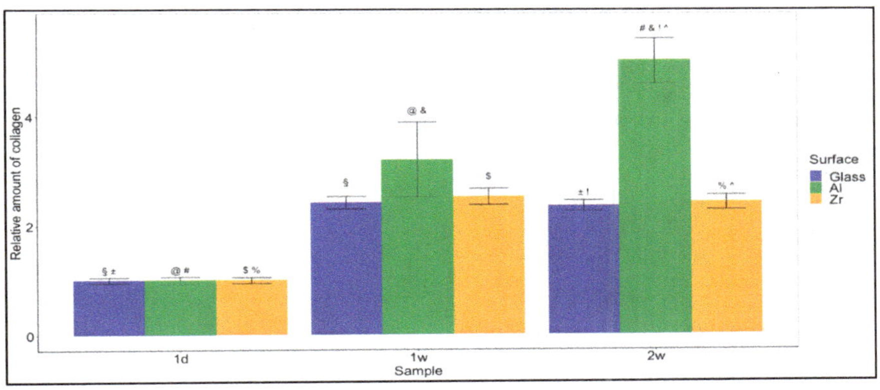

Figure 7. Collagen synthesis. Relative amount of collagen produced by cells grown on the different surfaces after 1 day, 1 week and 2 weeks. The results are presented as averages +/- standard errors. Statistically significant differences are indicated using different symbols. §—$p < 0.05$ between 1 day and 1 week glass; ±—$p < 0.05$ between 1 day and 2 week glass; @@@—$p < 0.001$ between 1 day and 1 week Al; ###—$p < 0.001$ between 1 day and 2 week Al; $$—$p < 0.01$ between 1 day and 1 week Zr; %—$p < 0.05$ between 1 day and 2 week Zr; &&&—$p < 0.001$ between 1 week and 2 week Al; !!!—$p < 0.001$ between 2 week glass and 2 week Al; ^^^—$p < 0.001$ between 2 week Al and 2 week Zr. Other statistically significant differences (those between both different materials and different timepoints) are not presented.

Other statistically significant differences were observed between Al surfaces after 1 day, 1 week and 2 weeks with $p < 0.001$ between all periods. As well as Zr surfaces after 1 day and 1 week with a $p < 0.05$.

The results show promise for tissue engineering applications, because they show a significant increase in the collagen produced in the cells grown on these surfaces. This means that the cells had not only attached to the surfaces, but they are comfortable and viable enough to start remodelling their environment via collagen synthesis.

4. Conclusions

We believe that the materials investigated in this work may be useful in constructing tissue engineered grafts of the future. This study provides a direct comparison between two materials that can be structured for laser two-photon polymerisation, assessing their support of primary rat myogenic cell growth. The differences between Zr and Al surfaces could potentially be attributed to a difference in adsorption of collagen, which in turn affects cellular adhesion strength, proliferation and the rate of extracellular matrix remodelling, but the results to support this are insignificant, suggesting a more thorough investigation into this process should be carried out.

An additional in vitro study should be performed in the future in order to evaluate how the adhesion-associated protein expression and activation changes over longer periods of time (several days or a couple of weeks) and whether that impacts cellular adhesion, viability and metabolism in a significant way.

Future studies in tissue engineering are likely to focus on integrating several materials with different properties in order to precisely guide cellular behaviour as well as investigating the role that artificial 3D niches with different geometries might play in affecting cellular differentiation and tissue maturation. As such, an extra option in the form of Al-containing hybrid organometallic polymer, presented in this work, is a choice, judging by its biocompatibility demonstrated in this study. In conjunction with its other advantages, like great 3D structuring capabilities, we believe the materials described in this work have potential to add value for the field of biomedical engineering.

Supplementary Materials:
Figure S1: A curve of collagen I content vs Sirius Red absorbance at 550 nm. An average of three independent experiments with n = 3 in each is presented ± standard deviation. Regression analysis was performed, showing that the dependence is linear in the measurement range (0–100 µg) with $R^2 = 0.994$ and $f(x) = 0.0199x-0.0322$, Figure S2: Characterization of rat skeletal muscle-derived progenitor cells. Fluorescent microscope images of cells immunostained for CD45, CD34, c-kit and Myf5. The scale bars correspond to 100 microns, Figure S3: Representative images of cells grown on glass, Al and Zr surfaces after 24, 48, 72 and 96 h. The scale bars correspond to 100 microns, Figure S4: Flow cytometry data on cellular viability on Glass, Al and Zr surfaces after 24, 48, 72 and 96 h. The results are presented as average absorbance +/- standard deviation, derived from 3 independent experiments with n = 3 per experiment. No statistically significant differences were found between the surfaces, Figure S5: Images of the Western blot membranes, showing the expression levels of p-Akt (Ser473), p-Akt (Thr308), Akt and FAK proteins. Tubulin was used as a reference. From left to right: 1st membrane: ladder, Al(4 h), Al(24 h), Zr(4 h), Zr(24 h), Glass(4 h), Glass(24 h), Al(4 h), Al(24 h), Zr(4 h); 2nd membrane: ladder, Zr(24 h), Glass(4 h), Glass(24 h), Al(4 h), Al(24h), Zr(4 h), Zr(24 h), Glass(4 h), Glass(24 h).

Author Contributions: Conceptualization, E.B. and D.B.; methodology, E.B., S.J.B., J.G.H., D.B.; formal analysis, all authors; investigation, all authors; resources, J.G.H. and D.B.; data curation, D.B.; writing—original draft preparation, all authors; writing—review and editing, E.B., J.G.H. and D.B.; supervision, J.G.H. and D.B.; project administration, D.B.; funding acquisition, J.G.H. and D.B.

References

1. Cima, L.G.; Vacanti, J.P.; Vacanti, C.; Ingber, D.; Mooney, D.; Langer, R. Tissue Engineering by Cell Transplantation Using Degradable Polymer Substrates. *J. Biomech. Eng.* **1991**, *113*, 143–151. [CrossRef] [PubMed]
2. Langer, R.; Vacanti, J. Advances in tissue engineering. *J. Pediatr. Surg.* **2016**, *51*, 8–12. [CrossRef] [PubMed]
3. Crapo, P.M.; Gilbert, T.W.; Badylak, S.F. An overview of tissue and whole organ decellularization processes. *Biomaterials* **2011**, *32*, 3233–3243. [CrossRef] [PubMed]

4. Melchels, F.P.W.; Domingos, M.A.N.; Klein, T.J.; Malda, J.; Bartolo, P.J.; Hutmacher, D.W. Additive manufacturing of tissues and organs. *Prog. Polym. Sci.* **2012**, *37*, 1079–1104. [CrossRef]
5. Raimondi, M.T.; Eaton, S.M.; Nava, M.M.; Laganà, M.; Cerullo, G.; Osellame, R. Two-photon laser polymerization: From fundamentals to biomedical application in tissue engineering and regenerative medicine. *J. Appl. Biomater. Funct. Mater.* **2012**, *10*, 55–65. [CrossRef]
6. Selimis, A.; Mironov, V.; Farsari, M. Direct laser writing: Principles and materials for scaffold 3D printing. *Microelectron. Eng.* **2014**, *132*, 83–89. [CrossRef]
7. Kufelt, O.; El-Tamer, A.; Sehring, C.; Schlie-Wolter, S.; Chichkov, B.N. Hyaluronic acid based materials for scaffolding via two-photon polymerization. *Biomacromolecules* **2014**, *15*, 650–659. [CrossRef]
8. Ovsianikov, A.; Deiwick, A.; Van Vlierberghe, S.; Dubruel, P.; Möller, L.; Drager, G.; Chichkov, B. Laser fabrication of three-dimensional CAD scaffolds from photosensitive gelatin for applications in tissue engineering. *Biomacromolecules* **2011**, *12*, 851–858. [CrossRef]
9. Ovsianikov, A.; Malinauskas, M.; Schlie, S.; Chichkov, B.; Gittard, S.; Narayan, R.; Löbler, M.; Sternberg, K.; Schmitz, K.P.; Haverich, A. Three-dimensional laser micro- and nano-structuring of acrylated poly(ethylene glycol) materials and evaluation of their cytoxicity for tissue engineering applications. *Acta Biomater.* **2011**, *7*, 967–974. [CrossRef]
10. Thiel, M.; Fischer, J.; Von Freymann, G.; Wegener, M. Direct laser writing of three-dimensional submicron structures using a continuous-wave laser at 532 nm. *Appl. Phys. Lett.* **2010**, *97*, 221102. [CrossRef]
11. Rekštyte, S.; Jonavičius, T.; Malinauskas, M. Direct laser writing of microstructures on optically opaque and reflective surfaces. *Optics Laser Eng.* **2014**, *53*, 90–97. [CrossRef]
12. Rekštytė, S.; Kaziulionytė, E.; Balčiūnas, E.; Kaškelytė, D.; Malinauskas, M. Direct laser fabrication of composite material 3D microstructured scaffolds. *J. Laser Micro Nanoeng.* **2014**, *9*, 25–30. [CrossRef]
13. Liu, X.; Miller, A.L.; Park, S.; George, M.N.; Waletzki, B.E.; Xu, H.; Terzic, A.; Lu, L. Two-Dimensional Black Phosphorus and Graphene Oxide Nanosheets Synergistically Enhance Cell Proliferation and Osteogenesis on 3D Printed Scaffolds. *ACS Appl. Mater. Interfaces* **2019**, *11*, 23558–23572. [CrossRef] [PubMed]
14. Zein, I.; Hutmacher, D.W.; Tan, K.C.; Teoh, S.H. Fused deposition modeling of novel scaffold architectures for tissue engineering applications. *Biomaterials* **2002**, *23*, 1169–1185. [CrossRef]
15. Lin, M.; Firoozi, N.; Tsai, C.T.; Wallace, M.B.; Kang, Y. 3D-printed flexible polymer stents for potential applications in inoperable esophageal malignancies. *Acta Biomater.* **2019**, *83*, 119–129. [CrossRef] [PubMed]
16. Olakanmi, E.O.; Cochrane, R.F.; Dalgarno, K.W. A review on selective laser sintering/melting (SLS/SLM) of aluminium alloy powders: Processing, microstructure, and properties. *Prog. Mater. Sci.* **2015**, *74*, 401–477. [CrossRef]
17. Farsari, M.; Chichkov, B.N. Materials processing: Two-photon fabrication. *Nat. Photonics* **2009**, *3*, 450–452.
18. Wang, X.; Lin, M.; Kang, Y. Engineering Porous β-Tricalcium Phosphate (β-TCP) Scaffolds with Multiple Channels to Promote Cell Migration, Proliferation, and Angiogenesis. *ACS Appl. Mater. Interfaces* **2019**, *11*, 9223–9232. [CrossRef]
19. Reignier, J.; Huneault, M.A. Preparation of interconnected poly(ε{lunate}-caprolactone) porous scaffolds by a combination of polymer and salt particulate leaching. *Polymer (Guildf.)* **2006**, *47*, 4703–4717.
20. Farsari, M.; Vamvakaki, M.; Chichkov, B.N. Multiphoton polymerization of hybrid materials. *J. Opt.* **2010**, *12*, 124001. [CrossRef]
21. Balčiūnas, E.; Baldock, S.J.; Dreižė, N.; Grubliauskaitė, M.; Coultas, S.; Rochester, D.L.; Valius, M.; Hardy, J.G.; Baltriukienė, D. 3D printing hybrid organometallic polymer-based biomaterials via laser two-photon polymerization. *Polym. Int.* **2019**, *68*, 1928–1940. [CrossRef]
22. Malinauskas, M.; Gaidukevičiute, A.; Purlys, V.; Žukauskas, A.; Sakellari, I.; Kabouraki, E.; Candiani, A.; Gray, D.; Pissadakis, S.; Gadonas, R.; et al. Direct laser writing of microoptical structures using a Ge-containing hybrid material. *Metamaterials* **2011**, *5*, 135–140. [CrossRef]
23. Sakellari, I.; Gaidukeviciute, A.; Giakoumaki, A.; Gray, D.; Fotakis, C.; Farsari, M.; Vamvakaki, M.; Reinhardt, C.; Ovsianikov, A.; Chichkov, B.N. Two-photon polymerization of titanium-containing sol-gel composites for three-dimensional structure fabrication. *Appl. Phys. A Mater. Sci. Process.* **2010**, *100*, 359–364. [CrossRef]
24. Kabouraki, E.; Giakoumaki, A.N.; Danilevicius, P.; Gray, D.; Vamvakaki, M.; Farsari, M. Redox multiphoton polymerization for 3D nanofabrication. *Nano Lett.* **2013**, *13*, 3831–3835. [CrossRef] [PubMed]

25. Ovsianikov, A.; Viertl, J.; Chichkov, B.; Oubaha, M.; MacCraith, B.; Sakellari, I.; Giakoumaki, A.; Gray, D.; Vamvakaki, M.; Farsari, M.; et al. Ultra-low shrinkage hybrid photosensitive material for two-photon polymerization microfabrication. *ACS Nano* **2008**, *2*, 2257–2262. [CrossRef] [PubMed]
26. Malinauskas, M.; Baltriukiene, D.; Kraniauskas, A.; Danilevicius, P.; Jarasiene, R.; Sirmenis, R.; Zukauskas, A.; Balciunas, E.; Purlys, V.; Gadonas, R.; et al. In vitro and in vivo biocompatibility study on laser 3D microstructurable polymers. *Appl. Phys. A Mater. Sci. Process.* **2012**, *108*, 751–759. [CrossRef]
27. Käpylä, E.; Sorkio, A.; Teymouri, S.; Lahtonen, K.; Vuori, L.; Valden, M.; Skottman, H.; Kellomäki, M.; Juuti-Uusitalo, K. Ormocomp-Modified glass increases collagen binding and promotes the adherence and maturation of human embryonic stem cell-derived retinal pigment epithelial cells. *Langmuir* **2014**, *30*, 14555–14565. [CrossRef]
28. Thingiverse. Available online: https://www.thingiverse.com (accessed on 24 October 2019).
29. Tullberg-Reinert, H.; Jundt, G. In situ measurement of collagen synthesis by human bone cells with a sirius red-based colorimetric microassay: Effects of transforming growth factor beta2 and ascorbic acid 2-phosphate. *Histochem. Cell Biol.* **1999**, *112*, 271–276. [CrossRef]
30. Mercille, S.; Massie, B. Induction of apoptosis in nutrient-deprived cultures of hybridoma and myeloma cells. *Biotechnol. Bioeng.* **1994**, *44*, 1140–1154. [CrossRef]
31. Raimondi, M.T.; Eaton, S.M.; Laganà, M.; Aprile, V.; Nava, M.M.; Cerullo, G.; Osellame, R. Three-dimensional structural niches engineered via two-photon laser polymerization promote stem cell homing. *Acta Biomater.* **2013**, *9*, 4579–4584. [CrossRef]
32. Murphy, W.L.; McDevitt, T.C.; Engler, A.J. Materials as stem cell regulators. *Nat. Mater.* **2014**, *13*, 547–557. [CrossRef] [PubMed]
33. Lamont, A.C.; Restaino, M.A.; Kim, M.J.; Sochol, R.D. A facile multi-material direct laser writing strategy. *Lab Chip* **2019**, *19*, 2340–2345. [CrossRef] [PubMed]
34. Malinauskas, M.; Rekštyte, S.; Lukoševičius, L.; Butkus, S.; Balčiunas, E.; Pečiukaityte, M.; Baltriukiene, D.; Bukelskiene, V.; Butkevičius, A.; Kucevičius, P.; et al. 3D microporous scaffolds manufactured via combination of fused filament fabrication and direct laser writing ablation. *Micromachines* **2014**, *5*, 839–858. [CrossRef]
35. Ber, S.; Torun Köse, G.; Hasirci, V. Bone tissue engineering on patterned collagen films: An in vitro study. *Biomaterials* **2005**, *26*, 1977–1986. [CrossRef]
36. Ramanathan, G.; Singaravelu, S.; Muthukumar, T.; Thyagarajan, S.; Perumal, P.T.; Sivagnanam, U.T. Design and characterization of 3D hybrid collagen matrixes as a dermal substitute in skin tissue engineering. *Mater. Sci. Eng. C* **2017**, *72*, 359–370. [CrossRef]
37. Antoine, E.E.; Vlachos, P.P.; Rylander, M.N. Review of Collagen I Hydrogels for Bioengineered Tissue Microenvironments: Characterization of Mechanics, Structure, and Transport. *Tissue Eng. Part B Rev.* **2014**, *20*, 683–696. [CrossRef]
38. Sarker, B.; Hum, J.; Nazhat, S.N.; Boccaccini, A.R. Combining collagen and bioactive glasses for bone tissue engineering: A review. *Adv. Healthc. Mater.* **2015**, *4*, 176–194. [CrossRef]
39. Dong, C.; Lv, Y. Application of collagen scaffold in tissue engineering: Recent advances and new perspectives. *Polymers (Basel).* **2016**, *8*, 42. [CrossRef]
40. Ying, P.; Jin, G.; Tao, Z. Competitive adsorption of collagen and bovine serum albumin - Effect of the surface wettability. *Colloids Surfaces B Biointerfaces* **2004**, *33*, 259–263. [CrossRef]
41. Nieponice, A.; Soletti, L.; Guan, J.; Hong, Y.; Gharaibeh, B.; Maul, T.M.; Huard, J.; Wagner, W.R.; Vorp, D.A. In Vivo Assessment of a Tissue-Engineered Vascular Graft Combining a Biodegradable Elastomeric Scaffold and Muscle-Derived Stem Cells in a Rat Model. *Tissue Eng. Part A* **2010**, *16*, 1215–1223. [CrossRef]
42. Shi, N.; Chen, S.Y. Mechanisms simultaneously regulate smooth muscle proliferation and differentiation. *J. Biomed. Res.* **2014**, *28*, 40–46. [PubMed]
43. Reddig, P.J.; Juliano, R.L. Clinging to life: Cell to matrix adhesion and cell survival. *Cancer Metastasis Rev.* **2005**, *24*, 425–439. [CrossRef] [PubMed]
44. Lee, D.Y.; Li, Y.S.J.; Chang, S.F.; Zhou, J.; Ho, H.M.; Chiu, J.J.; Chien, S. Oscillatory flow-induced proliferation of osteoblast-like cells is mediated by $\alpha v\beta 3$ and $\beta 1$ integrins through synergistic interactions of focal adhesion kinase and Shc with phosphatidylinositol 3-kinase and the Akt/mTOR/p70S6K pathway. *J. Biol. Chem.* **2010**, *285*, 30–42. [CrossRef] [PubMed]
45. Martini, M.; De Santis, M.C.; Braccini, L.; Gulluni, F.; Hirsch, E. PI3K/AKT signaling pathway and cancer: An updated review. *Ann. Med.* **2014**, *46*, 372–383. [CrossRef]

11

Effect of Surfactant type and Sonication Energy on the Electrical Conductivity Properties of Nanocellulose-CNT Nanocomposite Films

Sanna Siljander [1,*], Pasi Keinänen [1], Anna Räty [1], Karthik Ram Ramakrishnan [1], Sampo Tuukkanen [2], Vesa Kunnari [3], Ali Harlin [3], Jyrki Vuorinen [1] and Mikko Kanerva [1]

1 Laboratory of Materials Science, Tampere University of Technology, FI-33720 Tampere, Finland; pasi.keinanen@tut.fi (P.K.); anna.raty@tut.fi (A.R.); karthik.ramakrishnan@tut.fi (K.R.R.); jyrki.vuorinen@tut.fi (J.V.); mikko.kanerva@tut.fi (M.K.)

2 BioMediTech, Tampere University of Technology, FI-33720 Tampere, Finland; sampo.tuukkanen@tut.fi

3 VTT Research Center, FI-02044 Espoo, Finland; vesa.kunnari@vtt.fi (V.K.); ali.harlin@vtt.fi (A.H.)

* Correspondence: sanna.siljander@tut.fi

Abstract: We present a detailed study on the influence of sonication energy and surfactant type on the electrical conductivity of nanocellulose-carbon nanotube (NFC-CNT) nanocomposite films. The study was made using a minimum amount of processing steps, chemicals and materials, to optimize the conductivity properties of free-standing flexible nanocomposite films. In general, the NFC-CNT film preparation process is sensitive concerning the dispersing phase of CNTs into a solution with NFC. In our study, we used sonication to carry out the dispersing phase of processing in the presence of surfactant. In the final phase, the films were prepared from the dispersion using centrifugal cast molding. The solid films were analyzed regarding their electrical conductivity using a four-probe measuring technique. We also characterized how conductivity properties were enhanced when surfactant was removed from nanocomposite films; to our knowledge this has not been reported previously. The results of our study indicated that the optimization of the surfactant type clearly affected the formation of freestanding films. The effect of sonication energy was significant in terms of conductivity. Using a relatively low 16 wt. % concentration of multiwall carbon nanotubes we achieved the highest conductivity value of 8.4 S/cm for nanocellulose-CNT films ever published in the current literature. This was achieved by optimizing the surfactant type and sonication energy per dry mass. Additionally, to further increase the conductivity, we defined a preparation step to remove the used surfactant from the final nanocomposite structure.

Keywords: nanocellulose; carbon nanotubes; nanocomposite; conductivity; surfactant

---

1. Introduction

Conductive composite materials with micrometer and nanoscale fillers, like metallic powders, carbon black, graphite and carbon fibers, are used in many applications, such as antistatic films and electromagnetic interference (EMI) shielding. Electrical conductivity of 0.01 S/cm or higher is required for the composite to be considered conductive, while materials with lower conductivity can be used as antistatic and semiconducting materials. One of the drawbacks with most fillers is that the filler content ratio needs to be as high as 50 wt. % to achieve the percolation threshold (i.e., the critical concentration of filler that corresponds to the sharp rise of conductivity). However, this high filler content ratio might lead to a decrease in the resultant composite's mechanical properties [1,2]. Nanomaterials, such as carbon nanotubes (CNTs) and graphene, play a role in the development of future composite materials. For example, CNTs and graphene have been used to toughen matrix polymers [3], to adjust barrier properties of nanocomposite films [4], and to form hierarchical reinforcements [5]. It is possible to attain the percolation threshold in the insulating polymer matrix at a low CNT concentration due to their excellent electrical, mechanical and thermal properties.

Individual CNTs are part of a group of the strongest and most conductive nanomaterials known [6]. Additionally, CNTs can carry higher current density than any other known material, with its highest measured value being 109 A/cm^2 [7,8]. However, to obtain an ideal conductive network, the carbon nanotubes have to be well separated and homogenous dispersion should be maintained in the final product. Without efficient dispersion, filler aggregates act as defect sites, which leads to lower mechanical performance [9,10]. As the most abundant polymer on earth, cellulose is a promising and well-known material that can be used as a matrix in nanocomposites.

Cellulose is environmentally conscious, low-cost, strong, dimension-stable, non-melting, non-toxic and is a non-metal matrix. The interest towards nanoscale cellulose has increased during the past few years because of its inherent properties, including its good mechanical properties, which are better than those of the respective source biomass material [11]. Cellulose-based micro-/nanofibrils (MFC/NFC) can be extracted from various types of plant fibers using mechanical forces, chemical treatments, enzymes or combinations of these. The most typical approach, however, is to apply wood pulp and mechanical methods such as homogenization, microfluidization, microgrinding and cryocrushing. Finally, after fibrillation, the width of NFC is typically between 5 and 20 nm, with a length of several micrometers. Nanocellulose (NFC) has hydroxyl groups in its structure and is therefore associated with high aspect ratio and strong hydrogen bonds formed between nanocellulose fibers [12]. These bonds enhance mechanical properties and enable the formation of free standing films. A combination of CNTs and cellulose I provides a conductive nanocomposite network. CNT-cellulose composites have been reported to be used as supercapacitor electrodes [13,14], electromagnetic interference shielding devices [15], chemical vapor sensors [16], water sensors [17,18], and pressure sensors [19].

There are different manufacturing methods for the fabrication of CNT-cellulose nanocomposites, but all the methods typically include (1) a phase of dispersing CNTs into a solution, and (2) an impregnation phase into the cellulose substrates (e.g., paper, filter paper) [15,16,20–23]. Alternatively, the dispersion can be used as a wet component with bacterial cellulose [24,25], with cellulose I and regenerated cellulose fibers [13,18,26] or in an aerogel form [17]. The processing of nanocellulose in an aqueous medium is the most common way due to its tendency to react with water, and strong affinity to itself and hydroxyl group containing materials [12]. Chen et al. [27] showed that NFCs and CNTs can form a three-dimensional conductive network structure in a gel-film morphology to achieve high electrical conductivity.

The properties of the nanocellulose-CNT composites are affected by the quality of CNT dispersion, amount of structural and oxidative defects in the graphitic structure of the CNTs, the aspect ratio of the CNTs after the disaggregate treatment, the strength of the matrix, and the interactions between the CNTs and the cellulose matrix. [28] The key challenge in numerous industrial applications is to achieve uniform and stable CNT dispersion. The homogenization phase is vital to maximize the excellent mechanical, electrical and thermal properties of the CNTs and the eco-friendly, strong and low-cost nanocellulose matrix. This is particularly important in the case of submicron- or nanometer-sized particles. In these scales, the surface chemistry plays an important role, managing the particle dispersion within the final product [29]. CNT dispersions are challenging because as the surface area of particles increases, the attractive forces between the aggregates [29] and the high aspect ratio enable the entanglement and bundling of CNTs [30]. There are two phenomena that affect CNT dispersions: nanotube morphology and the forces between the tubes. Entanglement of CNTs occurs due to tube morphology, as well as molecular forces, high aspect ratio, and high flexibility. Dispersing these entangled aggregates is difficult without damaging the nanotubes. Both CNT and aggregate size are expected to play a crucial role in the achieved level of electrical conductivity [31].

Two typical dispersion methods for CNTs include high shear mixing and pure sonication [13,15,16,19–21,24,32]. Sonication is based on ultrasonic waves that generate microscopic bubbles or inertial cavitation, which produces a shearing action. This results in liquid and suspended particles becoming intensely agitated. Another common technique is to use a centrifuge in one of the processing steps to extract the unwanted agglomerates from the supernatant, but this additional phase takes time and effort and affects the concentration of dispersed particles in the dispersion. In general, sonication is superior to shear mixing, especially for low-viscosity systems [33], where conventional mixing does not create high enough strain rates to disintegrate the CNT aggregates.

Another issue in the manufacturing of films using NFC is the shrinkage and distortion of the structure because of faster evaporation rate on surface than the mass transport of moisture within the material. When strong enough gradient occurs, film distortions emerge because of local stresses [34,35].

One widely used method for CNT dispersion is the non-covalent method. In this method, chemical moieties are adsorbed onto the surface of CNTs, the CNTs are non-covalently dispersed in a water medium, and the resultant mixture is sonicated in the presence of the moieties, namely surfactants. Surfactants are a group of organic compounds that have a hydrophilic head and a hydrophobic tail, and they are commonly used as detergents, wetting agents, emulsifiers, foaming agents and dispersants. The advantage of the non-covalent method lies in the fact that it does not deteriorate the electronic structure of the CNTs' graphitic shells, maintaining their high electrical conductivity. Good dispersion can be achieved by having a mixture of both nanocellulose and carbon nanotubes with the help of surfactants, as the surfactants lower the interfacial free energy between the particles. Table 1 lists information about surfactants and their properties used in this study.

Table 1. Surfactants used in this study.

Product Name	**Triton™ x-100**	**Pluronic® F-127**	**CTAB**
Type	Non-ionic	Non-ionic Polymeric	Cationic
Name	Octylphenol Ethoxylate	Poloxamer	Hexadecyltri-methylammonium bromide
Chemical Structure	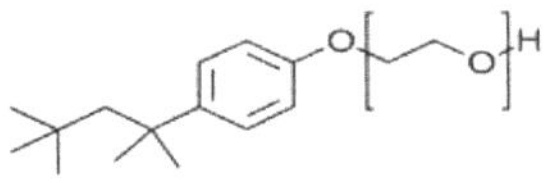	$H(OCH_2CH_2)_x(OCH_2CH(CH_3))_y(OCH_2CH_2)_zOH$	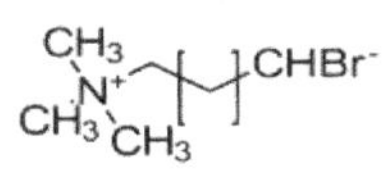
Critical Micelle Concentration	0.2–0.9 mM (20–25) °C	950–1000 ppm (25 °C)	0.92 mM (20–25) °C
HLB value	13.5	22	10

In the current literature, there are several different types of surfactants used for dispersing nanocellulose and carbon nanotubes. Choosing a surfactant type for effective dispersion of nanotubes through surfactant adsorption is complicated, as the results in the published literature often give contradictory results. For example, some researchers [29] have suggested that ionic surfactants are preferable for creating aqueous dispersions. However, the non-ionic surfactant Triton X-100 was shown to be a better surfactant than the anionic surfactant SDS, which was attributed to the π-π stacking ability of the former. The quality of the NFC-CNT dispersion is dependent on the nature of the surfactant, the concentration and the type of interactions between the surfactant and dispersing particles [36]. It has been stated that, for dispersing CNTs it is preferable for the surfactant to have a relatively high HLB (hydrophilic-lipophilic balance) value [29]. This assumption was proven false in our previous study [37]. Not only are the surfactant's nature and energy carried into the dispersed system, but the concentration of the surfactant also has a crucial role in the dispersion process [38]. Too high a surfactant concentration may negatively affect conductivity properties by blocking off the charge transport through the CNT network [39]. In addition, a low surfactant concentration can cause re-aggregation, because a sufficient amount is required to cover CNT surfaces to prevent re-aggregation [39,40]. It has been shown that an efficient CNT dispersion is only possible when the surfactant concentration is above the critical micelle concentration (CMC) value [41–44]. In some cases, the surfactant concentration is reported to be higher than the (CMC), but no micelle structures are observed in the dispersion. Presumably, most of the surfactant has been adsorbed onto the surface of the CNTs [40]. In other cases, surfactants can prefer surfactant-surfactant interactions over spreading on the CNT surface [45]. It has also been reported that dispersing agents can form stable dispersions below and equal to their CMC limit [46–49]. Moreover, it has been noted that commonly, the best results can be reached with a concentration of 0.5 CMC and that any further increase in the concentration of surfactant has only a minor effect [48].

The ISO 14887:2000(E) standard can be used to determine prospective dispersing agents for both cellulose and carbon. We can categorize nanocellulose and CNTs as solids. In that case, when using water as liquid, the category of suitable dispersing agent would be a poly ethylene-oxide (PEO)/alcohol for CNT and PEO/poly propylene oxide (PPO) copolymer for nanocellulose. The standard also provides information about commercial surfactants that fall into the mentioned categories. PEO/PPO copolymer is a suitable surfactant for nanocellulose. The standard denotes that a commercial equivalent is Pluronic®. In the case of CNTs, one example of alkyl phenoxy PEO ethanol dispersing agent is Triton™.

The typical approach to the manufacturing of conductive cellulose-CNT films has been to increase CNT weight percentage without optimizing the dispersion procedure or the used surfactants. Also, the effect of the particular ratios of the cellulose, CNT and surfactant toward each other has not been fully investigated. Even though ultrasonication is widely used for the dispersion and stabilization of CNTs, there is not a standard procedure for the sonication process, and different research groups have applied different sonication treatments to their samples. Sonication can cause chemical functionalization but it can also cause defects and breakage of CNTs [1,50–52]. This will further affect the performance of CNT-based materials and their applications. It has been found in the current literature that sonication parameters such as sonicator type, sonication time and temperature control vary significantly, with reported sonication times ranging from 2 min with tip sonication to 20 h for bath sonication. Dassios et al. [53] attempted to optimize the sonication parameters for the dispersion of MWCNTs in an aqueous solution. Two critical questions concerning the homogeneity of aqueous suspensions of carbon nanotubes by ultrasonic processing were identified; namely, the dependence of dispersion quality on the duration and intensity of sonication and the identification of the appropriate conditions for retaining the highly desirable initial aspect ratio of the free-standing tubes in the dispersed state. Fuge et al. [54] studied the effect of different ultrasonication parameters (time, amplitude) on undoped and nitrogen-doped MWCNTs in aqueous dispersions and found a nearly linear decrease of the arithmetic mean average in MWCNT length with increasing ultrasonication time.

The aim of this study was to optimize the conductivity of NFC-CNT nanocomposite films using a minimum amount of processing steps (e.g., without centrifugal processing of dispersion or pressing of the film), materials and chemicals. In this paper, NFC and multiwall carbon nanotubes (MWCNT) were used to prepare composite films and study the effect of the sonication energy and surfactant type on the electrical conductivity of the nanocomposite. In addition, we investigated the removal of the surfactant from the nanocomposites and the subsequent effect on the electrical conductivity. To our knowledge this is a novel approach and has not been reported previously. The conductivity properties of the nanocomposites were studied as a function of the used sonication energy amount, as well as with and without the presence of surfactant.

2. Results

The impact of sonication energy on electrical conductivity was one of the processing parameters with the highest interest in this study. This was due to the lack of previous research in the current literature. Also, our results show that the surfactant type and sonication energy play a major role in achieving excellent conductivity. In addition to the previously mentioned parameters, removal of surfactant can enhance conductivity values toward levels never seen or reported.

Overall, the shelf-life of the sonicated dispersion samples was significantly long, since samples remained unchanged before the film preparation. Also, sedimentation was not detected, based on the fact that conductivity values were at the same level when measured from both sides of the nanocomposite films. The appearance of the sonicated dispersion samples was identical; however, the consistency and visually inspected viscosity varied with increasing sonication energy. This was observed with Triton X-100 and Pluronic F-127 samples but not in cetyl trimethylammonium bromide (CTAB) surfactant-containing dispersions.

2.1. Conductivity of NFC-CNT Nanocomposite Films

Electrical conductivity of the even and uniform centrifugally cast films was measured using the four-probe measuring technique. With this method, it is possible to minimize the contact resistances and thus provide more accurate conductivity measurements than for the commonly used two-terminal measurement. The sheet resistances of prepared and cut nanocomposite films (size 30 mm × 30 mm) were measured using a four-point probe setup made in-house and a multimeter (Keithley 2002, Tektronix, Inc., Beaverton, OR, USA) in four-wire mode. The probes were placed in line, with equal 3 mm spacing. The four-probe setup is described elsewhere in detail [55]. The conductivity measurements were carried out using a 1 mA current and voltage was measured. Measurements were taken before and after removal of surfactant.

The selection of the most functional surfactant was an important aspect in this study. This selection was determined based on the sheet resistance measurements. The effect of different surfactants and sonication energy on conductivity is shown in Figure 1. From the conductivity diagrams, the effect of surfactant type can be visually observed and estimated.

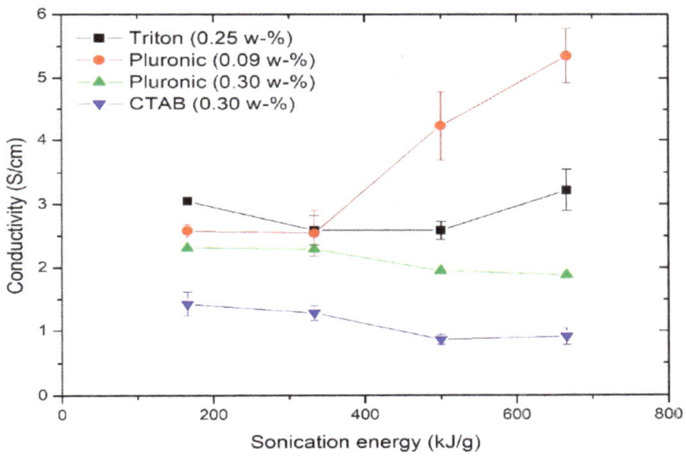

Figure 1. Conductivity of nanocomposite films processed using surfactants Triton X-100, Pluronic F-127 and cetyl trimethylammonium bromide. Concentration in weight percentages.

According to the standard, our assumption was that non-ionic surfactants would be the most promising surfactants. This was clearly the case, since the films made with surfactants Triton X-100 and Pluronic F-127 outperformed the films made with ionic surfactant CTAB.

Visual observations made with CTAB aqueous dispersion samples after sonication indicated that these samples did not gelate even with a higher amount of sonication energy per dry mass (666 kJ/g). This suggests that the dispersion process may not have been entirely successful, since samples had different consistencies and visually separate particles. The ionic surfactant (CTAB) was used to manufacture films at a 1 to 1 ratio of dry mass content of NFC and CNT. The conductivity of films processed using CTAB decreased as the sonication increased from almost 1.5 S/cm to less than 0.90 S/cm. The conductivity diagram of these films was different in its nature; the highest values were measured with the lowest amount of sonication energy.

Based on the standard Pluronic F-127, surfactant should be compatible with cellulosic materials. The first set of Pluronic F-127 nanocomposite films were done with a 1 to 1 ratio to dry mass content (0.30 wt. %). Results show that conductivity is decreasing as a function of sonication energy. Based on this finding, another set of films was manufactured using a surfactant concentration below the CMC limit (0.09 wt. %). Conductivity results for this set of samples show higher conductivity values than films manufactured using a surfactant concentration higher than the CMC value (0.30 wt. %). Using Pluronic F-127 surfactant, the highest conductivity for nanocomposite films was achieved using a sonication energy of 666 kJ/g. When comparing values of films below and above CMC value the difference is sensational 5.36 S/cm (0.09 wt. %) versus 1.88 S/cm (0.30 wt. %).

For Triton, the highest conductivity value of 3.37 S/cm was achieved with 666 kJ/g sonication energy. It should be noted that almost the same conductivity result (3.02 S/cm) was achieved using just 166 kJ/g of sonication energy.

2.2. *Effect of Surfactant Removal*

Conductivity measurements were also carried out after the removal of the surfactant used in the dispersing phase. Triton X-100 and Pluronic F-127 films were acetone treated and CTAB films were treated with ethanol. It can be clearly seen in Figure 2 that removal of surfactant has a strong effect. Removal of surfactant from films made with CTAB increased the conductivity significantly; the maximum conductivity was 3.02 S/cm for 166 kJ/g sonication energy. However, the films expressed a decrease in conductivity at sonication energy similar to the films with the surfactants present. For Pluronic F-127 films (Figure 2c) made below the CMC limit of the surfactant, the removal of surfactant did not have a significant effect on the conductivity. Even though there was no clear trend, the film with surfactant had somewhat higher conductivity than the one where surfactant was not present. For Pluronic F-127 (0.30 wt. %), the shape of the diagram differed from other previous sets. The films initially exhibited a decrease in conductivity as a function of increasing sonication energy, and the highest values were measured for the samples sonicated at the least energy, but also for the highest amount of sonication energy. For the samples with high conductivity, the removal of the non-ionic surfactant increased the conductivity. The highest value obtained from these measurements was 2.88 S/cm for a sonication energy of 166 kJ/g. A dramatic increase was observed in conductivity values of films manufactured with surfactant Triton X-100 (Figure 2a): the conductivity increases from approximately 3.0 S/cm to a value of 8.42 S/cm when the non-ionic surfactant was removed (sample was sonicated at 666 kJ/g). It should be noted that the films containing the surfactant did not exhibit as strong a sensitivity to increasing sonication energy (as those without surfactant).

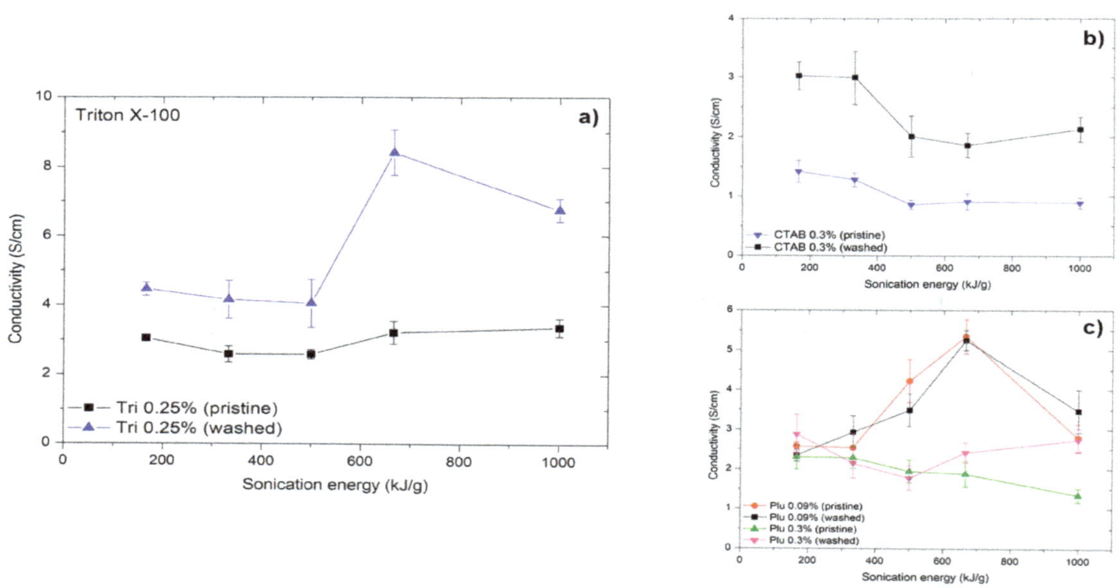

Figure 2. Conductivity of nanocomposite films before and after (**a**) Triton X-100 surfactant removal, (**b**) CTAB surfactant removal, and (**c**) Pluronic surfactant removal.

It is well known that surfactants can plasticize the structure of composites and interfere with conductivity properties by situating themselves at the interface between the conductive particles and matrix. This phenomenon was demonstrated when the properties of the nanocomposite films were compared in this study. Firstly, there was a clear increase in the conductivity of the films processed using the Triton X-100 and CTAB surfactants due to the removal of the surfactant. The diagrams (pristine vs. washed) were similar in their trend, and a clear increase in terms of conductivity was observed. When using the surfactant Pluronic F-127 for processing, a clear conclusion could not be made because the conductivity diagrams did not show a corresponding, monotonic trend due to surfactant removal. However, the results showed that, when the surfactant is present in the film structure, the effect of interference by Pluronic F-127 (concentration below CMC) on the electrical conductivity is at its minimum.

2.3. Comparison to Previous Results

When comparing our nanocomposite film's conductivity results to previous studies, we found that our results were superior to reported values. In Figure 3 are illustrated electrical conductivity results from studies that have used native NFC and manufactured homogenous nanocomposites from it. For non-ionic surfactants, the highest conductivity value found was 0.022 S/cm at a 10 wt. % CNT loading [17]. In our study, the highest value was 8.42 S/cm after removing Triton X-100 and, likewise, 5.35 S/cm with Pluronic F-127 still present in the film.

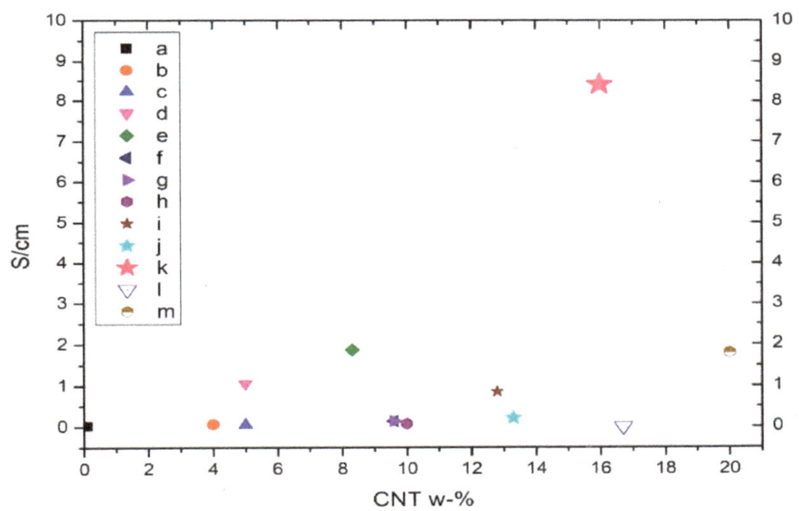

Figure 3. Comparison of obtained electrical conductivity of NFC-CNT nanocomposite films from the current literature. Pink star (letter k) refers to our data (Triton X-100), while other letters refer to a [9], b [56], c [57], d [58], e [22], f [9], g [24], h [57], i [59], j [15], l [32] and m [27].

Huang et al. [57] reported the results of a multiphase process which was used to accomplish a conductivity of 0.072 S/cm using MWCNT-doping at 10 wt. % and 0.056 S/cm with 5 wt. % doping with cotton linters and CTAB as a surfactant. CTAB surfactant was also used with bacterial nanocellulose and CNTs, where the conductivity was 0.027 S/cm (MWCNT 0.1 wt. %) [9]. Also, Yoon et al. [24] used bacterial cellulose as a matrix and obtained conductivity of 0.14 S/cm with 9.6 wt. % MWCNT loading. Electrical conductivity of TEMPO-oxidized cellulose films with 16.7 wt. % concentration of MWCNTs was 0.001 S/cm, which is lower than the conductive material limit [32]. For chitosan-cellulose-CNT membranes, Xiao et al. [56] accomplished conductivity of 0.062 S/cm with a 4 wt. % content of MWCNTs. By using comparable materials, but by applying a filtering method, Yamakawa et al. [58] obtained a 1.05 S/cm electrical conductivity with a 5 wt. % MWCNT loading and

Chen et al. 1.8 S/cm with 20 wt. % MWCNT. They were able to increase the conductivity to a value of 5.02 S/cm using a chemical alkali treatment.

In addition, studies about manufacturing conductive cellulose composites via coating cellulosic filter paper with a CNT dispersion have revealed rather good results, but the consistency of the materials is not homogeneous—not exactly an integral composite. Lee et al. [15] achieved conductivity of 1.11 S/cm using 13.3 wt. % MWCNT. Mondal et al. [59] reported conductivity values after using a dipping method, and they reached 0.85 S/cm with a 12.8 wt. % carbon nanofiber (CNF) content. Fugetsu et al. [22] manufactured conductive cellulose-based composites using a traditional paper making process with 8.32 wt. % CNT concentration and, finally, a conductivity of 1.87 S/cm was obtained.

2.4. Characterization of Nanocomposite Structure

The surface and cross-section of the films processed using surfactant Triton X-100 was studied with SEM. Images were taken with Zeiss ULTRAPlus scanning electron microscope (SEM). The effect of sonication as well as the removal of surfactant were studied with the SEM images shown in Figure 4. Two samples were specifically chosen for this inspection: 166 kJ/g and 666 kJ/g sonication energy films containing surfactant (Figure 4a,c) and after removal of surfactant Triton X-100 by washing them in acetone (Figure 4b,d).

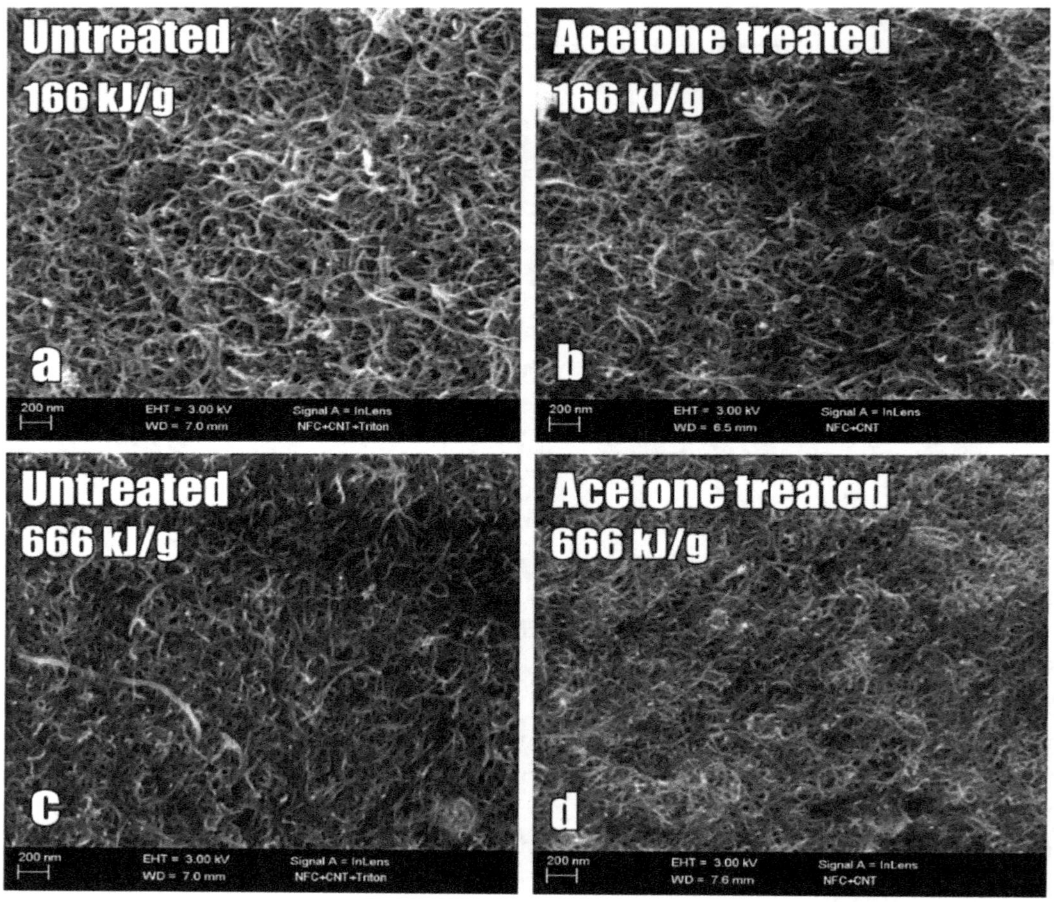

Figure 4. SEM imaging of NFC-CNT nanocomposite films surface (166 and 666 kJ/g) containing surfactant (**a**,**c**) and after removal of surfactant Triton X-100 by washing them in acetone (**b**,**d**).

In the top left (Figure 4a) image, some clusters of CNTs are present in the 166 kJ/g sonication energy sample, but not in the higher sonication energy 666 kJ/g sample. Both films have abundant amount of CNTs in the surface. Here, the 166 kJ/g film has a more porous structure than the 666 kJ/g film. In addition, the CNTs form a more consistent network in the 666 kJ/g film after washing the surfactant away (Figure 4d).

The SEM images of the sonicated samples in Figure 4 showed that there were clusters present in the 166 kJ/g sonicated film, while the higher sonicated energy sample did not have similar kinds of clusters. This indicates that, for lower sonication energies, non-dispersed particles remain in the films. This is not preferred, since the purpose is to achieve good dispersion of all the particles in the dispersion and in the films manufactured. This is an indication that the sonication process and amount of energy used affect the extent of the dispersion of particles.

The through-thickness structure of the films was also generally studied using polished cross-sections of the films embedded in epoxy. The cross-section in Figure 5 shows the CNT ends (bright contrast spots) and their even distribution in the film (500 kJ/g), (Triton system) through the thickness. A slightly layered structure can be observed and concluded as a result of dispersion flow during the casting.

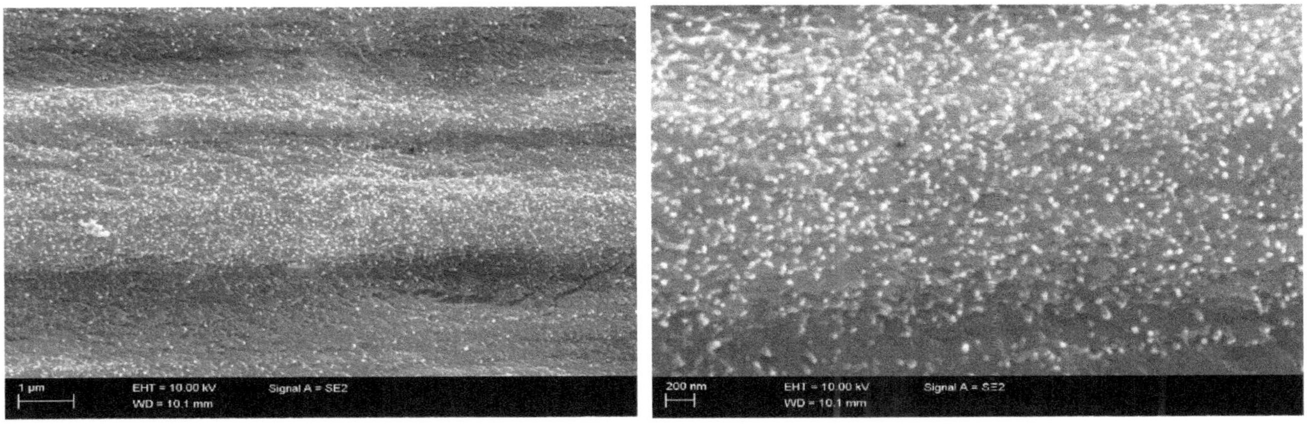

Figure 5. SEM imaging of the nanocomposite film (Triton X-100, 500 kJ/g) cross-section when embedded in epoxy: **Left** side: overall structure; **Right** side: magnification in the center of the film.

3. Discussion

Ultrasonication is a widely used process to manufacture aqueous CNT dispersions. However, how much it changes the properties of dispersed particles and the medium is often overlooked. It is known that sonication can, for example, generate hydrogen peroxide from water, degrade carbon nanotubes and ultimately destroy them. Therefore, it is important that the sonication process is optimized in terms of time and power. It also needs to be noted that the dispersion process is not linear, but follows an S-curve where temporal development of dispersion quality is related to quantity of un-dispersed solid.

Due to the re-agglomeration tendency of carbon nanotubes, it is necessary to use dispersion agents, i.e., surfactants, in manufacturing aqueous dispersions. If these dispersions are later used in conductive films, it is preferable to remove the surfactant to improve CNT network formation. Although carbon nanotubes are excellent conductors, CNT networks are not. This is due their high intertubular contact resistance. The contact points act essentially as a tunneling junction for electrons that is very sensitive to distance. The efficacy of surfactants is based on acting as a spacer between tubes, so any additional distance in a conductive network is detrimental to the conductivity itself.

4. Materials and Methods

Three-component systems containing nanocellulose, carbon nanotubes and surfactants are used in the strong, ecologically conscious nanocomposite films of this study. The CNTs add functionality to the nanocellulose matrix and the surfactant enables percolation network to build and maximize conductivity properties.

In this study, the nanocellulose (NFC) production was based on mechanical disintegration of bleached hardwood kraft pulp (BHKP). First, dried commercial BHKP produced from birch was soaked in water at approximately 1.7 wt. % concentration and dispersed using a high-shear Ystral dissolver for 10 min at 700 rpm. The chemical pulp suspension was predefined in a Masuko grinder (Supermasscolloider MKZA10-15J, Masuko Sangyo Co., Tokyo, Japan) at 1500 rpm and fluidized with six passes through a Microfluidizer (Microfluidics M-7115-30, Microfluidics Corp., Newton, MA, USA) using 1800 MPa pressure. The final material appearance of NCF was a viscous and translucent gel.

Multiwall carbon nanotubes (MWCNT, Nanocyl 7000, Nanocyl SA., Sambreville, Belgium) were purchased from Nanocyl Inc. and the product was used in the state it was in when received. This type of nanotubes is produced via catalytic chemical vapor deposition (CCVD). Concentration of CNTs was kept constant at 16 wt. % in the nanocomposite films, so the effects of sonication energy and surfactant type to the conductivity properties are more visible.

Three surfactants were chosen based on their ionic nature and standard: non-ionic Triton X-100 and Pluronic F-127, and anionic cetyl triammonium bromide (CTAB). Surfactants were purchased from Sigma-Aldrich (Merck KGaA, Darmstadt, Germany). The surfactants were diluted in deionized water to form solutions with variating dissolutions (1, 2.5, 10 wt. %).

Preparation of NFC-CNT Aqueous Dispersion

The NFC and CNT were sonicated simultaneously and after sonication no centrifuge was used so that the preparation of aqueous dispersions could be achieved using a minimum amount of processing phases. NFC-CNT aqueous dispersions with a total volume of 80 mL were prepared. One set contained NFC (0.25 wt. %), CNTs (0.05 wt. %), deionized water and one of the selected surfactants (Triton X-100, 0.25 wt. %, Pluronic F-127, 0.09 wt. % and 0.3 wt. % and CTAB 0.30 wt. %). Details about preparation produce are showed as Figure 6.

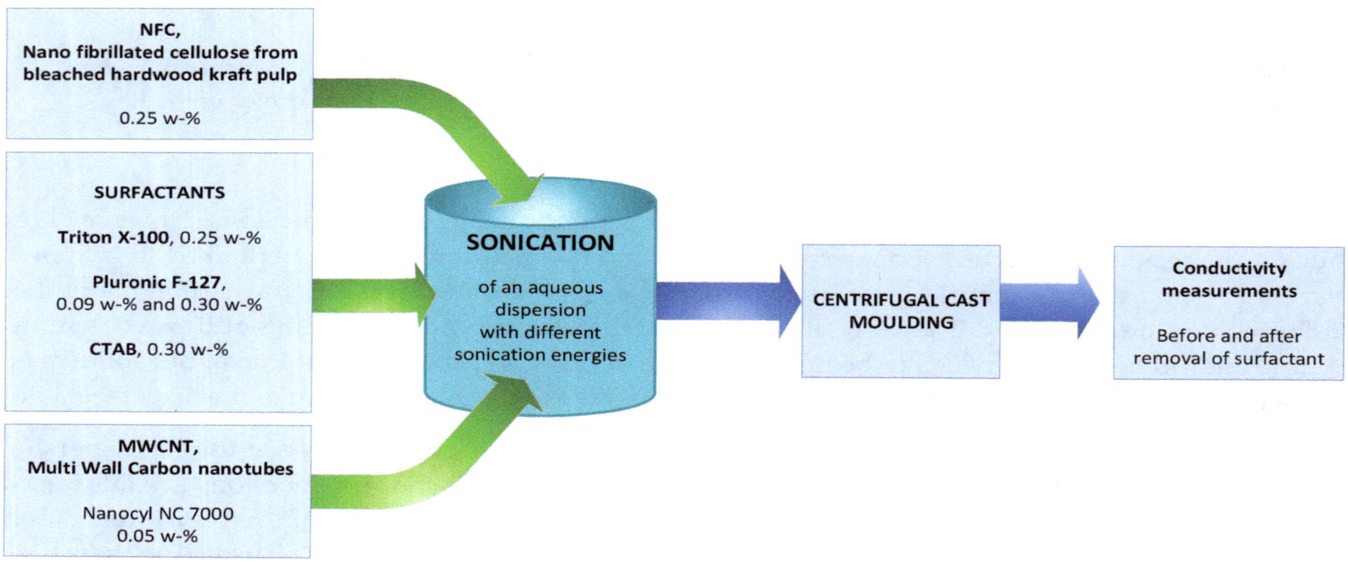

Figure 6. Preparation procedure of NFC-CNT dispersion and nanocomposite films.

The total dry mass for all the dispersions was 0.24 g. The sonication of the dispersion samples was performed with a tip horn (ø 12.7 mm) sonicator Q700 (QSonica LLC., Newton, CT, USA) in 100 mL glass beakers. The sonication amplitude of vibration (50%) was kept constant. The power output remained between 60 and 70 W for every sonication. The system included a water bath to keep samples cool during the sonication so that temperature would not rise above 30 °C. The water bath was cooled by circulating cooling glycerol through a chiller (PerkinElmer C6 Chiller, PerkinElmer Inc., Waltham, MA, USA). Samples were sonicated for four different amounts of energies per dry mass, respectively 166, 333, 500 and 666 kJ/g, which corresponded to energies of 40, 80, 120 and 160 kJ. Unsonicated samples manufactured using all three surfactants were not homogenous, and this is why film formation was unsuccessful and not analyzed.

5. Conclusions

The typical approach to the manufacturing of conductive cellulose-CNT films has been to increase CNT weight percentage without optimizing the dispersion procedure. In this study, NFC and multiwall carbon nanotubes (MWCNT) were used to prepare composite films using a minimum number of processing phases (e.g., no centrifugal dispersion or pressing of the film were used), materials and chemicals. The amount of CNTs was 0.05 wt. % in dispersion and 16 wt. % in the film after the evaporation of water in ratio to dry mass content of NFC and CNT. The effect of surfactant type (Triton X-100, Pluronic F-127 and CTAB) and sonication energy on the electrical conductivity of NFC-CNT nanocomposite films was investigated to identify optimal processing conditions for high conductivity of the nanocomposite. A conductivity of 5.36 S/cm was achieved by using Pluronic F-127 surfactant and 666 kJ/g of sonication energy. In addition, removal of the surfactant from film and its effect on the electrical conductivity was studied. A dramatic increase in conductivity values from approximately 3.0 S/cm to a value of 8.42 S/cm was observed for films manufactured with surfactant Triton X-100. Conductivity diagrams of the nanocomposite films show that sonication affects the electrical performance of the films. SEM images of sonicated samples showed that the films sonicated at 166 kJ/g have a more porous structure than the films sonicated at higher energy. The imaging also showed that the CNTs form a more consistent network with a combination of high sonication energy and surfactant removal. It can be concluded that the following parameters significantly affect the conductivity of NFC-CNT nanocomposite films:

(a) Surfactant type
(b) Surfactant concentration
(c) Sonication energy
(d) Removal of the used surfactant
(e) Film processing technique

To summarize, we manufactured nanocomposite films with exemplary conductivity in comparison to reported research and this was achieved by optimizing processing parameters and materials. Further research on the surfactant types and concentration can lead to better dispersion of the CNTs and therefore even higher conductivity.

Author Contributions: S.S. and P.K. conceived and designed the experiments; S.S. and A.R. performed the experiments and analyzed the data; V.K. and A.H. contributed materials; S.S., A.R., P.K., S.T. and K.R.R. wrote the paper. M.K. and J.V. coordinated the project aims in accordance to publication specific actions and delegation.

Acknowledgments: This work was funded by Tekes (Finnish Funding Agency for Innovation) through a strategic opening entitled Design Driven Value Chains in the World of Cellulose (DWoC 2.0). We acknowledge the contributions of Jarmo Laakso and Essi Sarlin for SEM imaging.

References

1. Huang, J.C. Carbon black filled conducting polymers and polymer blends. *Adv. Polym. Technol.* **2002**, *21*, 299–313. [CrossRef]
2. Ma, P.C.; Siddiqui, N.A.; Marom, G.; Kim, J.K. Dispersion and functionalization of carbon nanotubes for polymer-based nanocomposites: A review. *Compos. Part A Appl. Sci. Manuf.* **2010**, *41*, 1345–1367. [CrossRef]

3. Pereira, C.; Nóvoa, P.J.R.O.; Calard, V.; Forero, S.; Hepp, F.; Pambaguian, L. Characterization of Carbon Nanotube Papers Infused with Cyanate-Ester Resin. In Proceedings of the ICCM International Conference on Composite Materials, Edinburgh, UK, 27–31 July 2009.
4. Layek, R.K.; Das, A.K.; Park, M.J.; Kim, N.H.; Lee, J.H. Enhancement of physical, mechanical, and gas barrier properties in noncovalently functionalized graphene oxide/poly(vinylidene fluoride) composites. *Carbon N. Y.* **2015**, *81*, 329–338. [CrossRef]
5. Palola, S.; Sarlin, E.; Kolahgar Azari, S.; Koutsos, V.; Vuorinen, J. Microwave induced hierarchical nanostructures on aramid fibers and their influence on adhesion properties in a rubber matrix. *Appl. Surf. Sci.* **2017**, *410*, 145–153. [CrossRef]
6. Hamedi, M.M.; Hajian, A.; Fall, A.B.; Hkansson, K.; Salajkova, M.; Lundell, F.; Wgberg, L.; Berglund, L.A. Highly conducting, strong nanocomposites based on nanocellulose-assisted aqueous dispersions of single-wall carbon nanotubes. *ACS Nano* **2014**, *8*, 2467–2476. [CrossRef] [PubMed]
7. Tuukkanen, S.; Streiff, S.; Chenevier, P.; Pinault, M.; Jeong, H.J.; Enouz-Vedrenne, S.; Cojocaru, C.S.; Pribat, D.; Bourgoin, J.P. Toward full carbon interconnects: High conductivity of individual carbon nanotube to carbon nanotube regrowth junctions. *Appl. Phys. Lett.* **2009**, *95*, 113108. [CrossRef]
8. Haghi, A.K.; Thomas, S. *Carbon Nanotubes: Theoretical Concepts and Research Strategies for Engineers*; Apple Academic Press: Waretown, NJ, USA, 2015.
9. Jung, R.; Kim, H.-S.; Kim, Y.; Kwon, S.-M.; Lee, H.S.; Jin, H.-J. Electrically Conductive Transparent Papers Using Multiwalled Carbon Nanotubes. *J. Polym. Sci. Part B Polym. Phys.* **2008**, *46*, 1235–1242. [CrossRef]
10. Haghi, A.K.; Zaikov, G.E. *Advanced Nanotube and Nanofiber Materials*; Nova Science Publishers, Inc.: New York, NY, USA, 2012; ISBN 978-1-62-081201-3.
11. Hoeng, F.; Denneulin, A.; Bras, J. Use of nanocellulose in printed electronics: A review. *Nanoscale* **2016**, *8*, 13131–13154. [CrossRef] [PubMed]
12. Gardner, D.J.; Oporto, G.S.; Mills, R.; Samir, M.A.S.A. Adhesion and surface issues in cellulose and nanocellulose. *J. Adhes. Sci. Technol.* **2008**, *22*, 545–567. [CrossRef]
13. Kuzmenko, V.; Naboka, O.; Haque, M.; Staaf, H.; Göransson, G.; Gatenholm, P.; Enoksson, P. Sustainable carbon nanofibers/nanotubes composites from cellulose as electrodes for supercapacitors. *Energy* **2015**, *90*, 1490–1496. [CrossRef]
14. Lehtimäki, S.; Tuukkanen, S.; Pörhönen, J.; Moilanen, P.; Virtanen, J.; Honkanen, M.; Lupo, D. Low-cost, solution processable carbon nanotube supercapacitors and their characterization. *Appl. Phys. A Mater. Sci. Process.* **2014**, *117*, 1329–1334. [CrossRef]
15. Lee, T.W.; Lee, S.E.; Jeong, Y.G. Carbon nanotube/cellulose papers with high performance in electric heating and electromagnetic interference shielding. *Compos. Sci. Technol.* **2016**, *131*, 77–87. [CrossRef]
16. Yun, S.; Kim, J. Multi-walled carbon nanotubes-cellulose paper for a chemical vapor sensor. *Sens. Actuators B Chem.* **2010**, *150*, 308–313. [CrossRef]
17. Qi, H.; Liu, J.; Pionteck, J.; Pötschke, P.; Mäder, E. Carbon nanotube-cellulose composite aerogels for vapour sensing. *Sens. Actuators B Chem.* **2015**, *213*, 20–26. [CrossRef]
18. Qi, H.; Mäder, E.; Liu, J. Unique water sensors based on carbon nanotube-cellulose composites. *Sens. Actuators B Chem.* **2013**, *185*, 225–230. [CrossRef]
19. Wang, M.; Anoshkin, I.V.; Nasibulin, A.G.; Korhonen, J.T.; Seitsonen, J.; Pere, J.; Kauppinen, E.I.; Ras, R.H.A.; Ikkala, O. Modifying native nanocellulose aerogels with carbon nanotubes for mechanoresponsive conductivity and pressure sensing. *Adv. Mater.* **2013**, *25*, 2428–2432. [CrossRef] [PubMed]
20. Oya, T.; Ogino, T. Production of electrically conductive paper by adding carbon nanotubes. *Carbon N. Y.* **2008**, *46*, 169–171. [CrossRef]
21. Hu, L.; Choi, J.W.; Yang, Y.; Jeong, S.; La Mantia, F.; Cui, L.-F.; Cui, Y. Highly conductive paper for energy-storage devices. *Proc. Natl. Acad. Sci. USA* **2009**, *106*, 21490–21494. [CrossRef] [PubMed]
22. Fugetsu, B.; Sano, E.; Sunada, M.; Sambongi, Y.; Shibuya, T.; Wang, X.; Hiraki, T. Electrical conductivity and electromagnetic interference shielding efficiency of carbon nanotube/cellulose composite paper. *Carbon N. Y.* **2008**, *46*, 1256–1258. [CrossRef]
23. Imai, M.; Akiyama, K.; Tanaka, T.; Sano, E. Highly strong and conductive carbon nanotube/cellulose composite paper. *Compos. Sci. Technol.* **2010**, *70*, 1564–1570. [CrossRef]
24. Yoon, S.H.; Jin, H.J.; Kook, M.C.; Pyun, Y.R. Electrically conductive bacterial cellulose by incorporation of carbon nanotubes. *Biomacromolecules* **2006**, *7*, 1280–1284. [CrossRef] [PubMed]

25. Toomadj, F.; Farjana, S.; Sanz-Velasco, A.; Naboka, O.; Lundgren, P.; Rodriguez, K.; Toriz, G.; Gatenholm, P.; Enoksson, P. Strain sensitivity of carbon nanotubes modified cellulose. *Procedia Eng.* **2011**, *25*, 1353–1356. [CrossRef]
26. Liu, Y.; Liu, D.; Ma, Y.; Sui, G. Characterization and properties of transparent cellulose nanowhiskers-based graphene nanoplatelets/multi-walled carbon nanotubes films. *Compos. Part A Appl. Sci. Manuf.* **2016**, *86*, 77–86. [CrossRef]
27. Chen, C.; Mo, M.; Chen, W.; Pan, M.; Xu, Z.; Wang, H.; Li, D. Highly conductive nanocomposites based on cellulose nanofiber networks via NaOH treatments. *Compos. Sci. Technol.* **2018**, *156*, 103–108. [CrossRef]
28. Hilding, J.; Grulke, E.; George Zhang, Z.; Lockwood, F. Dispersion of Carbon Nanotubes in Liquids. *J. Dispers. Sci. Technol.* **2003**, *24*, 1–41. [CrossRef]
29. Vaisman, L.; Wagner, H.D.; Marom, G. The role of surfactants in dispersion of carbon nanotubes. *Adv. Colloid Interface Sci.* **2006**, *128–130*, 37–46. [CrossRef] [PubMed]
30. Rastogi, R.; Kaushal, R.; Tripathi, S.K.; Sharma, A.L.; Kaur, I.; Bharadwaj, L.M. Comparative study of carbon nanotube dispersion using surfactants. *J. Colloid Interface Sci.* **2008**, *328*, 421–428. [CrossRef] [PubMed]
31. Bai, J.B.; Allaoui, A. Effect of the length and the aggregate size of MWNTs on the improvement efficiency of the mechanical and electrical properties of nanocomposites—Experimental investigation. *Compos. Part A Appl. Sci. Manuf.* **2003**, *34*, 689–694. [CrossRef]
32. Salajkova, M.; Valentini, L.; Zhou, Q.; Berglund, L.A. Tough nanopaper structures based on cellulose nanofibers and carbon nanotubes. *Compos. Sci. Technol.* **2013**, *87*, 103–110. [CrossRef]
33. Huang, Y.Y.; Terentjev, E.M. Dispersion and rheology of carbon nanotubes in polymers. *Int. J. Mater. Form.* **2008**, *1*, 63–74. [CrossRef]
34. Gimåker, M.; Östlund, M.; Östlund, S.; Wågberg, L. Influence of beating and chemical additives on residual stresses in paper. *Nord. Pulp Pap. Res. J.* **2011**, *26*, 445–451. [CrossRef]
35. Baez, C.; Considine, J.; Rowlands, R. Influence of drying restraint on physical and mechanical properties of nanofibrillated cellulose films. *Cellulose* **2014**, *21*, 347–356. [CrossRef]
36. Rosen, M.J. *Surfactants and Interfacial Phenomena*; John Wiley & Sons, Inc.: Hoboken, NJ, USA, 2004; ISBN 978-0-47-147818-8.
37. Keinänen, P.; Siljander, S.; Koivula, M.; Sethi, J.; Vuorinen, J.; Kanerva, M. Optimized dispersion quality of aqueous carbon nanotube colloids as a function of sonochemical yield and surfactant/CNT ratio. *Heliyon* **2018**, in press.
38. Blanch, A.J.; Lenehan, C.E.; Quinton, J.S. Optimizing Surfactant Concentrations for Dispersion of Single-Walled Carbon Nanotubes in Aqueous Solution. *J. Phys. Chem. B* **2010**, *114*, 9805–9811. [CrossRef] [PubMed]
39. Yu, J.; Grossiord, N.; Koning, C.E.; Loos, J. Controlling the dispersion of multi-wall carbon nanotubes in aqueous surfactant solution. *Carbon N. Y.* **2007**, *45*, 618–623. [CrossRef]
40. Islam, M.F.; Rojas, E.; Bergey, D.M.; Johnson, A.T.; Yodh, A.G. High weight fraction surfactant solubilization of single-wall carbon nanotubes in water. *Nano Lett.* **2003**, *3*, 269–273. [CrossRef]
41. Utsumi, S.; Kanamaru, M.; Honda, H.; Kanoh, H.; Tanaka, H.; Ohkubo, T.; Sakai, H.; Abe, M.; Kaneko, K. RBM band shift-evidenced dispersion mechanism of single-wall carbon nanotube bundles with NaDDBS. *J. Colloid Interface Sci.* **2007**, *308*, 276–284. [CrossRef] [PubMed]
42. Sun, Z.; Nicolosi, V.; Rickard, D.; Bergin, S.D.; Aherne, D.; Coleman, J.N. Quantitative Evaluation of Surfactant-stabilized Single-walled Carbon Nanotubes: Dispersion Quality and Its Correlation with Zeta Potential. *J. Phys. Chem. C* **2008**, *112*, 10692–10699. [CrossRef]
43. Maillaud, L.; Zakri, C.; Ly, I.; Pénicaud, A.; Poulin, P. Conductivity of transparent electrodes made from interacting nanotubes. *Appl. Phys. Lett.* **2013**, *103*, 263106. [CrossRef]
44. Bai, Y.; Lin, D.; Wu, F.; Wang, Z.; Xing, B. Adsorption of Triton X-series surfactants and its role in stabilizing multi-walled carbon nanotube suspensions. *Chemosphere* **2010**, *79*, 362–367. [CrossRef] [PubMed]
45. Calvaresi, M.; Dallavalle, M.; Zerbetto, F. Wrapping nanotubes with micelles, Hemimicelles, and cylindrical micelles. *Small* **2009**, *5*, 2191–2198. [CrossRef] [PubMed]
46. Geng, Y.; Liu, M.Y.; Li, J.; Shi, X.M.; Kim, J.K. Effects of surfactant treatment on mechanical and electrical properties of CNT/epoxy nanocomposites. *Compos. Part A Appl. Sci. Manuf.* **2008**, *39*, 1876–1883. [CrossRef]
47. Bystrzejewski, M.; Huczko, A.; Lange, H.; Gemming, T.; Büchner, B.; Rümmeli, M.H. Dispersion and diameter separation of multi-wall carbon nanotubes in aqueous solutions. *J. Colloid Interface Sci.* **2010**, *345*, 138–142. [CrossRef] [PubMed]

48. Angelikopoulos, P.; Gromov, A.; Leen, A.; Nerushev, O.; Bock, H.; Campbell, E.E.B. Below the CMC. *J. Phys. Chem. C* **2010**, *114*, 2–9. [CrossRef]
49. Bonard, J.; Stora, T.; Salvetat, J.; Maier, F.; Stockli, T.; Duschl, C.; De Heer, W.A.; Forró, L.; Châtelain, A. Purification and Size-Selection of Carbon Nanotubes. *Adv. Mater.* **1997**, *9*, 827–831. [CrossRef]
50. Lu, P.; Hsieh, Y. Lo Multiwalled carbon nanotube (MWCNT) reinforced cellulose fibers by electrospinning. *ACS Appl. Mater. Interfaces* **2010**, *2*, 2413–2420. [CrossRef] [PubMed]
51. Rossell, M.D.; Kuebel, C.; Ilari, G.; Rechberger, F.; Heiligtag, F.J.; Niederberger, M.; Koziej, D.; Erni, R. Impact of sonication pretreatment on carbon nanotubes: A transmission electron microscopy study. *Carbon N. Y.* **2013**, *61*, 404–411. [CrossRef]
52. Yang, D.; Rochette, J.-F.; Sacher, E. Functionalization of Multiwalled Carbon Nanotubes by Mild Aqueous Sonication. *J. Phys. Chem. B* **2005**, *109*, 7788–7794. [CrossRef] [PubMed]
53. Dassios, K.G.; Alafogianni, P.; Antiohos, S.K.; Leptokaridis, C.; Barkoula, N.M.; Matikas, T.E. Optimization of sonication parameters for homogeneous surfactant assisted dispersion of multiwalled carbon nanotubes in aqueous solutions. *J. Phys. Chem. C* **2015**, *119*, 7506–7516. [CrossRef]
54. Fuge, R.; Liebscher, M.; Schröfl, C.; Oswald, S.; Leonhardt, A.; Büchner, B.; Mechtcherine, V. Fragmentation characteristics of undoped and nitrogen-doped multiwalled carbon nanotubes in aqueous dispersion in dependence on the ultrasonication parameters. *Diam. Relat. Mater.* **2016**, *66*, 126–134. [CrossRef]
55. Rajala, S.; Tuukkanen, S.; Halttunen, J. Characteristics of piezoelectric polymer film sensors with solution-processable graphene-based electrode materials. *IEEE Sens. J.* **2015**, *15*, 3102–3109. [CrossRef]
56. Xiao, W.; Wu, T.; Peng, J.; Bai, Y.; Li, J.; Lai, G.; Wu, Y.; Dai, L. Preparation, structure, and properties of chitosan/cellulose/multiwalled carbon nanotube composite membranes and fibers. *J. Appl. Polym. Sci.* **2013**, *128*, 1193–1199. [CrossRef]
57. Huang, H.D.; Liu, C.Y.; Zhang, L.Q.; Zhong, G.J.; Li, Z.M. Simultaneous reinforcement and toughening of carbon nanotube/cellulose conductive nanocomposite films by interfacial hydrogen bonding. *ACS Sustain. Chem. Eng.* **2015**, *3*, 317–324. [CrossRef]
58. Yamakawa, A.; Suzuki, S.; Oku, T.; Enomoto, K.; Ikeda, M.; Rodrigue, J.; Tateiwa, K.; Terada, Y.; Yano, H.; Kitamura, S. Nanostructure and physical properties of cellulose nanofiber-carbon nanotube composite films. *Carbohydr. Polym.* **2017**, *171*, 129–135. [CrossRef] [PubMed]
59. Mondal, S.; Ganguly, S.; Das, P.; Bhawal, P.; Das, T.K.; Nayak, L.; Khastgir, D.; Das, N.C. High-performance carbon nanofiber coated cellulose filter paper for electromagnetic interference shielding. *Cellulose* **2017**, *24*, 5117–5131. [CrossRef]

12

The Degradation Properties of MgO Whiskers/PLLA Composite In Vitro

Yun Zhao [1,2], Bei Liu [1], Hongwei Bi [3], Jinjun Yang [4], Wei Li [1,2], Hui Liang [1], Yue Liang [1], Zhibin Jia [1], Shuxin Shi [1] and Minfang Chen [1,2,*]

1. School of Materials Science and Engineering, Tianjin University of Technology, Tianjin 300384, China; yun_zhaotju@163.com (Y.Z.); liubei8679@163.com (B.L.); earlybird.184@126.com (W.L.); huiliang2014@126.com (H.L.); 15102260655@163.com (Y.L.); zhibin225225@163.com (Z.J.); 13752048706@163.com (S.S.)
2. Key Laboratory of Display Materials and Photoelectric Device (Ministry of Education), Tianjin University of Technology, Tianjin 300384, China
3. Tianjin Sannie Bioengineering Technology Co., Ltd., Tianjin 300384, China; bluewanli@126.com
4. School of Environmental Science and Safety Engineering, Tianjin University of Technology, Tianjin 300384, China; tjyjj_2014@tjut.edu.cn

* Correspondence: mfchentj@126.com

Abstract: In this study, composite films of stearic acid–modified magnesium oxide whiskers (Sa–w-MgO)/poly-L-lactic acid (PLLA) were prepared through solution casting, and the in vitro degradation properties and cytocompatibility of the composites with different whisker contents were investigated. The results showed that the degradation behavior of the composite samples depended significantly on the whisker content, and the degradation rate increased with the addition of MgO content. Furthermore, the degradation of the composites with higher contents of whiskers was influenced more severely by the hydrophilicity and pH value, leading to more final weight loss, but the decomposition rate decreased gradually. Furthermore, the pH value of the phosphate buffer solution (PBS) was obviously regulated by the dissolution of MgO whiskers through neutralization of the acidic product of PLLA degradation. The cytocompatibility of the composites also increased remarkably, as determined from the cell viability results, and was higher than that of PLLA at the chosen whisker content. This was beneficial for the cell affinity of the material, as it notably led to an enhanced biocompatibility of the PLLA, in favor of promoting cell proliferation, which significantly improved its bioactivity, as well.

Keywords: biopolymers composites; MgO whiskers; PLLA; in vitro degradation

1. Introduction

Over the last few decades, poly(L-lactide) (PLLA) has been given significant attention, owing to its advantages of good biocompatibility and processability, biodegradability, and bioresorbability, which are desirable in the biomedical field, especially for bone repair [1–4]. However, some drawbacks limit its wider application [5–10], particularly its degradation properties, which are an extremely important factor in bone repair, as accumulation of the lactic acid degradation product from PLLA can cause issues such as aseptic inflammation in vivo, hydrophobicity and lower biological activity that are unfavorable to the viability of the bone cell, and biodegradation that cannot be well controlled according to the requirement of bone formation.

To solve these problems, many researchers have attempted to improve PLLA properties through various methods [11–15]. Generally, a composite prepared by adding inorganic fillers is very promising, as it has the potential to enhance the mechanical and thermal properties, to enhance the hydrophilicity and bioactivity, and to reduce aseptic inflammation, because the fillers used have good bioactivity and similar constituents to the inorganic compounds in bone, such as calcium phosphate. However, these fillers, such as hydroxyapatite (HA) [4,7,11,13], β-tricalcium phosphate (β-TCP) [5,6,15], and bioactive glass [16,17], are usually slightly soluble [18], which tends to limit their impact on neutralizing the acidic environment induced by PLLA hydrolysis.

Magnesium oxide (MgO), another inorganic filler with good biocompatibility, and bioactive capacity that is not toxic, has attracted much attention [19–22] for polymer modification. In addition to the aforementioned advantages, the Mg^{2+} released from MgO dissolving in water is beneficial to the protein synthesized via the activation of many enzymes, which contribute to the excellent bioactivity of MgO. It has been reported that the dissolution of MgO in vitro prominently influences the solution pH through alkaline degradation [19]. Recently, the researchers utilized the different shapes of MgO [9,19–24] as fillers and introduced them to PLLA for preparing composites. The results mostly showed that the mechanical properties and thermal behavior of PLLA improved significantly with the presence of MgO fillers, which was also proved in our works [24]. However, few reports have focused on the effect of MgO contents on the degradation behavior of the composite during the decomposition process. Ma et al. [19] obtained the n-MgO/PLLA composite by adding n-MgO modified by poly(ε-caprolactone) into a PLLA matrix, and the reduction of pH was suppressed to a certain extent, which was also demonstrated by us [22]. Luo et al. [9] studied the in vitro degradation properties of the MgO whiskers (w-MgO)/PLLA composite; their work mainly focused on the effect of whiskers on PLLA degradation for a long decomposition period. It is worth mentioning that because such composites have the potential for application in bone repair, it is crucial to investigate their degradability and the cytocompatibility performance of the initial degradation stage, especially in terms of how it is affected by the filler contents, which is greatly important for the materials used, as implant inescapably contacted with bone cell and tissue [18]. Meanwhile, it was also noted that the change in the hydrophilicity of the PLLA matrix improved with the addition of fillers such as organic montmorillonite (OMMT), which prominently accelerated the hydrolytic degradation of PLLA, playing another key role in controlling the degradation process of PLLA [25,26]. As for MgO, they greatly impacted not only the hydrolysis behavior of PLLA, but also its degradation conditions, by controlling the pH. Therefore, the complex process in which the hydrolysis of PLLA could be facilitated can be controlled by varying the whisker content as well as be retarded by changes in the pH value, simultaneously [19,27]. Furthermore, it was reported that the hydrolysis process of PLLA was also favorable for promoting the growth of an ordered crystalline structure, due to the easier movement of the chains and segment under the degrading state [28,29]. Additionally, studying the degradation behavior in vitro can provide fundamental experimental results for the degrading properties of PLLA used in bone repair in vivo.

In this work, the MgO whiskers were modified using stearic acid, and the chemical bondings were obtained through the interaction between them, as demonstrated by our previous work [24]. Because stearic acid is effective in modifying inorganic fillers like hydroxyapatite [30] and TiO_2 [31] through changing the polarities, it was greatly helpful in improving the distribution of MgO whiskers and enhancing the interface of the PLLA composites. Then, this work undertook the detailed characterization of w-MgO/PLLA soaked in phosphate buffer solution (PBS) for different degrading periods and analyzed the influence of the whisker content on the composite degrading behaviors. The pH stability and cell affinity of the composite were also studied and evaluated by relative measurements and cell cultures to relate the changes in the composite degrading procedure to the material's biocompatibility and bioactivity.

2. Results

2.1. XRD Measurement

In order to investigate the effect of MgO whiskers on the properties of PLLA, such as its crystalline structure and crystallization behavior, the samples of PLLA and composites were tested after being soaked using XRD and DSC, and the results of XRD testing are shown in Figure 1 and Table 1. Figure 1 shows that the intensity of diffraction characteristic peaks of PLLA, for example with 2θ values of 17.76° and 18.68°, corresponding to the (110/200) and (203) diffraction planes, respectively, clearly strengthened gradually, probably due to the surface decomposition and increment of PLLA crystallinity during degrading periods. Meanwhile, a slight shift of these peaks was noted with an increase in the immersion time, indicating the generation and transformation of different types of crystalline PLLA in the degrading process [32], especially between the forms of α′ and α. Specifically, the strength of 2θ at 24.82° for the (206) diffraction plane generated by the α′ form became more intense after 28 days' immersion. Furthermore, these similar variations were also observed for the composite specimens, and the peak shifts were more prominent. This phenomenon can account for the accelerated degradation of the PLLA matrix, owing to the presence of MgO whiskers because of increased hydrophilicity [29]. Thus, a strong shift of the peaks was noted after 14 days. The little decrease in the 2θ value after 28 days of degradation seen for all composite samples, in comparison to that noted after 14 days, was attributed to the change in the relative content of the α′ and α forms caused by PLLA decomposition. Additionally, the characteristic peaks of MgO located at 43.5°, 62.5°, and 78.6°, corresponding to the (200), (220), and (222) diffraction planes, respectively, can also be found in the graphs of the composites, and PLLA3, with the highest content of 3 wt% MgO, showing the most significant peaks. However, the intensity of these peaks of MgO weakened gradually with progress in the experimental period: This change suggested a loss in the amount of MgO on the testing surface of composite films in the degrading process, showing no obvious peaks of the whiskers in the patterns of PLLA1, PLLA2, and PLLA3 for the 7 days in Figure 1.

Table 1. The 2θ values of samples on Day 0.

Sample	Diffraction Plane								
	(010)	(110/200)	(203)	(204)	(015)	(016)	(206)	(207)	(018)
	2θ(°)								
PLLA	14.69	17.76	18.68	21.36	22.45	23.78	24.82	26.62	28.71
PLLA1	14.63	16.48	18.96	21.36	22.28	23.72	24.87	26.60	28.78
PLLA2	14.63	16.67	19.02	21.39	22.43	23.83	24.09	26.64	28.93
PLLA3	14.74	16.67	19.06	21.43	22.34	23.79	24.94	26.64	28.82

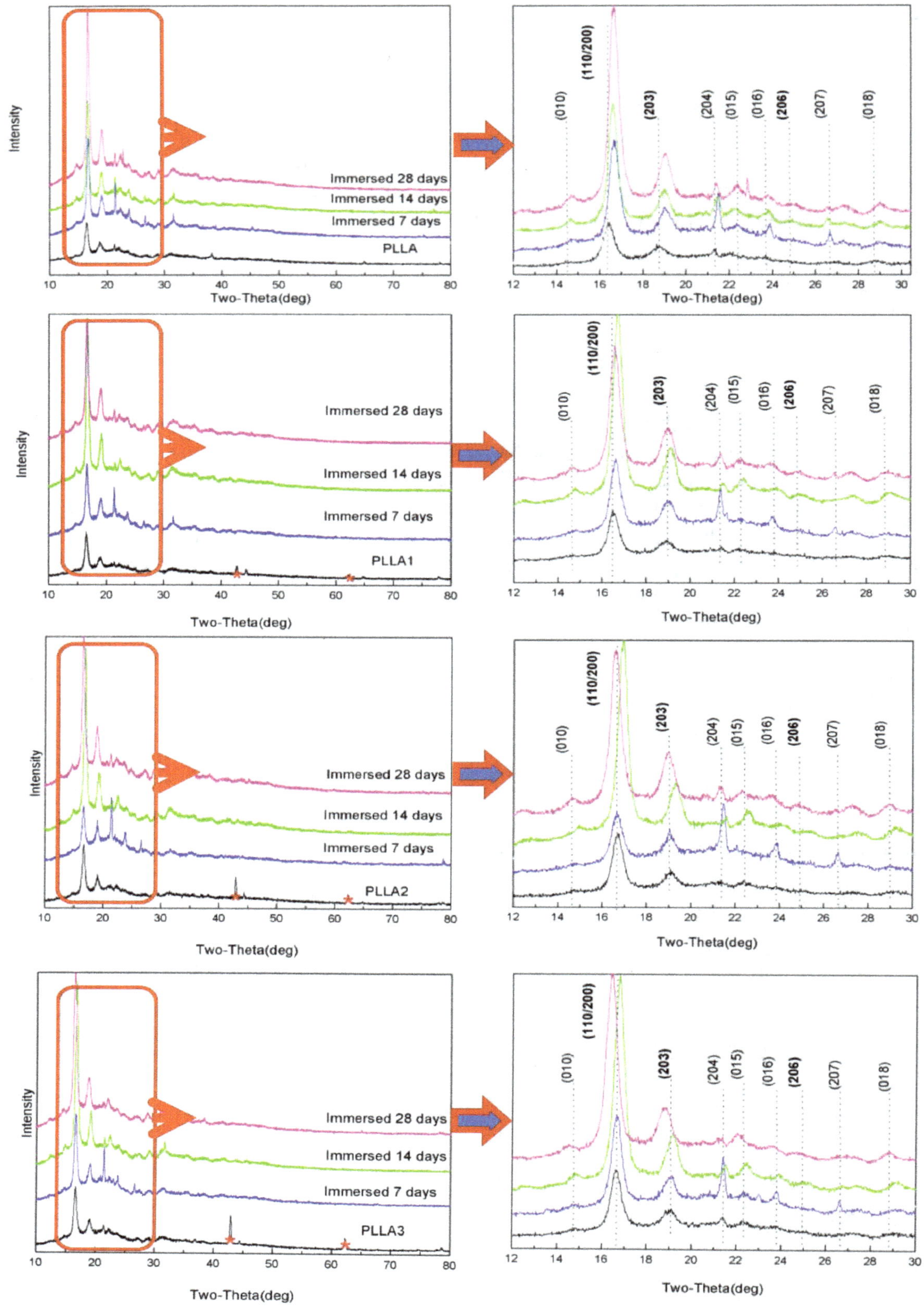

Figure 1. X-ray powder diffraction (XRD) patterns of the poly(L-lactide) (PLLA) and composite films at different degrading times and under partial enlargement.

2.2. DSC Analysis

The DSC data in the PLLA show the endothermic peaks of T_g and exothermic peaks of T_{cc} for the neat and composite materials for Day 0 in Figure 2. During the degradation process, the exothermic peaks observed around T_g and T_{cc} for the neat PLLA samples disappeared gradually, and meanwhile one weak exothermic peak was seen at the left side of the main endothermic peak, which was due to the $\alpha' \rightarrow \alpha$ transition [28] in the substrate from 7 days' to 28 days' immersion. Furthermore, it can be seen in Table 2 that the T_{mc} of the PLLA sample increased when it was immersed initially, and a small variation was observed in the value over the course of the experiment. This suggested an increased crystallinity of the specimen at the beginning of the procedure, caused by enhancement of the chain mobility, and particularly an increase in the amount of α form crystals, which was consistent with the XRD observation where a larger value of 2θ corresponding to the (110)/(200) crystal plane was obtained for PLLA after 28 days' immersion. Compared to the control, the composites exhibited different crystalline behaviors. Because the whiskers effectively promoted PLLA bulk crystallization, there were obvious exothermic peaks of $\alpha' \rightarrow \alpha$ transitions observed for the composite samples on Day 0. Double fusion peaks and a prominent reduction in the final T_{mc} value were observed after 28 days, indicating that the composites decomposed faster than the control sample under the same degrading procedure. Moreover, the PLLA3 sample, after 28 days' immersion, showed double fusion peaks similar to the ones shown by the PLLA1 and PLLA2 samples after 14 days, suggesting that an increased whisker amount was beneficial to inhibiting the degrading extent of the composites. This was also demonstrated by the T_{mc} value of PLLA1, 164.6 °C, which was the lowest among those exhibited by all specimens.

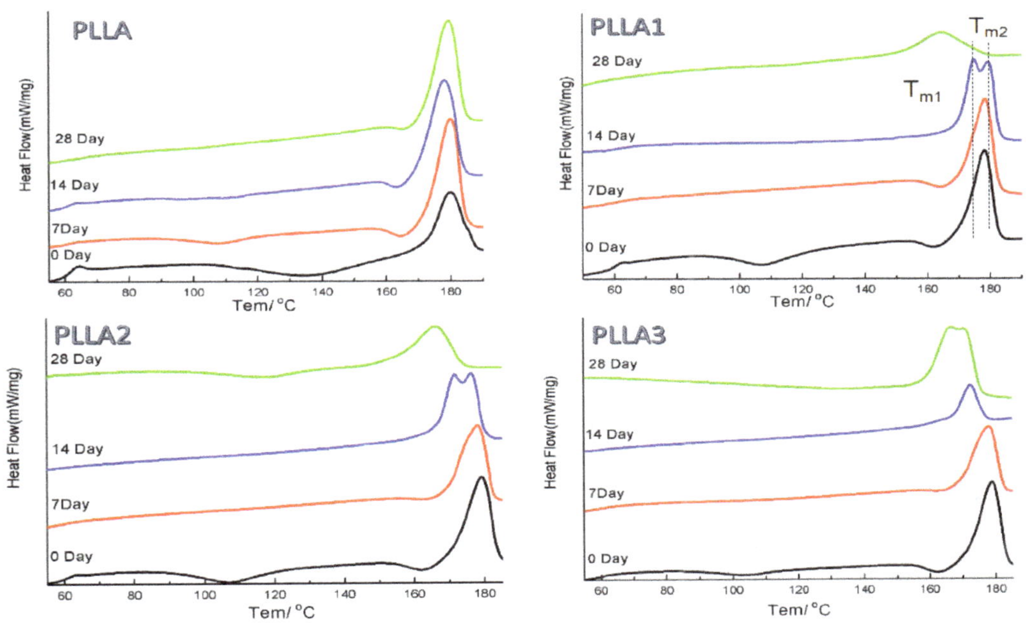

Figure 2. Secondary heating curves obtained by differential scanning calorimetry (DSC) for the samples after soaking in PBS for different degrading times.

Table 2. Melting points of the samples used in DSC measurements.

Degradation Time	T_m/°C			
	PLLA	PLLA1	PLLA2	PLLA3
0	179.8	178.2	179.2	179.1
7	180.0	178.5	178.1	177.9
14	178.6	174.9/179.4	171.4/176.4	172.4
28	179.6	164.6	166.3	166.8/170.4

2.3. Surface and Fracture Morphologies of Sample Films after Degradation

The surface and fracture morphologies (treated by liquid nitrogen for brittle fracture) of the sample films immersed in PBS for various degradation periods were obtained and are shown in Figure 3. It is obvious that the surfaces of all specimens degraded for 7 days became a little rougher in comparison to their initial smooth state, and it was also observed that the control film had a larger rough region than the composite ones, indicating an increase in the crystallinity of the composite on addition of MgO whiskers. This roughness observed after 7 days was probably due to the generation of the α′ form under the degradation, as shown in Figure 3a, of PLLA, PLLA1, and PLLA2, which had the bunch shape [33] because hydrolysis can facilitate matrix crystallinity through the movement of molecular chains [34,35] or reorganization and recrystallization in amorphous areas of the bulk [28]. The roughness was not obvious and not even observed after 14 days and 28 days, and a few small holes were found to be dispersed on the surface over the course of the sequential degrading process. This observation may be ascribed to the degradation of crystalline areas mentioned above with the progress of time [9], during which the rough surfaces disappeared.

To evaluate the effect of degradation on the inner structure of the films, the fracture morphology of the specimens after being treated for various degrading times are shown in Figure 3b. It shows that the samples had no obvious differences and still kept the dense microstructure for up to 28 days, and the fracture morphology of the composites still exhibited better toughness than neat PLLA. However, some micropores can be seen on the composites and can probably be attributed to the loss of the non-crystalline region during the PLLA hydrolysis degrading procedure, although more residual parts of the crystalline component were retained on them, which is consistent with Luo's results [9].

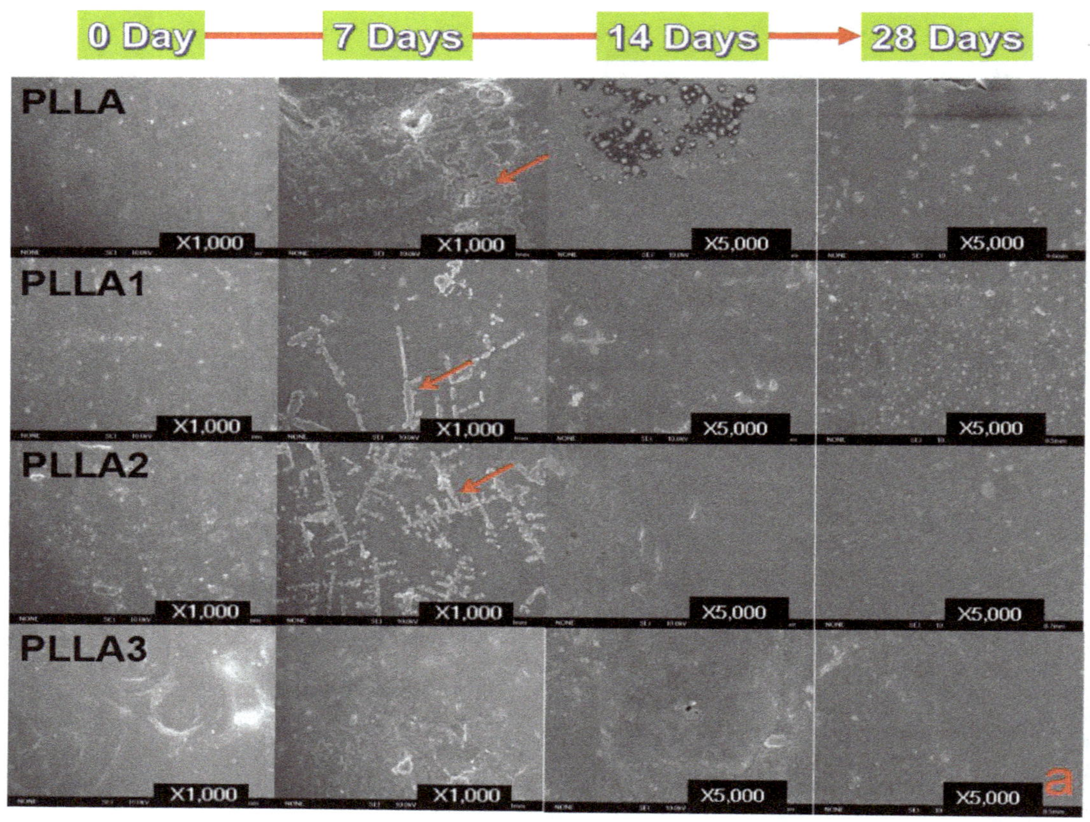

Figure 3. *Cont.*

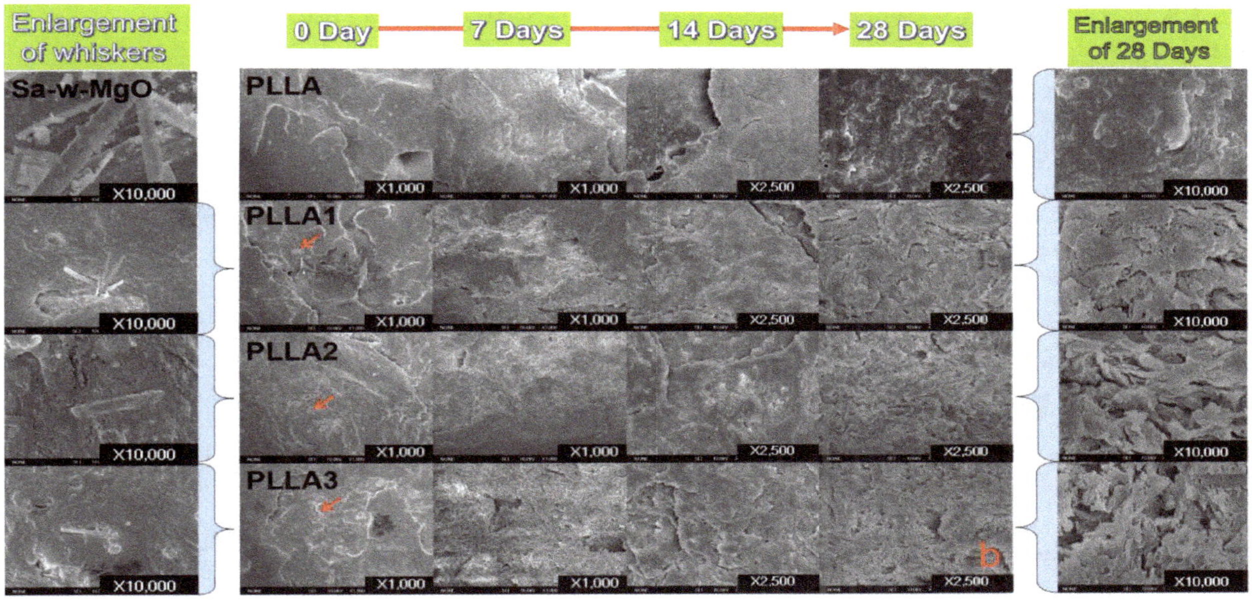

Figure 3. SEM micrographs of PLLA and composite films at different degradation times: (**a**) surface morphology (red arrows: crystalline region from the degradation); (**b**) fracture morphology (red arrows: MgO whiskers).

2.4. The pH Values of the Immersing Solution, the Weight Loss and Mechanical Properties of Sample Films after Degradation

The pH values of the PBS after the degradation of neat PLLA and composite films are shown in Figure 4. It shows that the pH changes of the film samples in the testing period were not obvious, as pH varied in the range of 7.0 to 7.5. With an increase in the degradation time, the pH value of the soaking solution in the presence of neat PLLA film decreased from 7.4 (Day 0) to 7.25 (Day 28), and this was ascribed to the acidic product after the degradation of neat PLLA. Meanwhile, a slight buffering effect can be seen, attributed to the OH- released after dissolution of MgO in the PBS, owing to the added MgO whiskers in PLLA1. Obviously, increasing the incorporated content of whiskers in PLLA2 and PLLA3 resulted in an effective buffering on the pH value of the PBS, as the pH values observed were higher than those of the former. This suggested that the higher the amount of MgO whiskers added, the higher the pH value obtained would be and the more notable the buffering effect on the degrading solution, caused by the release of OH- because of MgO dissolving. However, it should also be noted that the higher value of pH, like alkalinity condition, was also able to accelerate PLLA degradation, and the degradation was even faster than its acidic autocatalytic degradation [36,37]. Generally, the degradation rate of PLLA in ambient pH value follows the order alkaline > acid > neutral. Therefore, it is illustrated that the much higher content of whiskers was not favorable for further enhancing the stability of the pH value of the PBS during the initial degrading stage of PLLA-based composite films.

For further analysis of the whiskers affecting the degradation of the PLLA bulk, the decomposition rate of the samples over time was investigated through their weight losses, and the results are shown in Figure 5. The weight of the sample films was reduced gradually during the degrading procedure. The weight losses of the composites PLLA2 and PLLA3 were comparatively more obvious, suggesting that a composite with a higher content of MgO whiskers exhibited a faster degradation rate, due to the increase in the pH value of the PBS, owing to MgO dissolution. Then, the weights of the films changed slightly after 28 days, because the non-crystalline region was degraded by the hydrolysis process and the residual part that was highly crystalline could not be hydrolyzed in the PBS, owing to the tight and regular arrangement of its molecular chains. Furthermore, PLLA1 degraded more slowly compared with PLLA in the later stage, which was consistent with the pH results and could be accounted for by the buffering effect of MgO neutralizing the acid from PLLA hydrolysis.

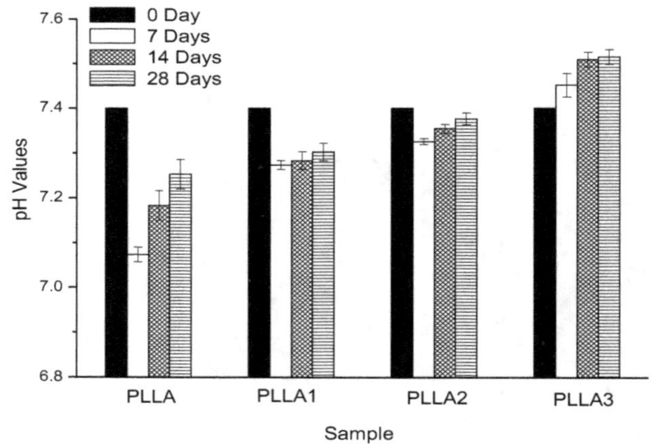

Figure 4. The pH value as a function of degradation time of the neat PLLA and composite films immersed in the phosphate buffer solution (PBS).

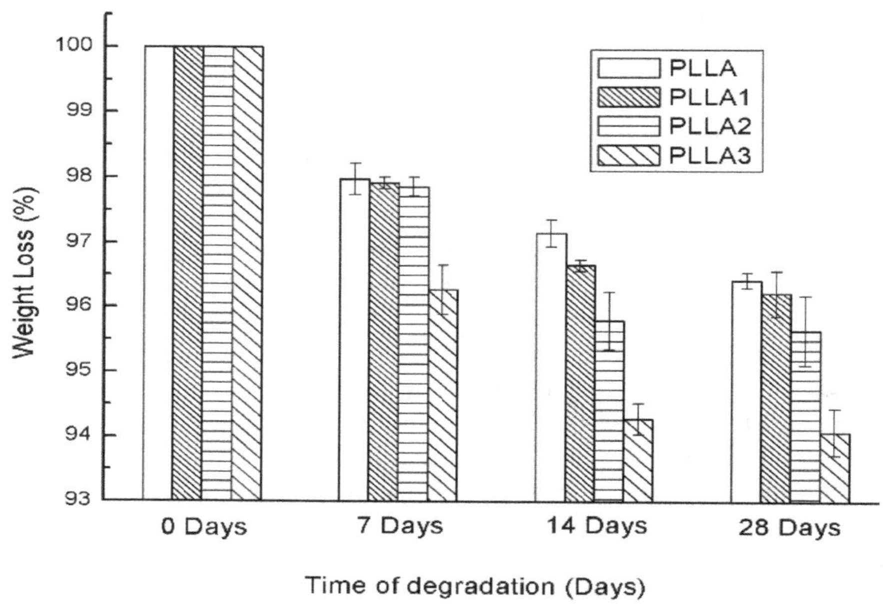

Figure 5. Weight losses as a function of degradation time of neat PLLA and composite films immersed in PBS.

The variation of mechanical properties of the samples during the degrading process is shown in Figure 6. Basically, the mechanical properties of PLLA and composite, including tensile strength, elastic modulus, and elongation at break, decreased as immersing time went. With increasing whisker content, the mechanical properties of the composite decreased notably; especially, the PLLA 3 with 3% whisker content exhibited a more obvious reduction in the tensile strength and Young modulus in comparison to pure PLLA. This deteriorated performance was probably due to the higher pH value of PLLA3 presented during the degradation. Meanwhile, it can be seen that the mechanical properties of composite were changed more greatly within the initially degrading stage, and this can probably be ascribed to the increasing hydrophilicity of composite induced by whiskers added [24].

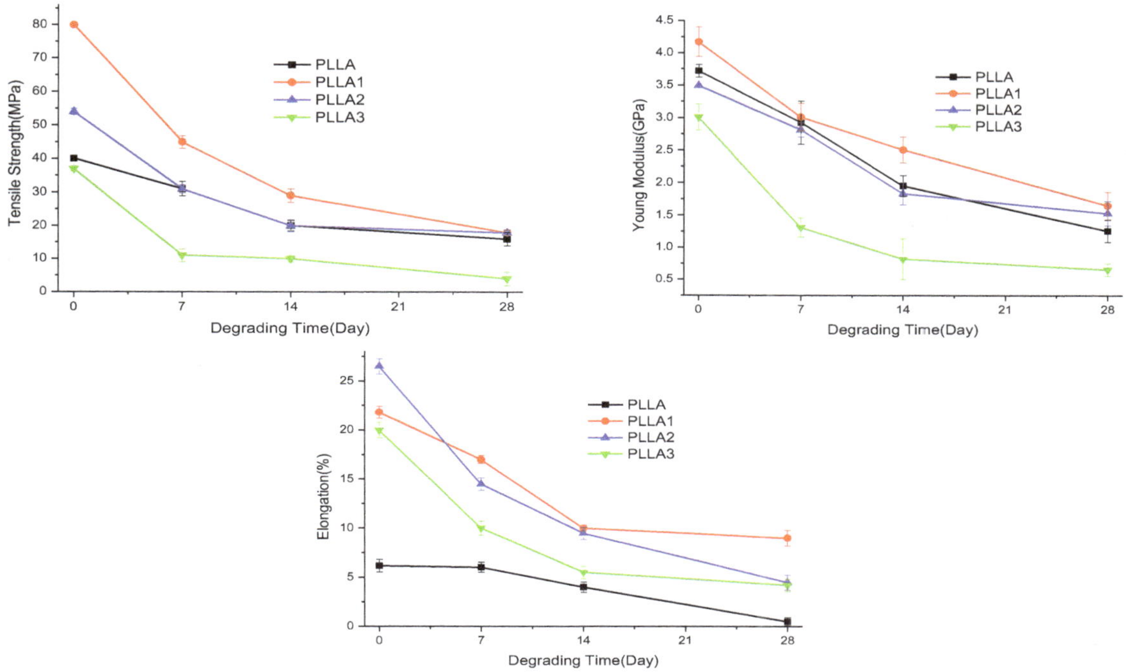

Figure 6. Mechanical properties of PLLA and composites at different degrading time.

2.5. *Cell Experiments Results of Sample Films with Different Whisker Contents*

The results of cell morphology observation and viability evaluations of the sample films are displayed in Figures 7 and 8. It can be seen that small cells exhibited a good shuttle shape with an adherent and spreading morphology, whereas floating cells with spherical or round shapes observed in Figure 7 indicate a shape change and activity loss of the cells. Compared to that of the negative control group, the amount of cells showing loss in activity in PLLA was comparatively higher, but contrarily a decrease in the comparative number of floating cells was also observed in the composite samples in Figure 7c–e, with the cell shape being maintained similar to that of the negative control group. The variations of the cell states suggested that cells can show good growth and proliferation in the degrading solution of the composite films, because of the beneficial bioactivity and biocompatibility of MgO whiskers.

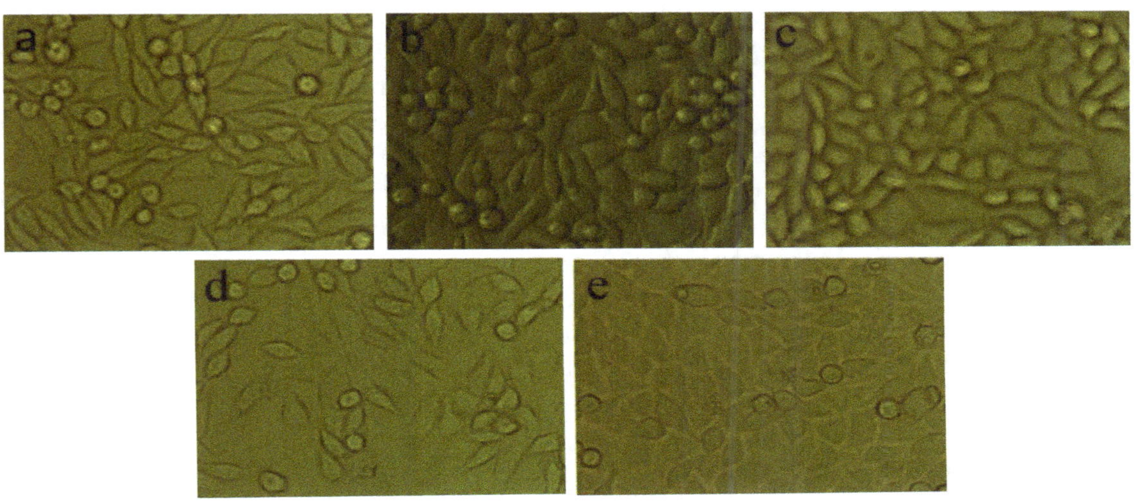

Figure 7. Morphology of the cell cultivated in leach liquor for 2 days (×100): (**a**) Negative control, (**b**) PLLA, (**c**) PLLA1, (**d**) PLLA2, (**e**) PLLA3.

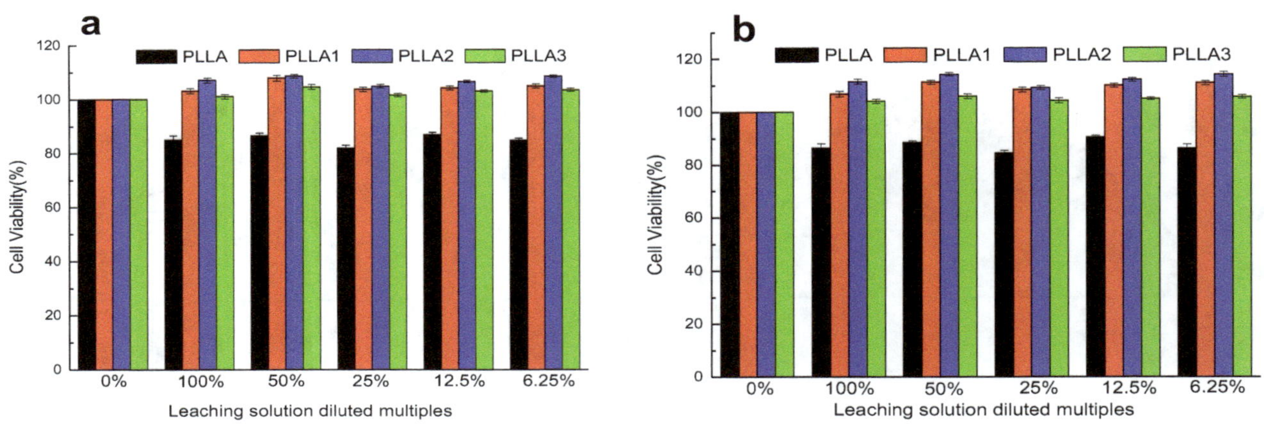

Figure 8. Cell viability of on PLLA, PLLA1, PLLA2, PLLA3 diluted 0, 2, 4, 8, and 16 times in leach liquor and culture for (**a**) 4 days and (**b**) 7 days, respectively.

With the addition of MgO whiskers, cell viability in the composite films increased significantly compared to that of neat PLLA, as shown in Figure 8. Consistent with the cell morphology results shown in Figure 7, it exhibited a relatively higher cell viability on the composite than on neat PLLA, especially in the presence of PLLA2, with a 2 wt% content of whiskers. On extending the culture time, the composite had a higher cell number, indicating prominent cell proliferation. The improved cell viability in the presence of MgO whiskers was probably due to the buffering effect of MgO whiskers on the PBS and significant Mg^{2+} bioactivity as a cofactor in the cell metabolism acting on the cells. The former is helpful in providing a relatively stable pH state, and the latter is beneficial for enhancing cell growth rate.

3. Discussion

PLLA, as a widely used biodegradable polymer for biomedical applications, usually shows different degradation properties, depending on the degrading environmental conditions. In addition to the impact of enzymes in the body on its decomposition, the behavior of bulk hydrolysis mainly accounts for its degradation in vivo, and the performance of the implant in the early degrading process is extremely important, in particular, because new bone is usually intensively active during the early recovery stage [29–32].

Generally, the features of this hydrolysis include two sections with the decomposition of the non-crystalline region and that of the crystalline part, in that order [9]. The results of the XRD patterns and DSC curves for the pristine PLLA in Figures 1 and 2 showed that the relative crystallinity of PLLA increased after degradation in PBS, as observed in Figure 3, demonstrating that the non-crystalline region in matrix was hydrolyzed first and, simultaneously, the crystalline structures of PLLA as α'-form were fabricated on inducing of hydrolysis through a chain or fragment movement in this process. On adding MgO whiskers, the crystallization degree, mechanical properties, and hydrophilicity of PLLA improved significantly, and the content of the whisker was undoubtedly also a key factor that affected the PLLA degrading properties [28,35]. Regarding the composite degradation behaviors, the initial change was similar to the degrading behaviors of the control PLLA, and the amorphous regions in the PLLA matrix were gradually destroyed, as shown by the corresponding changes in the patterns and curves of XRD and DSC. However, the hydrolysis rate of the PLLA bulk increased remarkably in the presence of whiskers, and the decomposing rate of the composites was prominently faster than that of the crystalline structure forming, which could be attributed to the hydrophilicity of the PLLA enhanced by the MgO whiskers. This hydrophilicity resulted in the notable acceleration and increment of water molecular seeping into the bulk (as shown in Figure 9), as reported previously [28]. As noted in Figures 1 and 2, this increase in the degradation was limited in the specimen of PLLA3, which showed a lower rate of hydrolysis in comparison to PLLA2 in 28 days, and PLLA3 degradation showed a tendency of accelerating and then retarding, depending on the variation of T_m, as shown in Table 2.

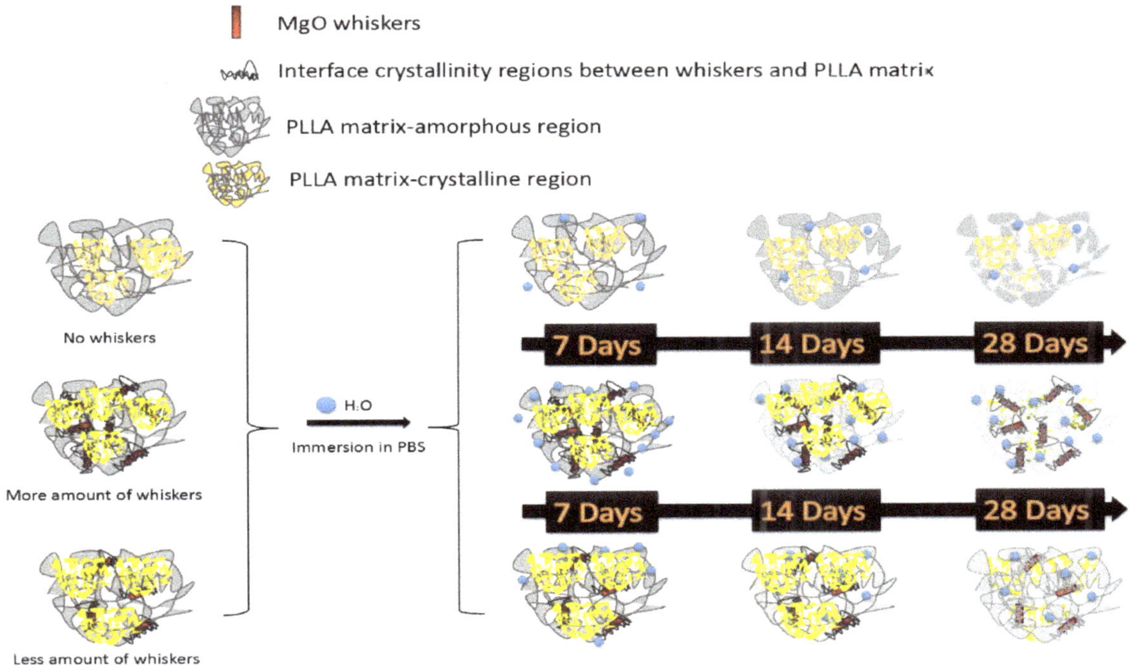

Figure 9. Diagram of in vitro degradation process for PLLA/magnesium oxide (MgO) whiskers composite films with different whisker contents.

This was different from previously reported studies [25,28] mentioning an increase in the degradation rate with an increase in the amount of inorganic fillers, such as carbon nanotube–organic montmorillonite. This was probably due to two aspects: one was that the effect of the whiskers' heterogeneous nucleation promoted the crystallization of PLLA, and an interface crystallinity [38] was also inclined to be introduced between the matrix and the whiskers, especially around the whisker regions, owing to the tight and regular packing of molecular chains. The numerous crystalline structures favorably prevented the water molecules permeating, in spite of the MgO whiskers increasing the hydrophilicity of the matrix, thereby inhibiting the water uptake of the materials. The other factor was the neutralizing effect of MgO whiskers that can effectively control the pH value of the PBS through their alkaline characteristics in solution. The MgO whiskers can impede the effect of acidic conditions and the autocatalysis caused by the PLLA hydrolysis, as a result of slowing down the degrading procedure of the PLLA. This hypothesis agrees with the SEM results of the surface and fracture morphologies of the samples, as shown in Figure 3. It should be noted that the influence of the dissolution of MgO on the pH of the PBS was more prominent for higher whisker contents. PLLA3 exhibited a higher degradation at the beginning of the immersion over 7 days, with a relatively higher weight loss, whereas the weight change was not obvious from 14 days to 28 days, which was similar to the mechanical properties in Figure 6. This was ascribed to the combined effects of the hydrophilicity and pH increase of the composites, which resulted in the fast loss of the amorphous region of the composites occurring initially. Then, the higher amount of OH- could control the pH value favorably, and this was beneficial in retarding the degrading course of the composites. Meanwhile, a small negative effect on cell viability of the PLLA3, as seen in Figures 7 and 8, was also probably due to the pH increase of the sample at the earlier stage, and this can also be detrimental to the films' biocompatibility and biological activity in terms of cell proliferation and viability. Simultaneously, the samples with 1 wt% and 2 wt% whisker contents were relatively more conductive for cell performance, and this was different from the results in previously reported studies, which selected a weight content of 5 wt% for the whiskers for investigating the effect of MgO whiskers on cell behaviors. Compared to the results above, the composites of our study showed faster degradation and beneficial pH regulation during the PLLA hydrolysis. The difference in the optimal whisker content in the studies performed was probably due to the type of surface modifiers of MgO whiskers. Therefore, the surface modifiers of whiskers can be an effective approach to controlling the degradation behaviors of the PLLA, and specific studies will be conducted on the modifier as a factor on composite material performance in following studies.

4. Materials and Methods

4.1. Materials

Powdered PLLA (M_w = 417,000) was purchased from Daigang Biological Engineering Co., Ltd. (Jinan, China), and MgO whiskers were prepared in the laboratory according to the procedure described in the literature [24]. All other agents used were of analytical grade and did not require further treatment.

4.2. Preparation of MgO Whiskers and Their Modification

Using the procedure described in Reference [24], MgO whiskers were prepared according to a sequence of steps based on the chemical synthesis of a whisker precursor (magnesium carbonate hydrate) and its calcination. Na_2CO_3 (0.6 M) was added dropwise into an equal volume of $MgCl_2$ (0.6 M) and stirred for 20 min. The mixture was aged at room temperature for 10 h and then filtered, washed, and dried at 80 °C for 3–4 h. The precursor was calcined at 750 °C for 4 h (the heating rate was 5 °C/min). The final product of this process was MgO whiskers (w-MgO).

Mixtures of 0.5 g of whiskers and 50 mL of ethanol were placed in three-necked flasks, which were subjected to ultrasonication in a bath to achieve full dispersion of the mixtures. The mixtures were then heated to 45 °C under reflux condensation. Stearic acid (Sa, 0.005 g) was dissolved into 20 mL of ethanol, added dropwise into each MgO suspension, and allowed to react for 1 h. The mixtures were then centrifuged, washed, and dried. The resultant product was denoted as Sa–w-MgO.

4.3. Preparation of Sa–w-MgO/PLLA Composite Films

PLLA/Sa–w-MgO composites with three different Sa–w-MgO contents (1, 2, and 3 wt%) were prepared through the solution casting method. The detailed procedure has been described in our previous reports [24]. Detailed information on the samples prepared is given in Table 3.

Table 3. Detailed information of the samples.

Abbreviation	Sample	Whisker Content (*w*/*w*)
PLLA	PLLA	0
PLLA1	1 wt% Sa–w-MgO/PLLA	1/100
PLLA2	2 wt% Sa–w-MgO/PLLA	2/100
PLLA3	3 wt% Sa–w-MgO/PLLA	3/100

4.4. Degradation of the Composite Material in PBS

We carried out in vitro degradation of pure PLLA and composite samples of the dimensions 1 cm × 1 cm with a thickness of 1 mm by immersing them in 10 mL of a phosphate buffer solution in a rocking water bath at 37 °C for 7, 14, and 28 days. The specimens of each material were removed from PBS at the end of these periods. After rinsing with deionized water and removing the surface water using filter paper, the samples were weighed and then dried in vacuum at 40 °C to a constant weight. The weight losses were calculated using the following equation:

$$W_{Loss}(\%) = \frac{m_0 - m_d}{m_0} \times 100\% \quad (1)$$

where W_{Loss} is the average degrading rate, and m_0 and m_d are the initial and final weights (after drying to constant weight), respectively. When the degradation was complete, the pH value of the PBS was determined by averaging the results of 3 independent measurements, obtained from three identical samples for each content. The process of film preparation and degradation is displayed in Figure 10.

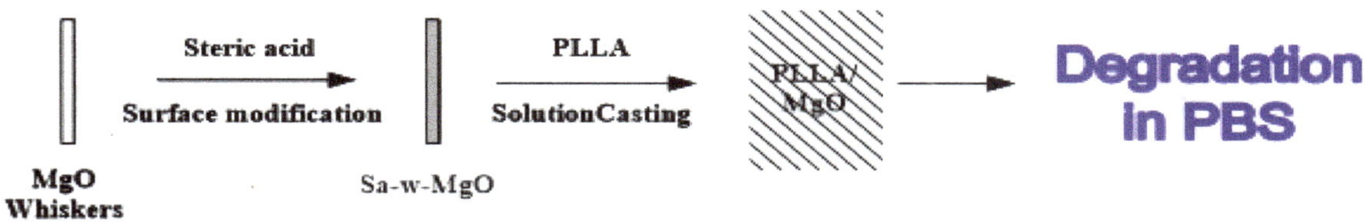

Figure 10. Schematic illustration of the preparation and degradation of the composite.

4.5. Characterization

X-ray powder diffraction (XRD, Rigaku D/max/2500PC, Tokyo, Japan) was performed to obtain XRD patterns of the samples, using Cu Kα radiation with λ = 1.5418 Å, operating at 40 kV/100 mA, with a scanning speed of 8°/min.

The morphology of the samples was characterized by field-emission scanning electron microscopy (FESEM, JOEL 6700F, Osaka, Japan, operating at 10 kV).

Differential scanning calorimetry (DSC) analyses were performed with a SETRAM SETSYS EVOLUTION16/18 (DSC 200F3, Netzsch Co., Selb, Germany). The samples, weighing approximately 5–8 mg each, were sealed in an aluminum pan, heated under nitrogen flow from room temperature to 220 °C at a heating rate of 20 °C/min, isothermally conditioned at 220 °C for 2 min, cooled to 0 °C, and then reheated to 220 °C at a heating rate of 10 °C/min.

The Figure 11 shows the detailed information of the samples' shape used in mechanical testing. The measurements were taken in tension using a universal testing machine (Instron 5845, Boston, MA, USA) at an extension rate of 0.5 mm/min. There were three replicates of each sample tested, and the average and standard deviation of each set of replicates were calculated.

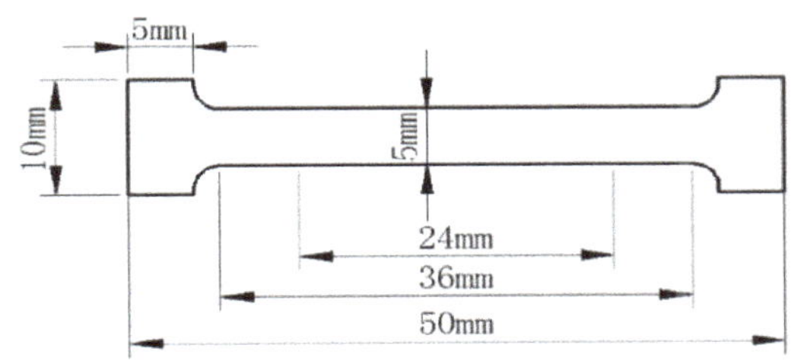

Figure 11. Detailed information of the samples used in mechanical testing.

4.6. Cytotoxicity Testing

4.6.1. Preparation of Material Extracts

With the ratio of film samples to culture medium set at 1 cm^2/mL, the disinfected pristine PLLA and composite samples (1 cm × 1 cm) were immersed in an RPMI-1640 solution (Gibco, Grand Island, NY, USA) under a humidified atmosphere with 5% CO_2 at 37 °C for 3 days to get the extraction medium. Prior to cell seeding, extracts were sterile-filtered through 0.2 mL syringe filters and then diluted by 2, 4, 8, and 16 times. The control groups involved the use of 10% fetal calf serum and 90% 1640 medium as negative controls.

4.6.2. Culture of Murine Fibroblast Cells (L929)

Murine fibroblast cells (L929) were used to evaluate the in vitro cytotoxicity of the sample films. L-929 cells (Dingguo, Tianjin, China) were cultured in RPMI-1640 medium, supplemented with 10% fetal bovine serum (FBS) (Hyclone, Logan, UT, USA), 100 U/mL penicillin, and 100 U/mL streptomycin in a humidified incubator at 95% relative humidity and 5% CO_2 at 37 °C. Then, the cells were cultured through the enzyme (Tryosin-EDTA, Sorabio, Beijing, China) digesting method.

4.6.3. Cell Behavior and Viability

(1) After well sterilized, the pure PLLA and composite films were fixed in the 24-well flat-bottomed cell culture plates. Then, 10^4 cells/500 μL medium were added in each well with the plates incubated for 2 days at 37 °C/5% CO_2, and cell morphology was consequently observed by optical microscopy (Nikon ECLIPSE Ti inverted microscope, Tokyo, Japan).

(2) We incubated 24-well flat-bottomed cell culture plates with 10^4 cells/500 μL medium in each well for 24 h to allow attachment. After the growth states were confirmed, the medium in the experimental group was removed and 500 μL extracts with various concentrations (dilutions of 0, 2, 4, 8, and 16 times) were placed in cell culture plates. The medium in the negative control group was replaced with 500 μL RPMI-1640 and 10% fetal calf serum, and at least 3 samples from each group were tested to confirm reproducibility. After incubating the cells in a humidified atmosphere for 4 and 7 days at 37 °C/5% CO_2, 500 μL of 3-(4,5-dimethyl-2-thiazolyl)-2,5-diphenyl-2-*H*-tetrazolium bromide (MTT) (sigma, Burlington, MA, USA) (5 mg/mL) was added to each well and the samples were incubated with MTT in the incubator at 37 °C for 3 h, and 350 μL DMSO (Sorabio, China) was added into each well. Finally, the absorbance was recorded by a multimode detector on a microplate reader (Epoch, BioTek, Winooski, VT, USA) at a wavelength of 570 nm, and the OD value of each group was statistically analyzed to calculate cell viability. The cell viability were calculated using the following equation:

$$\text{Cell viability}(\%) = \frac{OD_{\text{Experiment}}}{OD_{\text{Negative control}}} \times 100\% \quad (2)$$

where $OD_{\text{Experiment}}$ is the OD value from the sample films and $OD_{\text{Negative control}}$ is the OD value from the negative control group, respectively.

5. Conclusions

In this study, the composites of PLLA/MgO whiskers with different whisker contents were soaked in PBS for relatively short treating periods to investigate their initial in vitro degrading properties while used as implants in the context of bone repair. It was demonstrated that an increase in the whisker content can accelerate the degradation rate of the PLLA matrix. The degrading procedure of composites with a higher content of whiskers was influenced more significantly by the hydrophilicity and pH value. As the results showed, the degradation of the composite began with loss of the non-crystalline region prior to destruction of the crystalline part, and the content of MgO whiskers present was a significant factor that influenced the degrading procedure. The cell morphology observation and cell viability measurements suggested that, in the presence of MgO whiskers, the composite showed higher cell viability and led to cells with flattened and narrow rhombus shapes, which enhanced the bioactivity of the PLLA matrix significantly. This PLLA/MgO biocomposite can be potentially utilized for bone repair with better performance.

Author Contributions: Conceptualization: M.C., J.Y. and H.B.; Data curation: B.L., Y.Z. and H.L.; Investigation: Y.L.; Resources: B.L. and S.S.; Validation: Z.J.; Writing: Y.Z., W.L., M.C.

Acknowledgments: The authors acknowledge the financial support for this work from Key projects of the Joint Foundation of the National Natural Science Foundation of China (U1764254), Tianjin Natural Science Foundation (No. 17JCQNJC03100), and the National Natural Science Foundation of China (No. 51501129).

References

1. Alsberg, E.; Kong, H.J.; Hirano, Y.; Smith, M.K.; Albeiruti, A.; Mooney, D.J. Regulating bone formation via controlled scaffold degradation. *J. Dent. Res.* **2003**, *82*, 903–908. [CrossRef] [PubMed]
2. Kakinoki, S.; Uchida, S.; Ehashi, T.; Murakami, A.; Yamaoka, T. Surface Modification of Poly(L-lactic acid) Nanofiber with Oligo(D-lactic acid) Bioactive-Peptide Conjugates for Peripheral Nerve Regeneration. *Polymers* **2011**, *3*, 820–832. [CrossRef]
3. Rogina, A.; Pribolšan, L.; Hanžek, A.; Gómez-Estrada, L.; Ferrer, G.G.; Marijanović, I.; Ivanković, M.; Ivanković, H. Macroporous poly(lactic acid) construct supporting the osteoinductive porous chitosan-based hydrogel for bone tissue engineering. *Polymer* **2016**, *98*, 172–181. [CrossRef]
4. Wang, X.J.; Song, G.J.; Lou, T. Fabrication and characterization of nano composite scaffold of poly(L-lactic acid)/hydroxyapatite. *J. Mater. Sci.* **2010**, *21*, 183–188. [CrossRef] [PubMed]
5. Yang, Y.F.; Zhao, Y.H.; Tang, G.W.; Li, H.; Yuan, X.Y.; Fan, Y.B. In vitro degradation of porous poly(L-lactide-co-glycolide)/β-tricalcium phosphate (PLGA/β-TCP) scaffolds under dynamic and static conditions. *Polym. Degrad. Stab.* **2008**, *93*, 1838–1845. [CrossRef]
6. Liao, L.; Chen, L.; Chen, A.Z.; Pu, X.M.; Kang, Y.Q.; Yao, Y.D.; Liao, X.M.; Huang, Z.B.; Yin, G.F. Preparation and characteristics of novel poly-L-lactide/β-calcium metaphosphate fracture fixation composite rods. *J. Mater. Res.* **2007**, *22*, 3324–3329. [CrossRef]
7. Shikinami, Y.; Matsusue, Y.; Nakamura, T. The complete process of bioresorption and bone replacement using devices made of forged composites of raw hydroxyapatite particles/poly L-lactide (F-u-HA/PLLA). *Biomaterials* **2005**, *26*, 5542–5551. [CrossRef] [PubMed]
8. Li, X.; Chu, C.L.; Liu, L.; Liu, X.K.; Bai, J.; Guo, C.; Xue, F.; Lin, P.H.; Chu, P.K. Biodegradable poly-lactic acid based-composite reinforced unidirectionally with high-strength magnesium alloy wires. *Biomaterials* **2015**, *49*, 135–144. [CrossRef] [PubMed]
9. Wen, W.; Zou, Z.P.; Luo, B.H.; Zhou, C.R. In vitro degradation and cytocompatibility of g-MgO whiskers/PLLA composites. *J. Mater. Sci.* **2017**, *52*, 2329–2344. [CrossRef]
10. Nampoothiri, K.M.; Nair, N.R.; John, R.P. An overview of the recent developments in polylactide (PLA) research. *Bioresour. Technol.* **2010**, *101*, 8493–8501. [CrossRef] [PubMed]
11. Jiang, L.Y.; Xiong, C.D.; Jiang, L.X.; Xu, L.J. Effect of hydroxyapatite with different morphology on the crystallization behavior, mechanical property and in vitro degradation of hydroxyapatite/poly(lactic-co-glycolic) composite. *Compos. Sci. Technol.* **2014**, *93*, 61–67.
12. Cifuentes, S.C.; Gavilán, R.; Lieblich, M.; Benavente, R.; González-Carrasco, J.L. In vitro degradation of biodegradable polylactic acid/magnesium composites: Relevance of Mg particle shape. *Acta Biomater.* **2016**, *32*, 348–357. [CrossRef] [PubMed]
13. Shalumon, K.T.; Sheu, C.; Fong, Y.T.; Liao, H.T.; Chen, J.P. Microsphere-Based Hierarchically Juxtapositioned Biphasic Scaffolds Prepared from Poly(Lactic-co-Glycolic Acid) and Nanohydroxyapatite for Osteochondral Tissue Engineering. *Polymers* **2016**, *8*, 429. [CrossRef]
14. Kothapalli, C.R.; Shaw, M.T.; Wei, M. Biodegradable HA-PLA 3-D porous scaffolds: Effect of nano-sized filler content on scaffold properties. *Acta Biomater.* **2005**, *1*, 653–662. [CrossRef] [PubMed]
15. Loher, S.; Reboul, V.; Brunner, T.J.; Simonet, M.; Dora, C.; Neuenschwander, P.; Stark, W.J. Improved degradation and bioactivity of amorphous aerosol derived tricalcium phosphate nanoparticles in poly(lactide-co-glycolide). *Nanotechnology* **2006**, *17*, 2054–2061. [CrossRef]
16. Yun, H.; Kim, S.; Park, E.K. Bioactive glass-poly(ε-caprolactone) composite scaffolds with 3 dimensionally hierarchical pore networks. *Mater. Sci. Eng. C* **2016**, *31*, 198–205. [CrossRef]
17. Fernandes, J.S.; Gentile, P.; Martins, M.; Neves, N.M.; Miller, C.; Crawford, A.; Pires, R.A.; Hatton, P.; Reis, R.L. Reinforcement of poly-L-lactic acid electrospun membranes with strontium borosilicate bioactive glasses for bone tissue engineering. *Acta Biomater.* **2016**, *44*, 168–177. [CrossRef] [PubMed]
18. Johnson, A.J.W.; Herschler, B.A. A review of the mechanical behavior of CaP and CaP/polymer composites for applications in bone replacement and repair. *Acta Biomater.* **2011**, *7*, 16–30. [CrossRef] [PubMed]
19. Ma, F.Q.; Lu, X.L.; Wang, Z.M.; Sun, Z.J.; Zhang, F.F.; Zheng, Y.F. Nanocomposites of poly(L-lactide) and surface modified magnesia nanoparticles: Fabrication, mechanical property and biodegradability. *J. Phys. Chem. Solids* **2011**, *72*, 111–116. [CrossRef]

20. Kum, C.H.; Cho, Y.; Seo, S.H.; Joung, Y.K.; Ahn, D.J.; Han, D.K. A poly(lactide) stereocomplex structure with modified magnesium oxide and its effects in enhancing the mechanical properties and suppressing inflammation. *Small* **2014**, *10*, 3783–3794. [CrossRef] [PubMed]
21. Kum, C.H.; Cho, Y.; Joung, Y.K.; Choi, J.; Park, K.; Seo, S.H.; Park, Y.S.; Ahn, D.J.; Han, D.K. Biodegradable poly(L-lactide) composites by oligolactide-grafted magnesium hydroxide for mechanical reinforcement and reduced inflammation. *J. Mater. Chem. B* **2013**, *1*, 2764–2772. [CrossRef]
22. Yang, J.J.; Cao, X.X.; Zhao, Y.; Wang, L.; Liu, B.; Jia, J.P.; Liang, H.; Chen, M.F. Enhanced pH stability, cell viability and reduced degradation rate of poly(L-lactide)-based composite in vitro: Effect of modified magnesium oxide nanoparticles. *J. Biomater. Sci. Polym. Ed.* **2017**, *28*, 486–503. [CrossRef] [PubMed]
23. Wen, W.; Luo, B.; Qin, X.; Li, C.; Liu, M.; Ding, S.; Zhou, C. Strengthening and toughening of poly(L-lactide) composites by surface modified MgO whiskers. *Appl. Surf. Sci.* **2015**, *332*, 215–223. [CrossRef]
24. Zhao, Y.; Liu, B.; You, C.; Chen, M.F. Effects of MgO whiskers on mechanical properties and crystallization behavior of PLLA/MgO composites. *Mater. Des.* **2016**, *89*, 573–581. [CrossRef]
25. Chen, H.M.; Chen, J.W.; Chen, J.; Yang, J.H.; Huang, T.; Zhang, N.; Wan, Y. Effect of organic montmorillonite on cold crystallization and hydrolytic degradation of poly(L-lactide). *Polym. Degrad. Stab.* **2012**, *97*, 2273–2283. [CrossRef]
26. Paul, M.A.; Delcourt, C.; Alexandre, M.; Degée, P.; Monteverde, F.; Dubois, P. Polylactide/montmorillonite nanocomposites: Study of the hydrolytic degradation. *Polym.Degrad. Stab.* **2005**, *87*, 535–542. [CrossRef]
27. Fukushima, K.; Tabuani, D.; Dottori, M.; Armentano, I.; Kenny, J.M.; Gamino, G. Effect of temperature and nanoparticle type on hydrolytic degradation of poly(lactic acid) nanocomposites. *Polym. Degrad. Stab.* **2011**, *96*, 2120–2129. [CrossRef]
28. Chen, H.M.; Feng, C.X.; Zhang, W.B.; Yang, J.H.; Huang, T.; Zhang, N.; Wang, Y. Hydrolytic degradation behavior of poly(L-lactide)/carbon nanotubes nanocomposites. *Polym. Degrad. Stab.* **2013**, *98*, 198–208. [CrossRef]
29. Bose, S.; Tarafder, S. Calcium phosphate ceramic systems in growth factor and drug delivery for bone tissue engineering: A review. *Acta Biomater.* **2012**, *8*, 1401–1421. [CrossRef] [PubMed]
30. Li, Y.B.; Weng, W.J. Surface modification of hydroxyapatite by stearic acid: Characterization and in vitro behaviors. *J. Mater. Sci. Mater. Med.* **2008**, *19*, 19–25. [CrossRef] [PubMed]
31. Zhang, L.; Chen, L.; Wan, H.Q.; Chen, J.M.; Zhou, H.D. Synthesis and Tribological Properties of Stearic Acid-Modified Anatase (TiO_2) Nanoparticles. *Tribol. Lett.* **2011**, *41*, 409–416. [CrossRef]
32. Barry, M.; Pearce, H.; Cross, L.; Tatullo, M.; Gaharwar, A.K. Advances in Nanotechnology for the Treatment of Osteoporosis. *Curr. Osteoporos. Rep.* **2016**, *14*, 87–94. [CrossRef] [PubMed]
33. Hutmacher, D.W. Scafolds in tissue engineering bone and cartilage. *Biomaterials* **2000**, *21*, 2529–2543. [CrossRef]
34. Rezwan, K.; Chen, Q.Z.; Blaker, J.J.; Boccaccini, A.R. Biodegradable and bioactive porous polymer/inorganic composite scaffolds for bone tissue engineering. *Biomaterials* **2006**, *27*, 3413–3431. [CrossRef] [PubMed]
35. Chen, H.M.; Shen, Y.; Yang, J.H.; Huang, T.; Zhang, N.; Wang, Y.; Zhou, Z.W. Molecular ordering and α'-form formation of poly(L-lactide) during the hydrolytic degradation. *Polymer* **2013**, *54*, 6644–6653. [CrossRef]
36. Tsuji, H.; Nakahara, K. Poly(L-lactide). IX. Hydrolysis in Acid Media. *J. Appl. Polym. Sci.* **2002**, *86*, 186–194. [CrossRef]
37. Li, S.M. Hydrolytic degradation characteristics of aliphatic polyesters derived from lactic and glycolic acids. *J. Biomed. Mater. Res.* **1999**, *48*, 342–353. [CrossRef]
38. Ning, N.Y.; Fu, S.R.; Zhang, W.; Chen, F.; Wang, K.; Deng, H.; Zhang, Q.; Fu, Q. Realizing the enhancement of interfacial interaction in semicrystalline polymer/filler composites via interfacial crystallization. *Prog. Polym. Sci.* **2012**, *37*, 1425–1455. [CrossRef]

13

One-Minute Synthesis of Size-Controlled Fucoidan-Gold Nanosystems: Antitumoral Activity and Dark Field Imaging

Ricardo J. B. Pinto [1,*], Daniela Bispo [1], Carla Vilela [1], Alexandre M. P. Botas [2], Rute A. S. Ferreira [2], Ana C. Menezes [3], Fábio Campos [3], Helena Oliveira [3], Maria H. Abreu [4], Sónia A. O. Santos [1] and Carmen S. R. Freire [1,*]

1 Department of Chemistry, CICECO—Aveiro Institute of Materials, University of Aveiro, 3810-193 Aveiro, Portugal; d.bispo@ua.pt (D.B.); cvilela@ua.pt (C.V.); santos.sonia@ua.pt (S.A.O.S.)

2 Phantom-G, Department of Physics, CICECO—Aveiro Institute of Materials, University of Aveiro, 3810-193 Aveiro, Portugal; a.botas@ua.pt (A.M.P.B.); rferreira@ua.pt (R.A.S.F.)

3 Department of Biology & CESAM, University of Aveiro, 3810-193 Aveiro, Portugal; catarinamenezes@msn.com (A.C.M.); f.m.c@ua.pt (F.C.); holiveira@ua.pt (H.O.)

4 ALGAplus—Prod. e Comerc. De Algas e Seus Derivados, Lda., 3830-196 Ílhavo, Portugal; helena.abreu@algaplus.pt

* Correspondence: r.pinto@ua.pt (R.J.B.P.); cfreire@ua.pt (C.S.R.F.)

Abstract: Gold nanoparticles (AuNPs) are one of the most studied nanosystems with great potential for biomedical applications, including cancer therapy. Although some gold-based systems have been described, the use of green and faster methods that allow the control of their properties is of prime importance. Thus, the present study reports a one-minute microwave-assisted synthesis of fucoidan-coated AuNPs with controllable size and high antitumoral activity. The NPs were synthesized using a fucoidan-enriched fraction extracted from *Fucus vesiculosus*, as the reducing and capping agent. The ensuing monodispersed and spherical NPs exhibit tiny diameters between 5.8 and 13.4 nm for concentrations of fucoidan between 0.5 and 0.05% (w/v), respectively, as excellent colloidal stability in distinct solutions and culture media. Furthermore, the NPs present antitumoral activity against three human tumor cell lines (MNT-1, HepG2, and MG-63), and flow cytometry in combination with dark-field imaging confirmed the cellular uptake of NPs by MG-63 cell line.

Keywords: gold nanoparticles; fucoidan; microwave irradiation; antitumoral activity; darkfield imaging

---

1. Introduction

Cancer, i.e., the abnormal growth and proliferation of cells, is one of the leading causes of mortality and morbidity worldwide. According to the World Health Organization (WHO), cancer was responsible for 9.9 million deaths in 2018 [1], and the number of cases is anticipated to increase by about 70% over the next two decades. Each cancer type involves a specific treatment procedure that embraces one or more modalities, namely surgery (to remove the tumor), radiotherapy, or chemotherapy. Despite the high cure rates observed when cancer is detected early and if appropriated treatment is provided, most of the presently employed therapies, particularly conventional chemotherapy, are associated with severe side effects, including hair loss, nausea and vomiting, pain, anemia, fertility issues, edema, among many others [2], that strongly affect the patient's quality of life. In recent decades, the design of more effective alternatives that allows a targeting action, with almost no impact on healthy tissues or organs, has received considerable attention [3]. Of these, nanosystems that combine a therapeutic effect and imaging properties, or that promote intertwined diagnosis and therapy, the so-called nanotheranostics, has opened new avenues for cancer-conquering [4–6].

Metal nanoparticles (NPs) are a class of nanomaterials with a panoply of biomedical and therapeutic applications, including cancer therapy and imaging of tumors [7,8]. In particular, gold (Au) nanoparticles owing to their unique physical and optical properties have attracted enormous interest in this realm during the past decades [9–12]. The colloidal stability of NPs in biological environments and their interactions with cells is strongly influenced by their surface properties, and thus distinct coating strategies of NPs, using small molecules, polymers, or lipids, have been described [13]. Moreover, this methodology could also target the reduction of the toxicity of NPs and improvement of biological functionalities, typically associated with the use of active biomolecules [13]. The synthesis of AuNPs using biomacromolecules, including alginate, starch, cellulose, chitosan, gelatin, collagen and fucoidan, among others, as reducing and stabilizing agents is a well-documented strategy to achieve these goals [14] and, at the same time, overcome the environmental effects of the conventional methodologies that commonly involve the use of harmful reducing agents.

Fucoidan is a natural occurring sulfated marine polysaccharide extracted from brown seaweeds that presents various biological properties, including antiangiogenic, antitumoral, and anti-inflammatory properties [15,16], and because of that has been widely investigated for the development of nanomaterials for biomedical applications [17]. However, only a few number of papers reported the combination of fucoidan and AuNPs for cancer treatment. One of the first studies in this topic involved the synthesis of AuNPs (44 nm average size) using sodium borohydride as the reducing agent and a fucoidan-mimetic glycopolymer as the capping agent [18]. The obtained AuNPs displayed excellent colloidal stability and selective cytotoxicity to human colon cancer cell line (HCT116). Afterward, Manivasagan et al. [19] produced biocompatible AuNPs (82 nm average size) by using naturally occurring fucoidan as the reducing and capping agent, avoiding the use of harmful reducing agents. This study also demonstrated the applicability of these NPs as a carrier for doxorubicin (DOX) and photoacoustic imaging of breast cancer tumors. In a follow-up study, this research team explored similar fucoidan-AuNPs for dual-chemo-photothermal treatment of eye tumors [20]. In another study, size-controlled fucoidan-AuNPs (15-80 nm) were produced by varying the concentration of fucoidan during the synthesis step [21]. These NPs showed anticancer effect against human oral squamous cell carcinoma (HSC3), and its surface modification and conjugation with DOX also improved their effect.

These studies clearly demonstrate the prospective of the partnership between AuNPs and fucoidan in cancer treatment. However, to achieve high antitumoral activities, viz., less than 20% cell viability, the conjugation with other chemotherapeutics (DOX), or the use of high dosages (up to 50 $\mu g\ mL^{-1}$ of NPs) was typically required. Moreover, some methodologies are somewhat time-consuming and laborious, e.g., up to 2 h for the synthesis of the AuNPs and 24 h for the conjugation with DOX. Additionally, the number of investigated tumor cell lines is limited, and the imaging properties of fucoidan-AuNPs have been only marginally explored. Thus, some essential traits still need to be tackled envisioning their scale-up production and broad application, viz., the establishment of fast and straightforward procedures for the synthesis of fucoidan-Au nanosystems with controllable size and morphology and improved antitumoral activity against different tumor cell lines.

As a developing heating tool, microwave irradiation has been shown to considerably reduce the reaction times and provide a uniform bulk heating that allows the synthesis of various nanomaterials, including AuNPs [22,23], with defined structures and narrow size distributions. However, to the best of our knowledge, this methodology has never been sightseen as a simple, time-saving approach to fabricate fucoidan-AuNPs for application in cancer therapy.

In this line, in the present study, we report for the first time a one-minute microwave-assisted synthesis of fucoidan-coated AuNPs with controllable size and high antitumoral activity. This is a pioneering achievement with respect to previous methodologies to produce Au-fucoidan NPs. The fucoidan-AuNPs were synthesized, using a fucoidan-enriched fraction extracted from *F. vesiculosus*, and characterized in terms of structure, colloidal stability, antitumoral activity against different cell lines (MNT-1, HepG2, and MG-63), and cellular uptake by flow cytometry and dark field imaging of

NPs. The antitumoral activity of Au-fucoidan NPs against these cell lines and their imaging properties by dark-field is also reported here for the first time.

2. Materials and Methods

2.1. Materials

Fucus vesiculosus was collected (January 2014) in Mindelo beach (41°18′38′′N, 8°,43′42′′W), Portugal. The biomass was washed with water to remove salts, epiphytes, and/or microorganisms and dried at 25 °C until reaching a total moisture content of 12–14%. Algae samples were transformed into flakes (1–2 mm) with a knife mill (Retsch SM100, Haan, Germany), packed and stored in airtight bags at the ALGAplus warehouse. The milled algae samples were then washed with a solvent mixture (1 g per 20 mL) of chloroform and methanol (2:1 v/v) under stirring for 20 min, centrifuged at 2500 rpm during 20 min, and dried at 40 °C in a vacuum drying oven.

Gold(III) chloride trihydrate (≥99.9% trace metals basis), dimethyl sulfoxide (DMSO), paraformaldehyde, and Triton X-100 were purchased from Sigma-Aldrich (St. Louis, MO, USA). Phosphate buffer saline (PBS, pH 7.4), Dulbecco's modified Eagle's medium (DMEM), fetal bovine serum (FBS), L-glutamine, penicillin, streptomycin and amphotericin B were supplied by Gibco® (Life Technologies, Grand Island, NY, USA), 3-(4,5-dimethylthiazol-2-yl)-2,5-diphenyltetrazolium bromide (MTT, 98%) was purchased from Sigma-Aldrich, and 4′,6-diamidino-2-phenylindole (DAPI)-containing Vectashield mounting medium was acquired from Vector Labs. Ultra-purified water (Type 1, 18.2 MΩ·cm at 25 °C) was obtained by a Simplicity® Water Purification System (Merck, Darmstadt, Germany). Human osteosarcoma cell line MG-63 was a kind gift by INEB, University of Porto, Portugal. The HepG2 cell line, a liver hepatocellular carcinoma cell line, was obtained from the European Collection of Authenticated Cell Cultures (ECACC, Salisbury, UK) and supplied by Sigma-Aldrich. MNT-1 cells were kindly provided by Doctor Manuela Gaspar (iMed.ULisboa, Lisbon, Portugal).

2.2. Microwave-Assisted Extraction (MAE) of Fucoidan from F. Vesiculosus

The MAE of fucoidan-rich fraction from *F. vesiculosus* followed the methodology described by Rodriguez-Jasso et al. [24]. About 0.4 g of macroalgae sample was suspended in water, with a solid-liquid ratio of 1:25 (w/v). The extraction was performed in a Monowave 300 (Anton Paar, Graz, Austria) equipment, at 172 °C, for 1 min. The samples were cooled on ice and then centrifuged at 4000 rpm during 5 min). The aqueous extract was mixed with a 1% (w/v) $CaCl_2$ aqueous solution, in a solid-liquid ratio of 1:1 (v/v), and maintained overnight at 4 °C, to precipitate alginate. This was separated by filtration, and ethanol was added to the filtrate (1:2, v/v) and maintained at 4 °C for 8 h. The fucoidan-rich fraction (MWF) was obtained after centrifugation (4000 rpm, 5 min) and dried at room temperature.

2.3. Microwave-Assisted Synthesis of Fucoidan-AuNPs

The microwave (MW)-assisted synthesis of fucoidan-AuNPs was carried out on a Monowave 300 equipment (Anton Paar, firmware version 2.0). Total of 145 μL of $HAuCl_4$ $3H_2O$ solution (17.2 mM) was added to microwave vials with 5 mL of the three distinct concentrations of the fucoidan-rich fraction (0.5, 0.1, and 0.05% w/v). The mixtures were heated at 120 °C for 1 min. After the reaction, the obtained fucoidan-AuNPs were centrifuged for 1 h at 15,000 rpm and 4 °C, sonicated and washed three times with ultra-purified water, and finally stored at 4 °C until usage.

2.4. Structural Characterization of the Fucoidan-Enriched Fraction and Fucoidan-AuNPs

The sulfate content of the fucoidan-enriched fraction was determined by elemental analysis. The fucoidan-rich fraction was grounded with a mortar and analyzed (about 2–3 mg) in a Leco TruSpec 630-200-200 CHNS elemental analyzer (LECO Corporation, St. Joseph, MI, USA), in order to assess the carbon (C), hydrogen (H), nitrogen (N), and sulfur (S) contents. The sulfate content of the sample was calculated by the following equation: sulfate content (%) = 3.22 × S (%), where S (%) is the S content, as proposed by several authors [25,26].

Total sugars were determined by the phenol-sulfuric acid method, following the methodology described by DuBois et al. [27] where D(+)-glucose was used as standard. An aqueous solution of the fucoidan-rich fraction (0.05% w/v) was prepared and diluted as necessary. The UV absorbance measurements were performed in a Shimadzu UV-1800 spectrophotometer (Shimadzu Corp., Kyoto, Japan) at $\lambda = 490$ nm. Triplicate measurements were carried out.

Optical spectra of the fucoidan-AuNPs were recorded by a Thermo Scientific Evolution 220 spectrophotometer (Thermo Fisher Scientific, Waltham, MA, USA) using 100 scans min^{-1} with a bandwidth of 2 nm and an integration time of 0.3 s.

The Fourier transform infrared (FTIR) spectra of fucoidan-enriched fraction extracted from *F. vesiculosus* and fucoidan-Au colloidal in the form of KBr pellet were collected by a Mattson 7000 spectrometer using 256 scans at a resolution of 4 cm^{-1} and with a signal gain of 1.

Transmission electron microscopy (TEM) images were obtained by a Hitachi H-9000 (Hitachi High-Technologies Corporation, Tokyo, Japan) operating at 300 kV. Scanning transmission electron microscopy (STEM) images were acquired by a field-emission gun (FEG) SEM Hitachi SU70 microscope operated at 15 kV. Samples for microscopy analysis were prepared by placing a washed colloid drop directly onto a carbon-coated copper grid and then allowing the solvent to evaporate. The average diameter of the NPs was determined by measuring over 100 NPs for each STEM image with the Fiji image processing software.

2.5. *Colloidal Stability of Fucoidan-AuNPs*

The colloidal stability of the fucoidan-AuNPs was evaluated in five different mediums, namely ultra-purified water, HCl (0.01 M, pH 2.1), NaOH (0.01 M, pH 12.0), PBS (pH 7.4), and DMEM. Typically, 100 μL of the fucoidan-AuNPs 0.1% (w/v) was added to the vials already with 2.9 mL of the distinct mediums. The colloidal suspensions were placed under mechanical stirring during 48 h at room temperature, and the UV-Vis spectra at specific times (0, 6, 24, and 48 h) were recorded. All assays were performed in triplicate. After 48 h, the NPs in each medium were centrifuged for 15 min at 15,000 rpm (4 °C), sonicated and washed three times with ultra-purified water and analyzed by STEM as previously described.

2.6. *Cell Culture*

The cell lines were cultured in DMEM supplemented with 10% fetal bovine serum, 2 mM L-glutamine, 100 U mL^{-1} penicillin, 100 μg mL^{-1} streptomycin, and 2.5 μg mL^{-1} amphotericin B. Cells were incubated in a humidified atmosphere of 5% CO_2 at 37 °C. Sub confluent cells were trypsinized with trypsin-EDTA (0.25% trypsin, 1 mM EDTA) when monolayers reached 70% confluence.

2.7. *Cytotoxicity Evaluation*

Cell viability was determined by the colorimetric MTT assay, which measures the formation of purple formazan in viable cells [28]. Cells were seeded in 96-well plates and after 24 h, medium was replaced with fresh medium containing fucoidan extract (0, 0.25, 0.5, 1.0, 2.5, and 5 mg mL^{-1}) or fucoidan-AuNPs (0, 2.5, 5.0, 10.0, 15.0, and 26.0 μg mL^{-1}). Cell viability was measured after 24, 48, and 72 h. After that, 50 μL of MTT reagent (1 mg mL^{-1}) in PBS was added to each well and incubated for 4 h at 37 °C, and 5% CO_2. The medium was then removed, and 150 μL of DMSO was added to each well for crystals solubilization. The optical density of the reduced MTT was measured at 570 nm in a microtiter plate reader (Synergy HT Multi-Mode, BioTeK instruments, Winooski, VT, USA), and the cell metabolic activity (MA, a usual marker for cell viability) was calculated as MA (%) = $((Abs_{sample} - Abs_{DMSO})/(Abs_{control} - Abs_{DMSO})) \times 100$. Three independent assays were performed with at least three technical replicates each and the results compared with the control (incubated with culture medium).

From the MTT results, the concentrations of 5 and 12 μg mL^{-1} of fucoidan-AuNPs were selected for the following assays.

2.8. Uptake Potential by Flow Cytometry

The uptake potential of fucoidan-AuNPs by MG-63 cells was assessed by flow cytometry (FCM), as previously described by Suzuki et al. [29] and Bastos et al. [30]. Briefly, cells were seeded in 6-well plates, and after fucoidan-AuNPs exposure for 24 h, they were trypsinized, collected to FCM tubes, and analyzed by FCM. Two parameters, namely forward scatter (FS), which gives information on the particle's size, and side scatter (SS), which provides information on the complexity of particles, were measured in an Attune® Acoustic Focusing Cytometer (Thermo Scientific, Waltham, MA, USA) equipped with a 488 nm laser. For each sample, 5000–20,000 cells were analyzed at a flow rate of about 300 cells s^{-1}.

For MTT assay and cellular uptake by flow cytometry, the statistical significance between control and exposed cells was performed by one-way ANOVA, followed by Dunnet and Dunn's method (as parametric and non-parametric test, respectively), using Sigma Plot 12.5 software (Systat Software Inc.).

2.9. Dark Field Imaging

MG-63 cells were grown on glass coverslips and cultured in the presence of 12 μg mL^{-1} fucoidan-AuNPs dispersed in culture medium for 24 h. Cells were fixed with a 4% paraformaldehyde in PBS for 10 min, permeabilized with a 0.1% Triton X-100/PBS solution. Following washes with PBS and deionized water, coverslips were mounted onto the glass slides with DAPI-containing Vectashield mounting medium.

The microscopic images were recorded using an Olympus BX51 microscope (50× objective) (Olympus, Tokyo, Japan) equipped with a digital CCD camera (Retiga 4000R, QImaging) used to capture the microphotographs. The dark field images were acquired under white light illumination by replacing the standard microscope condenser by the CytoViva enhanced dark field illumination system (CytoViva, Auburn, AL, USA). For the images under white light illumination and UV irradiation, a DC regulated illuminator (DC-950, Fiber-Lite) and a LED light (LLS-365, Ocean Optics, emission at 365 ± 25 nm) were used, respectively.

The hyperspectral images were recorded with a hyperspectral imaging system from CytoViva, accoupled to the Olympus BX51 microscope, that includes a digital camera (Pixelfly USB, PCO) coupled to a spectrograph (V10E 2/3'', Specim, 30 μm slit, nominal spectral range of 400–1000 nm and nominal spectral resolution of 2.73 nm). Each pixel field-of-view on the hyperspectral images corresponds to 258 × 258 nm^2 on the samples' plane. The hyperspectral scanning is vertical, and each image results from 696 lines, using an exposure time of 3 s for each line. All the hyperspectral data were acquired and analyzed using ENVI 4.8 software, and the spectra were corrected using the tool Calibration and Correction of the ENVI 4.8 software.

3. Results and Discussion

A fucoidan-rich fraction from *F. vesiculosus* was used as the reducing and capping agent in the green synthesis of antitumoral fucoidan-AuNPs for application in cancer therapy (Figure 1). MW technology was used for both the extraction of fucoidan from *F. vesiculosus* and the synthesis of the AuNPs, pursuing the establishment of a timesaving methodology to produce fucoidan-Au nanosystems with controllable size, morphology, and high antitumoral effect, as will be discussed in the following paragraphs.

The extraction yield obtained for the fucoidan-rich fraction was 5.2 ± 0.8% that is in good agreement with the published data for this algae specie under similar MAE conditions [31]. However, the degree of sulfation (5.42%) and total sugars content (13.7 ± 0.6%) are lower than those previously reported [31]. These differences are certainly associated with the natural variability of seaweed biomass [32]. The presence of fucoidan was further confirmed by FTIR analysis. The FTIR spectrum of the fucoidan-rich fraction (Figure 2) displays the typical absorption bands of fucoidans [31], namely a band at around 1260 cm^{-1} (asymmetric stretching of the O=S=O groups of sulfate esters, with the contribution of C–OH, C–C and C–O vibrations). Moreover, a band at 840 cm^{-1} (C–O–S bending associated with the axial substitution at C-4 position) and a shoulder at around 820 cm^{-1} (C–O–S bending associated with substitution at C-2 and C-3 equatorial positions of fucopyranosyl moieties) are also present [33]. The higher intensity of the band at 820 cm^{-1} suggests that the extracted fucoidan is mainly characterized by repeated units of disaccharides primarily composed of fucose residues with -OSO_3^- at C-2 and C-3 positions and with a single -OSO_3^- at C-2 position (Figure 2).

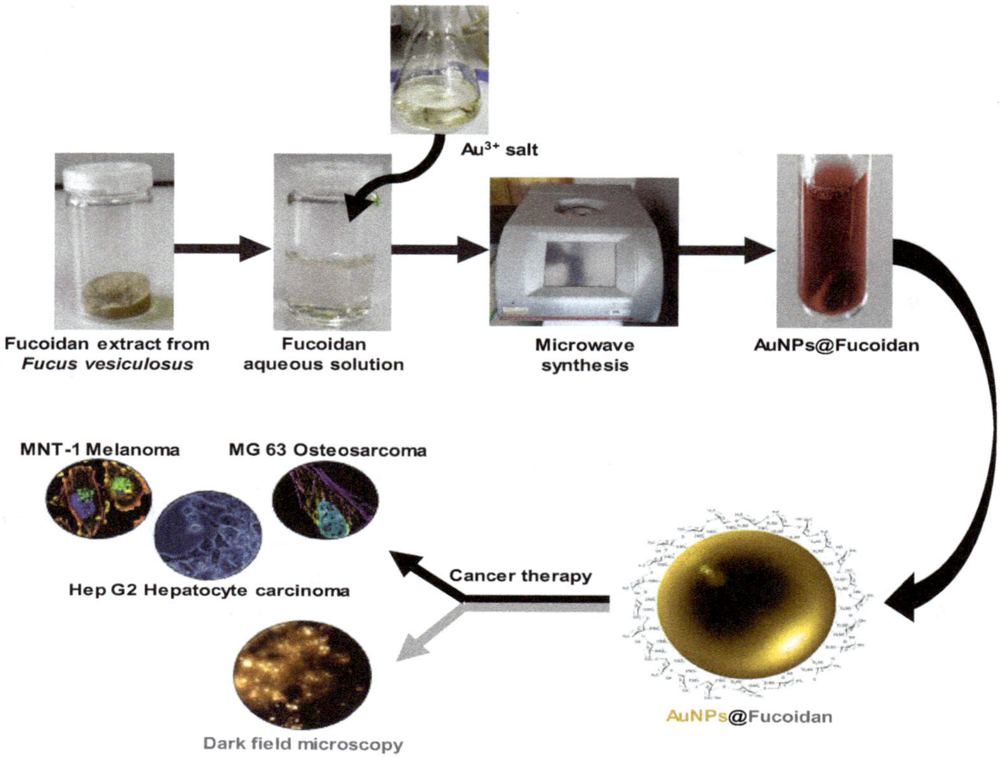

Figure 1. Schematic representation of the microwave irradiated synthesis of fucoidan-AuNPs (molecular structure of fucoidan) with antitumoral activity.

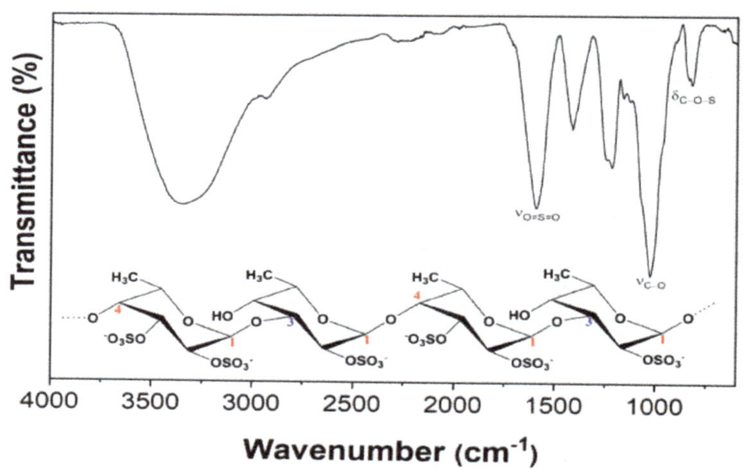

Figure 2. Fourier transform infrared (FTIR) spectrum of the fucoidan rich fraction extracted from *F. vesiculosus* (vibrational modes: ν = stretching, δ = bending).

3.1. *Structural and Morphological Characterization of the Fucoidan-AuNPs*

The MW-assisted synthesis of AuNPs, using the prepared fucoidan-rich fraction, as the reducing and capping agent, was achieved in only 1 min. Three different concentrations of fucoidan were tested, namely 0.05%, 0.1%, and 0.5% (w/v), aiming to produce AuNPs with an appropriated size and excellent colloidal stability. The color change of the Au solutions from yellowish to ruby red, for fucoidan concentrations of 0.1% and 0.5%, and to purplish for the fucoidan solution with 0.05%, is an early confirmation of the formation of the AuNPs (Figure 3A). The formation of a purple color in the case of the lowest fucoidan concentration (0.05% w/v) could be associated with the formation of larger and close particles due to the lower amount of fucoidan present at the surface of the AuNPs [21]. Additionally, the UV-vis spectra of these fucoidan-AuNPs colloids showed the typical surface plasmon resonance at around 520 nm for fucoidan concentrations of 0.1% and 0.5% (w/v). The position of these bands is indicative of the formation of small spherical NPs. The displacement of this band for higher wavelength values (around 550 nm) for the lowest fucoidan concentration (0.05%) is also in line with the formation of larger particles. Similar results were reported by Jang et al. [21] following a conventional solvothermal method but using considerably higher fucoidan concentrations (from 0.5% to 2.5% w/v).

FTIR analysis of the obtained fucoidan-AuNPs (Figure 3B) clearly confirmed the capping role of fucoidan, because of the occurrence of the typical absorption bands of fucoidan [31], as well as a correlation between the content of fucoidan in the surface of the NPs and its amount used in the synthesis.

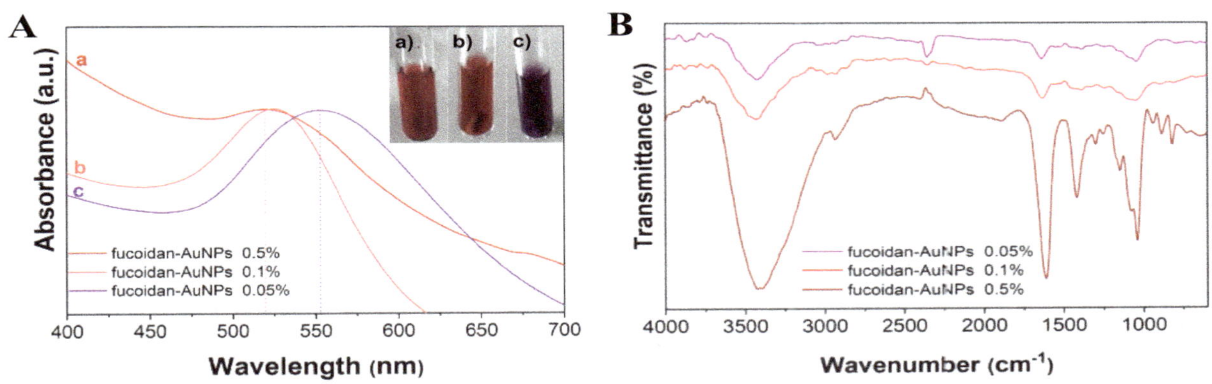

Figure 3. (**A**) UV-Vis and (**B**) FTIR spectra of AuNPs colloids obtained with different concentrations of fucoidan: a) 0.5%, b) 0.1%, and c) 0.05% w/v of fucoidan-rich fraction.

The STEM micrographs provided clear information about the shape and size of the fucoidan-AuNPs (Figure 4). All the obtained fucoidan-AuNPs were monodispersed and spherical, with average sizes of 5.8 ± 0.9 nm, 10.4 ± 1.4 nm, and 13.4 ± 3.0 nm for initial concentrations of fucoidan of 0.5%, 0.1%, and 0.05% (w/v), respectively. STEM images corroborate the increase of the NPs diameter with the decrease of the fucoidan concentration. It is also perceptible that for the fucoidan concentration of 0.05%, AuNPs are closer to each other, but still individualized, because of lower colloidal stability associated with an inferior content of fucoidan on the surface of the NPs. These results are actually remarkable because in conventional solvothermal synthesis, for concentrations equal and lower than 0.5% (w/v) of fucoidan, the AuNPs formed are unstable and aggregate during synthesis leading to large anisotropic particles [21]. These results are a good indication that the MW technology allows the rapid formation of stable spherical AuNPs with controllable size and using lower concentrations of fucoidan when compared with conventional methodologies. Fucoidan-AuNPs sample with an average diameter of 10.4 ± 1.4 nm was selected for the colloidal stability studies and biological evaluations because of the monodispersity and no visual aggregation of the colloidal suspension.

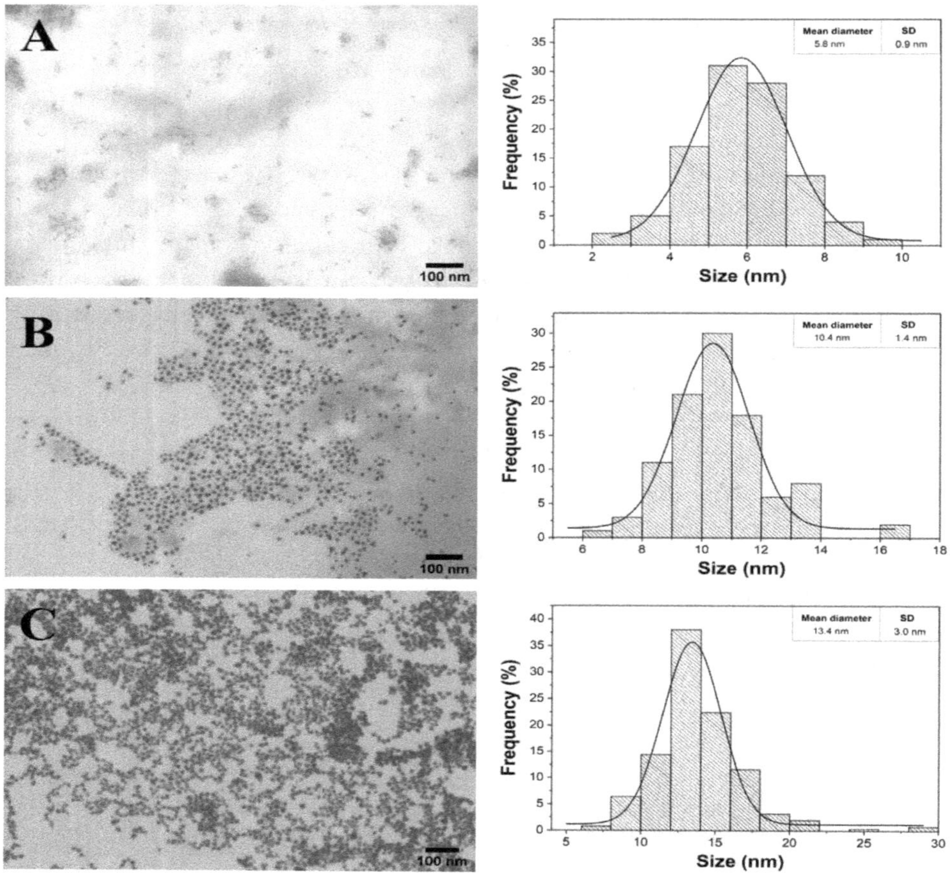

Figure 4. Scanning transmission electron microscopy (STEM) micrographs with the respective histograms of the size distribution of AuNPs colloids obtained with different concentrations of fucoidan: (**A**) 0.5%, (**B**) 0.1%, and (**C**) 0.05% w/v of the fucoidan-rich fraction.

3.2. Colloidal.Stability of the Fucoidan-AuNPs

The colloidal stability of the fucoidan-AuNPs (10.4 ± 1.4 nm average diameter) was investigated under different conditions, namely by using acid (pH 2.1) and basic (pH 12) solutions, PBS (pH 7.4), DMEM (culture medium), and ultra-purified water, at room temperature. The stability profile over 48 h was inspected by UV-Vis and STEM analysis. The observation of the fucoidan-AuNPs colloid over time (Figure S1), and for the different conditions allowed to conclude that, in general, these NPs are considerably stable because no color changes were perceived. Based on the relative absorbance maximum obtained in UV-Vis analysis (Figure 5A), it is evident that these fucoidan-AuNPs are highly stable (more than 90% of maximum absorption) in ultra-pure water, DMEM, and alkaline solution (pH 12), with no significant variations over time. Similar results were previously reported for fucoidan-AuNPs with 15–80 nm [21] and fucoidan-Au nanorods [34]. However, in PBS (pH 7.4) and acidic (pH 2.1) solutions, the maximum absorptions slightly decrease over time, reaching about 70% and 80%, respectively. However, the analysis of the colloids by UV-Vis only gives a rough indication of the stability of the NPs because the decrease in the maximum absorption could be associated with different causes. Possible reasons are the change of the refractive index of the surrounding medium and/or the distance between the AuNPs that causes changes in the $\lambda_{máx}$ of absorption (525 nm), as well as in the correspondent absorbance values [35].

To have a deep insight into the effect of the different studied conditions in the stability of the obtained colloids, in particular on their size and morphology, STEM analysis of the AuNPs after 48 h of incubation in the solutions mentioned above was also carried out. STEM micrographs (Figure 5B) confirmed that their morphology and size was not affected, but in acidic conditions, the NPs are much closer. These results indicate that under acidic conditions the fucoidan capping layer is weakened, certainly due to the protonation of sulfate groups and partial detachment from the surface of the AuNPs [36], leading to slightly less stable colloids.

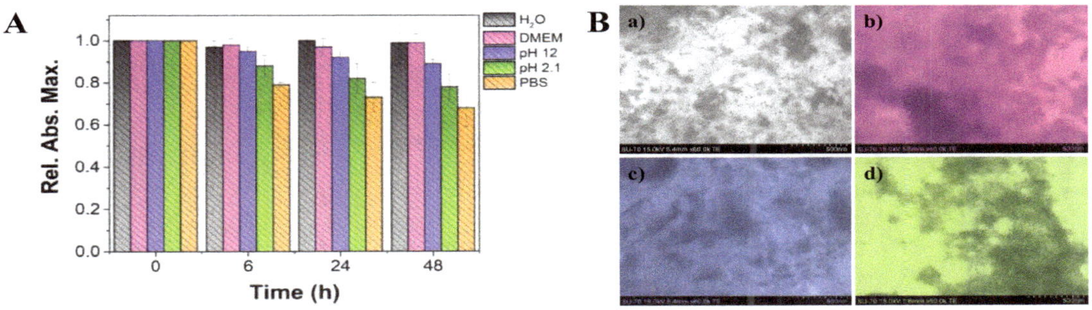

Figure 5. (**A**) Colloidal stability assay of fucoidan-AuNPs 0.1% (w/v) up to 48 h in distinct mediums: acid solution (pH 2.1), basic solution (pH 12), PBS, DMEM (culture medium), and ultra-purified water. (**B**) STEM images of fucoidan-AuNPs after 48 h immersed in the distinct mediums: (**a**) ultra-purified water, (b) DMEM, (c) NaOH (pH 12), and (d) HCl (pH 2.1). The color of micrographs (**b**), (**c**), and (**d**) were changed for visual guidance in order to match the color of the respective medium displayed in (A).

3.3. Cytotoxicity Assays of Fucoidan-AuNPs and Cellular Uptake by Flow Cytometry

In this study, the in vitro cytotoxicity of the fucoidan-enriched extract and fucoidan-AuNPs (10.4 ± 1.4 nm) was investigated against MNT-1 (pigmented human melanoma cells), HepG2 (human hepatocyte carcinoma), and MG-63 (human osteosarcoma) cell lines for 24, 48, and 72 h at concentrations ranging from 0 to 5 mg mL^{-1} (Figure 6). The fucoidan extract obtained in this study is not cytotoxic against the three cell lines tested, in the concentration range of 0.25–2.5 mg mL^{-1}, with cell viabilities higher than 90% in most cases. However, for a concentration of 5 mg mL^{-1}, a significant reduction of cell viability (up to 60%) was observed, particularly for 48 and 72 h of exposure. The antitumoral activity of fucoidan (and fucoidan-enriched extracts) is well documented, as well as its dependence on the source of fucoidan [15]. For instance, Manivasagan et al. [19] reported that fucoidan from *F. vesiculosus* inhibits the proliferation of human breast cancer cells with an inhibitory concentration of 35 μg mL^{-1} and Jang et al. [21] described cell viabilities of around 80% for cancer cells (HSC3) treated with 100 mg mL^{-1} of a commercial fucoidan.

The cytotoxic effect of the fucoidan-AuNPs was investigated for concentrations in the range of 2.5–26 μg mL^{-1}. In general, the NPs showed a dose-depend decrease in cell viability but with noticeable differences for distinct cell lines and exposure times. For the HepG2 cell line, it was observed a reduction of cell viability with the concentration of fucoidan-AuNPs reaching about 60% cell viability for 26 μg mL^{-1} of NPs. In this case, no time dependency effect was perceived for the three exposure times investigated. For the other cell lines, it was also observed a decrease in cell viability with the concentration of NPs, but with most pronounced reductions for 48 and 72 h of exposure. For example, for a concentration of 26 μg mL^{-1} of NPs and 72 h of exposure, cell viabilities of 32 and 10% were observed for MNT-1 and MG-63 cell lines, respectively. These results demonstrate the more significant antitumoral effect of fucoidan when combined with the AuNPs because considerably higher cell viability reduction was obtained for much lower concentrations of fucoidan-AuNPs when compared with the fucoidan-enriched extract (around 60% cell viability for 5 mg mL^{-1} of fucoidan). This behavior could be associated with the small size and high surface area of the AuNPs that result in extraordinary surface concentrations of fucoidan and higher interaction with the cells. Jang et al. [21] also reported a higher cell viability reduction for fucoidan-AuNPs comparatively with fucoidan when using the same concentration of 100 mg mL^{-1}.

The cellular uptake of fucoidan-AuNPs was only tested for the MG-63 cells, given the higher antitumoral activity of these nanosystems toward this cell line. According to Figure 7, both concentrations (5 and 12 μg mL^{-1}) of fucoidan-AuNPs induce an increase in side scatter (SS) intensity without change of forward scatter (FS) intensity of MG-63 cells, which means that particles are internalized by the cells.

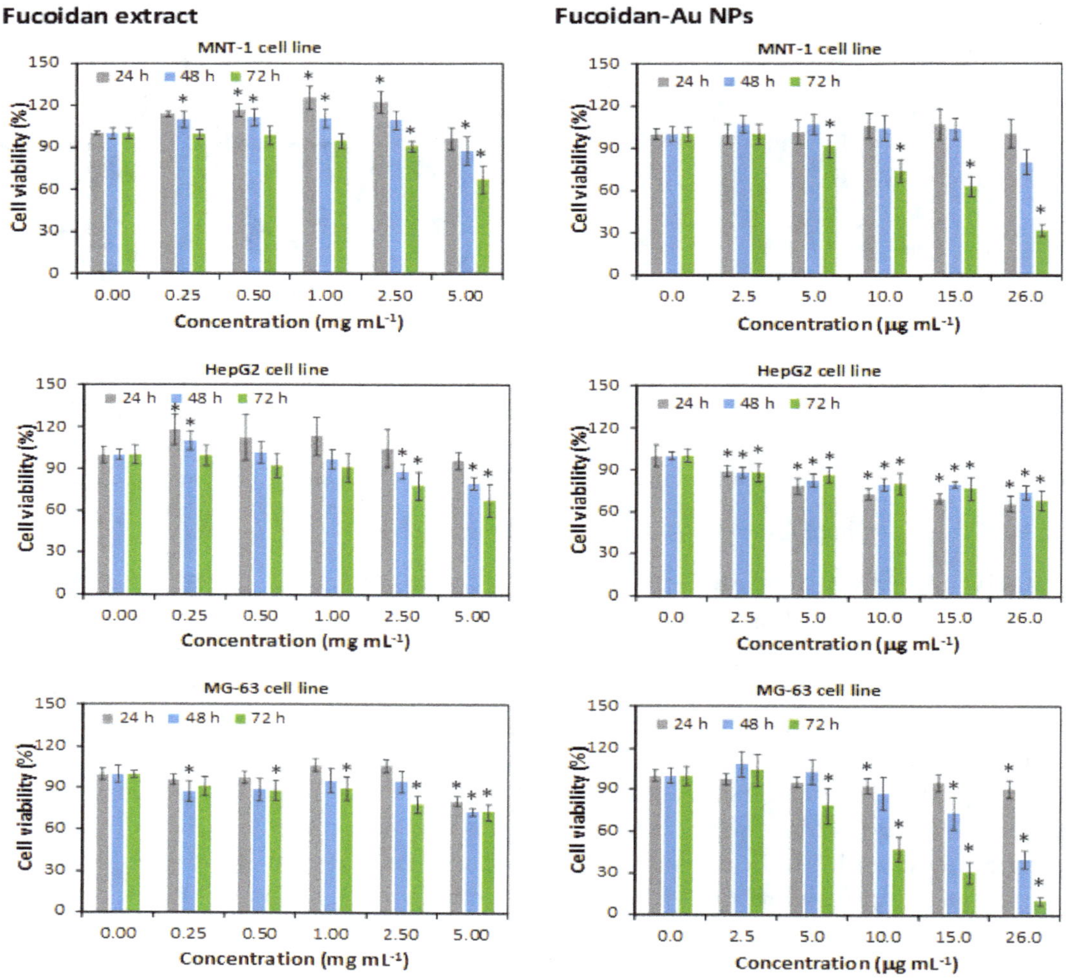

Figure 6. Viability measured by 3-(4,5-dimethylthiazol-2-yl)-2,5-diphenyltetrazolium bromide (MTT) assay after 24, 48, and 72 h exposure to fucoidan extract (left) and fucoidan-AuNPs (right). Values are the mean of nine replicates, and the error bars represent the standard deviation; the asterisk (*) denotes statistically significant differences to the control ($p < 0.05$).

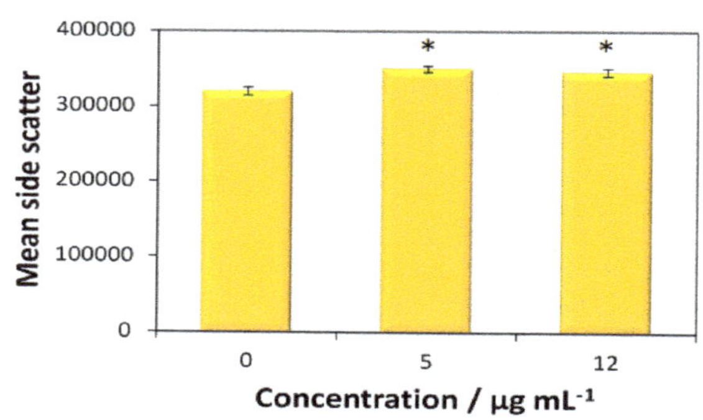

Figure 7. Uptake of fucoidan-AuNPs by MG-63 cells. The uptake was assessed by the side scattered light through flow cytometry after 24 h exposure to 5 and 12 μg mL^{-1} of fucoidan-AuNPs. Values are the mean of three replicates, and the error bars represent the standard deviation; the asterisk (*) denotes statistically significant differences to the control ($p < 0.05$).

3.4. Dark Field Imaging of MG-63 Cells Incubated with Fucoidan-AuNPs

Dark field microscopy of the MG-63 cells and those incubated with fucoidan-AuNPs are shown in Figure 8A,B, respectively. The darker regions in the dark field images are assigned to the cell's nucleus with a diameter of around 20 μm. The cells' nucleus identification is unequivocally confirmed, taking advantage of the fact that the cells were marked with a fluorescent stain (DAPI) that binds specifically to the regions of the nucleus. Thus, under UV irradiation, blue areas are discerned (Figure 8C,D), assigned to the emission spectra around 460 nm of the DAPI (Figure S2). The overlap between the dark field images and those acquired under UV the dark areas overlap that revealing blue emission (Figure 8E). This shows that dark field imaging under white light can be used to identify the nucleus of the cells, without the need to use a fluorescent stain.

Also, the dark field images of the MG-63 cells incubated with fucoidan-AuNPs also shows bright spots with diameter values between 1 and 10 μm. The light scattering from those regions was analyzed by hyperspectral microscopy. Figure 8G,H compare the hyperspectral images of the MG-63 cells and of those incubated with fucoidan-AuNPs, revealing that the bright spots in the images of the MG-63 cells incubated with fucoidan-AuNPs are characterized by a broad spectrum. In fact, it displays a low-relative intensity band in the same region as that found for the absorption of the Au-particles (Figure 3) and a more intense one in the red spectral region, that results from the light scattered by the fucoidan-AuNPs indicating the presence of fucoidan-AuNPs aggregates (Figure 8F, Figures S3 and S4). The larger dimension of those aggregates, when compared to STEM data (Figure 4), is due to the spatial resolution of the optical image, and we also note that the contribution of the guidance of the scattered photons from the particles for the larger bright spots cannot be excluded [37]. We note that some of those bright spots (marked with arrows in Figure 8H) are localized in the same coordinates of the plane in which was possible to detect the cells, suggesting the incorporation of the fucoidan-AuNPs in the MG-63 cells, that it is in line with the flow cytometry results.

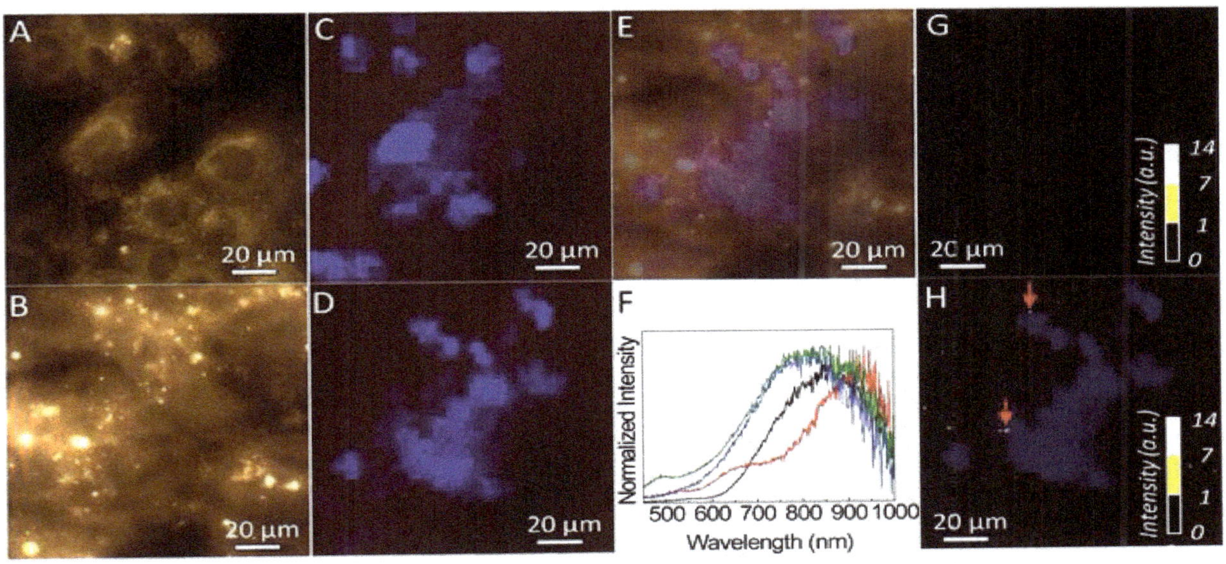

Figure 8. Optical images in dark field transmission mode under white light, of MG-63 cells, incubated (**A**) without and (**B**) with fucoidan-AuNPs. Optical images, in brightfield field reflectance mode under UV irradiation, of MG-63 cells incubated (**C**) without and (**D**) with fucoidan-AuNPs. (**E**) Show the overlay of (B) and (D). Spectra measured in several single pixels of the bright spots shown in the hyperspectral image shown in the (**F**). (**G**) and (**H**) show the hyperspectral images measured for the same sample and illumination conditions of (A) and (B), respectively. The color scale is based on the intensity of the spectra of each pixel at 750 nm. In (H), the hyperspectral image is superimposed with (D).

4. Conclusions

The one-minute microwave-assisted synthesis of fucoidan-coated gold nanoparticles (AuNPs) with controllable, high stability, and antitumoral activity is the first and foremost contribution of the present study, where fucoidan-AuNPs were synthesized by using a fucoidan-enriched fraction extracted from *F. vesiculosus*, as simultaneous reducing and capping agent. The resulting monodispersed and spherical fucoidan-AuNPs present very small diameters, namely 5.8 ± 0.9 nm, 10.4 ± 1.4 nm, and 13.4 ± 3.0 nm that depend on the initial concentrations of fucoidan: 0.5%, 0.1%, and 0.05% (w/v), respectively, together with excellent colloidal stability in acidic and basic solutions, ultra-purified water and culture media. The second innovative input of this study lies in the antitumoral activity of the fucoidan-AuNPs against three human tumor cell lines, namely MNT-1 (pigmented human melanoma), HepG2 (human hepatocyte carcinoma), and MG-63 (human osteosarcoma) cells. Moreover, the use of flow cytometry in combination with dark field imaging confirmed the cellular uptake of fucoidan-AuNPs by the MG-63 cell line, which demonstrates the antitumoral activity of these nanomaterials, and thus their potential for cancer therapy.

Supplementary Materials:
Figure S1: Digital photographs of fucoidan-AuNPs colloid over time in the different media. Figure S2: Hyperspectral images of MG-63 cells incubated (**A**) without and (**B**) with fucoidan-AuNPs, measured for the same sample and illumination conditions of Figure 8C,D of the manuscript. The color scale is based on the intensity of the spectra of each pixel at 460 nm. (**C**) Spectra measured in collection areas of 20 × 20 pixels of (A) (red line) and (B) (blue line). Figure S3: Spectra measured in several single pixels of the hyperspectral image measured for the same sample and illumination conditions of the Figure 8A of the manuscript, before (**A**) and after (**C**) correction. (**B**) Spectrum measured in an area of 20 × 20 pixels, of the Figure 8A of the manuscript used to perform the correction. The inset in (A) shows the normalized spectra for better comparison. Figure S4: Spectra measured in several single pixels of the hyperspectral image shown in Figure 8B of the manuscript, (**A**) before and (**C**) after correction using the spectrum shown in Figure S3B. The normalized spectra before correction are shown in (**B**) for better comparison.

Author Contributions: Conceptualization, R.J.B.P., S.A.O.S., and C.S.R.F.; investigation, R.J.B.P., D.B., C.V., A.M.P.B., A.C.M., and F.C.; resources, R.A.S.F., H.O., M.H.A., S.A.O.S., and C.S.R.F.; writing—original draft preparation, C.S.R.F.; writing—review and editing, R.J.B.P., D.B., C.V., A.M.P.B., R.A.S.F., A.C.M., F.C., H.O., M.H.A., S.A.O.S., and C.S.R.F.; supervision, C.S.R.F.; funding acquisition, R.A.S.F., H.O., S.A.O.S., and C.S.R.F. All authors have read and agreed to the published version of the manuscript.

References

1. World Health Organization. Available online: https://www.who.int/cancer/en/ (accessed on 6 June 2019).
2. Roy Chowdhury, M.; Schumann, C.; Bhakta-Guha, D.; Guha, G. Cancer nanotheranostics: Strategies, promises and impediments. *Biomed. Pharmacother.* **2016**, *84*, 291–304. [CrossRef] [PubMed]
3. Mitchell, M.J.; Jain, R.K.; Langer, R. Engineering and physical sciences in oncology: Challenges and opportunities. *Nat. Rev. Cancer* **2017**, *17*, 659–675. [CrossRef] [PubMed]
4. Chen, H.; Zhang, W.; Zhu, G.; Xie, J.; Chen, X. Rethinking cancer nanotheranostics. *Nat. Rev. Mater.* **2017**, *2*, 17024. [CrossRef] [PubMed]
5. Aftab, S.; Shah, A.; Nadhman, A.; Kurbanoglu, S.; Aysıl Ozkan, S.; Dionysiou, D.D.; Shukla, S.S.; Aminabhavi, T.M. Nanomedicine: An effective tool in cancer therapy. *Int. J. Pharm.* **2018**, *540*, 132–149. [CrossRef]
6. Silva, C.; Pinho, J.; Lopes, J.; Almeida, A.; Gaspar, M.; Reis, C. Current Trends in Cancer Nanotheranostics: Metallic, Polymeric, and Lipid-Based Systems. *Pharmaceutics* **2019**, *11*, 22. [CrossRef]
7. Arvizo, R.R.; Bhattacharyya, S.; Kudgus, R.A.; Giri, K.; Bhattacharya, R.; Mukherjee, P. Intrinsic therapeutic applications of noble metal nanoparticles: Past, present and future. *Chem. Soc. Rev.* **2012**, *41*, 2943–2970. [CrossRef]
8. Menon, J.U.; Jadeja, P.; Tambe, P.; Vu, K.; Yuan, B.; Nguyen, K.T. Nanomaterials for Photo-Based Diagnostic and Therapeutic Applications. *Theranostics* **2013**, *3*, 152–166. [CrossRef]
9. Li, W.; Chen, X. Gold nanoparticles for photoacoustic imaging. *Nanomedicine* **2015**, *10*, 299–320. [CrossRef]
10. Weber, J.; Beard, P.C.; Bohndiek, S.E. Contrast agents for molecular photoacoustic imaging. *Nat. Methods* **2016**, *13*, 639–650. [CrossRef] [PubMed]
11. Abadeer, N.S.; Murphy, C.J. Recent Progress in Cancer Thermal Therapy Using Gold Nanoparticles. *J. Phys. Chem. C* **2016**, *120*, 4691–4716. [CrossRef]

12. Her, S.; Jaffray, D.A.; Allen, C. Gold nanoparticles for applications in cancer radiotherapy: Mechanisms and recent advancements. *Adv. Drug Deliv. Rev.* **2017**, *109*, 84–101. [CrossRef] [PubMed]
13. Nam, J.; Won, N.; Bang, J.; Jin, H.; Park, J.; Jung, S.; Jung, S.; Park, Y.; Kim, S. Surface engineering of inorganic nanoparticles for imaging and therapy. *Adv. Drug Deliv. Rev.* **2013**, *65*, 622–648. [CrossRef] [PubMed]
14. Yadav, P.; Singh, S.P.; Rengan, A.K.; Shanavas, A.; Srivastava, R. Gold laced bio-macromolecules for theranostic application. *Int. J. Biol. Macromol.* **2018**, *110*, 39–53. [CrossRef] [PubMed]
15. Wu, L.; Sun, J.; Su, X.; Yu, Q.; Yu, Q.; Zhang, P. A review about the development of fucoidan in antitumor activity: Progress and challenges. *Carbohydr. Polym.* **2016**, *154*, 96–111. [CrossRef]
16. Fitton, J.; Stringer, D.; Karpiniec, S. Therapies from Fucoidan: An Update. *Mar. Drugs* **2015**, *13*, 5920–5946. [CrossRef]
17. Manivasagan, P.; Oh, J. Marine polysaccharide-based nanomaterials as a novel source of nanobiotechnological applications. *Int. J. Biol. Macromol.* **2016**, *82*, 315–327. [CrossRef]
18. Tengdelius, M.; Gurav, D.; Konradsson, P.; Påhlsson, P.; Griffith, M.; Oommen, O.P. Synthesis and anticancer properties of fucoidan-mimetic glycopolymer coated gold nanoparticles. *Chem. Commun.* **2015**, *51*, 8532–8535. [CrossRef]
19. Manivasagan, P.; Bharathiraja, S.; Bui, N.Q.; Jang, B.; Oh, Y.O.; Lim, I.G.; Oh, J. Doxorubicin-loaded fucoidan capped gold nanoparticles for drug delivery and photoacoustic imaging. *Int. J. Biol. Macromol.* **2016**, *91*, 578–588. [CrossRef]
20. Kim, H.; Nguyen, V.P.; Manivasagan, P.; Jung, M.J.; Kim, S.W.; Oh, J.; Kang, H.W. Doxorubicin-fucoidan-gold nanoparticles composite for dual-chemo-photothermal treatment on eye tumors. *Oncotarget* **2017**, *8*, 113719–113733. [CrossRef]
21. Jang, H.; Kang, K.; El-Sayed, M.A. Facile size-controlled synthesis of fucoidan-coated gold nanoparticles and cooperative anticancer effect with doxorubicin. *J. Mater. Chem. B* **2017**, *5*, 6147–6153. [CrossRef]
22. Arshi, N.; Ahmed, F.; Kumar, S.; Anwar, M.; Lu, J.; Heun Koo, B.; Gyu Lee, C. Microwave assisted synthesis of gold nanoparticles and their antibacterial activity against Escherichia coli (*E. coli*). *Curr. Appl. Phys.* **2011**, *11*, S360–S363. [CrossRef]
23. Bayazit, M.K.; Yue, J.; Cao, E.; Gavriilidis, A.; Tang, J. Controllable Synthesis of Gold Nanoparticles in Aqueous Solution by Microwave Assisted Flow Chemistry. *ACS Sustain. Chem. Eng.* **2016**, *4*, 6435–6442. [CrossRef]
24. Rodriguez-Jasso, R.M.; Mussatto, S.; Pastrana, L.; Aguilar, C.; Teixeira, J. Chemical composition and antioxidant activity of sulphated polysaccharides extracted from Fucus vesiculosus using different hydrothermal processes. *Chem. Pap.* **2014**, *68*, 203–209. [CrossRef]
25. Rioux, L.-E.; Turgeon, S.L.; Beaulieu, M. Rheological characterisation of polysaccharides extracted from brown seaweeds. *J. Sci. Food Agric.* **2007**, *87*, 1630–1638. [CrossRef]
26. Roger, O.; Kervarec, N.; Ratiskol, J.; Colliec-Jouault, S.; Chevolot, L. Structural studies of the main exopolysaccharide produced by the deep-sea bacterium Alteromonas infernus. *Carbohydr. Res.* **2004**, *339*, 2371–2380. [CrossRef] [PubMed]
27. DuBois, M.; Gilles, K.A.; Hamilton, J.K.; Rebers, P.A.; Smith, F. Colorimetric Method for Determination of Sugars and Related Substances. *Anal. Chem.* **1956**, *28*, 350–356. [CrossRef]
28. Twentyman, P.; Luscombe, M. A study of some variables in a tetrazolium dye (MTT) based assay for cell growth and chemosensitivity. *Br. J. Cancer* **1987**, *56*, 279–285. [CrossRef]
29. Suzuki, H.; Toyooka, T.; Ibuki, Y. Simple and Easy Method to Evaluate Uptake Potential of Nanoparticles in Mammalian Cells Using a Flow Cytometric Light Scatter Analysis. *Environ. Sci. Technol.* **2007**, *41*, 3018–3024. [CrossRef]
30. Bastos, V.; Ferreira-de-Oliveira, J.M.P.; Carrola, J.; Daniel-da-Silva, A.L.; Duarte, I.F.; Santos, C.; Oliveira, H. Coating independent cytotoxicity of citrate- and PEG-coated silver nanoparticles on a human hepatoma cell line. *J. Environ. Sci.* **2017**, *51*, 191–201. [CrossRef]
31. Rodriguez-Jasso, R.M.; Mussatto, S.I.; Pastrana, L.; Aguilar, C.N.; Teixeira, J.A. Microwave-assisted extraction of sulfated polysaccharides (fucoidan) from brown seaweed. *Carbohydr. Polym.* **2011**, *86*, 1137–1144. [CrossRef]
32. Ale, M.T.; Mikkelsen, J.D.; Meyer, A.S. Important determinants for fucoidan bioactivity: A critical review of structure-function relations and extraction methods for fucose-containing sulfated polysaccharides from brown seaweeds. *Mar. Drugs* **2011**, *9*, 2106–2130. [CrossRef] [PubMed]

33. Pielesz, A.; Biniaś, W. Cellulose acetate membrane electrophoresis and FTIR spectroscopy as methods of identifying a fucoidan in Fucus vesiculosus Linnaeus. *Carbohydr. Res.* **2010**, *345*, 2676–2682. [CrossRef] [PubMed]
34. Manivasagan, P.; Bharathiraja, S.; Santha Moorthy, M.; Oh, Y.O.; Song, K.; Seo, H.; Oh, J. Anti-EGFR Antibody Conjugation of Fucoidan-Coated Gold Nanorods as Novel Photothermal Ablation Agents for Cancer Therapy. *ACS Appl. Mater. Interfaces* **2017**, *9*, 14633–14646. [CrossRef] [PubMed]
35. Liz-Marzán, L.M. Nanometals. *Mater. Today* **2004**, *7*, 26–31. [CrossRef]
36. Santha Moorthy, M.; Subramanian, B.; Panchanathan, M.; Mondal, S.; Kim, H.; Lee, K.D.; Oh, J. Fucoidan- coated core–shell magnetic mesoporous silica nanoparticles for chemotherapy and magnetic hyperthermia-based thermal therapy applications. *New J. Chem.* **2017**, *41*, 15334–15346. [CrossRef]
37. Gonell, F.; Botas, A.M.P.; Brites, C.D.S.; Amorós, P.; Carlos, L.D.; Julián-López, B.; Ferreira, R.A.S. Aggregation-induced heterogeneities in the emission of upconverting nanoparticles at the submicron scale unfolded by hyperspectral microscopy. *Nanoscale Adv.* **2019**, *1*, 2537–2545. [CrossRef]

14

Morphological Study of Chitosan/Poly (Vinyl Alcohol) Nanofibers Prepared by Electrospinning, Collected on Reticulated Vitreous Carbon

Diana Isela Sanchez-Alvarado [1], Javier Guzmán-Pantoja [2], Ulises Páramo-García [1,*], Alfredo Maciel-Cerda [3], Reinaldo David Martínez-Orozco [1] and Ricardo Vera-Graziano [3]

1 Tecnológico Nacional de México/Instituto Tecnológico de Cd. Madero, Centro de Investigación en Petroquímica, Prol. Bahía de Aldhair y Av. De las Bahías, Parque de la Pequeña y Mediana Industria, Altamira, Tamaulipas 89600, Mexico; diana.sanchez@itsna.edu.mx (D.I.S.-A.); rd.martinez.orozco@gmail.com (R.D.M.-O.)

2 Instituto Mexicano del Petróleo, Gerencia de Transformación de Hidrocarburos, Eje Central Lázaro Cárdenas No. 152, Col. San Bartolo Atepehuacán, 07730 Ciudad de México, Mexico; jguzmanp@imp.mx

3 Instituto de Investigaciones en Materiales, Universidad Nacional Autónoma de México, Circuito Exterior S/N, Circuito de la Investigación Científica, Ciudad Universitaria, 04510 Ciudad de México, Mexico; macielal@unam.mx (A.M.-C.); graziano@unam.mx (R.V.-G.)

* Correspondence: uparamo@itcm.edu.mx

Abstract: In this work, chitosan (CS)/poly (vinyl alcohol) (PVA) nanofibers were prepared by using the electrospinning method. Different CS concentrations (0.5, 1, 2, and 3 wt %), maintaining the PVA concentration at 8 wt %, were tested. Likewise, the studied electrospinning experimental parameters were: syringe/collector distance, solution flow and voltage. Subsequently, the electrospun fibers were collected on a reticulated vitreous carbon (RVC) support for 0.25, 0.5, 1, 1.5, and 2 h. The morphology and diameter of the CS/PVA nanofibers were characterized by scanning electron microscopy (SEM), finding diameters in the order of 132 and 212 nm; the best results (uniform fibers) were obtained from the solution with 2 wt % of chitosan and a voltage, distance, and flow rate of 16 kV, 20 cm, and 0.13 mL/h, respectively. Afterwards, a treatment with an ethanolic NaOH solution was performed, observing a change in the fiber morphology and a diameter decrease (117 ± 9 nm).

Keywords: chitosan; PVA; nanofibers; electrospinning

1. Introduction

The electrospinning technique is an efficient method for the manufacture of micro- and nanofibers; it is also an efficient and simple method to produce homogenous fibers for various applications. This technique relies on multiple operating parameters such as applied voltage, collector type, temperature, relative humidity, flow rate, etc., as well as on intrinsic solution parameters like molecular weight, polymeric concentration, viscosity, surface tension, electrical conductivity, among others. The electrospinning process uses electrostatic forces to produce polymeric fibers. When high voltage is applied, polymer drops are subjected to instability, forming fibers when the surface tension of the polymer solution drop at the tip of the needle is overcome. A series of parameters must be optimized in order to produce fibers instead of drops [1–3].

Chitosan (CS) is a biopolymer derived from chitin, often obtained from the exoskeleton of crustaceans. It has very useful properties such as biocompatibility, biodegradability, and antimicrobial activity; however, the production of fibers of this polymer is very difficult due to its insolubility in organic solvents, its ionic character in solution and the formation of three-dimensional networks by strong hydrogen bonds [4–6].

In this sense, Li and Hsieh (2006) found that by combining an anionic polymer such as poly (vinyl alcohol) (PVA) and chitosan, which is a cationic polymer, the formation of fibers was improved, which is in contrast with the combination of polymers having the same charge [7]. PVA is a biodegradable polymer that may be used to immobilize chitosan [8]. Moreover, PVA can reduce the crystallinity of chitosan [9]. The functional groups of chitosan make hydrogen bonds with PVA which lead to form defect-free nanofibers [10].

Although CS and PVA can improve the production of fibers, the right proportion to have a stable solution has to be established. In this sense, several works studying different CS/PVA ratios have been reported [11–13]. To do so, the calculation of the thermodynamic interaction parameter (X) by using the Flory–Huggins theory is necessary. This parameter represents the energy associated with the blending of polymers; depending on its value, the interaction can be classified as favorable or unfavorable.In our case, we did not consider its calculation because the CS/PVA ratio selected to carry out the electrospinning experiments produced a homogeneous solution, which we could witness in the lab by observing a miscible behavior.

Paipitak et al. [14] used the electrospinning method to produce CS/PVA nanofibers, with a glass plate as collector, and study the effect of CS concentration on the formation of nanofibers. The morphology and diameter of the CS/PVA nanocomposite fibers were analyzed by SEM, finding a fiber diameter of 100 nm. The best CS solution concentration for fiber formation was 2 wt % due to the maximum fiber yield.

According to the points described above and in order to improve the efficiency of the electrospinning method and morphology of the obtained CS/PVA nanofibers, in the present work, reticulated vitreous carbon (RVC) was used as a collector due to its high electrical conductivity and minimal reactivity. RVC is currently used in electrochemical systems such as battery and fuel cells, metal ion removal, electroanalytical analysis, sensors and synthesis of organic compounds. As it can be seen, these applications are related to the promotion of turbulence aimed at increasing the mass transfer phenomena in reactors. To the best of our knowledge, in the present research work, RVC has been used for the first time as a fiber collector in the electrospinning process. The ultimate goal, after fundamental studies, is to produce CS/PVA nanofibers, establishing the best parameters of the electrospinning technique (voltage, drop-collector distance and volumetric flow), on a highly porous support to generate turbulence and adequate contact between the materials and a solution. The obtained results encouraged us to consider the experimental procedure featured in this work for its application in the reduction of sulfur compounds in FCC (Fluid Catalytic Cracking) gasoline, where the chitosan fibers could work as adsorbents of metals such as nickel and molybdenum in an oxidative desulfurization process. The results of this study will be part of a forthcoming paper.

2. Results and Discussions

2.1. Electrospinning Method: Experimental Parameters

Preliminary electrospinning processes were carried out using different operation parameters. Not all the analyzed electrospinning parameters showed good results since in some cases, the fibers were not uniform with large nodules incorporated into them or simply no fibers were observed.

The operation parameters used to produce fibers are shown in Table 1. In order to establish the best CS/PVA proportions, four of them were experimentally tested (0.5 wt % CS/8 wt % PVA, 1 wt % CS/8 wt % PVA, 2 wt % CS/8 wt % PVA, and 3 wt % CS/8 wt % PVA). As the most uniform nanofibers were obtained with 2 wt % CS/8 wt % PVA, the other three concentrations were discarded (data not shown). Then, the optimal electrospinning conditions were obtained with a 2 wt % CS/8 wt % PVA solution, and a flow rate, distance, and voltage of 0.13 mL/h, 20 cm, and 16 kV, respectively.

Table 1. Experimentally evaluated parameters using a 2 wt % CS/8 wt % PVA solution.

2 wt. CS/8 wt % PVA Solution	Image Id.	Flow Rate mL/h	Distance cm	Voltage kV	Observations *
Flow rate	a	0.12	15	14	Non-uniform fibers
	b	0.13	15	14	Non-uniform fibers with nodules
Voltage	c	0.13	15	15	Non-uniform fibers with nodules
	d	0.13	15	14	Non-uniform fibers
Tip–target distance	e	0.12	20	14	Non-uniform fibers
	f	0.12	25	14	Non-uniform fibers
	g	0.13	20	16	Non-uniform fibers
	h	0.12	25	16	Non-uniform fibers
	i	0.13	20	14	Non-uniform fibers
	j	0.13	25	14	Non-uniform fibers
	k	0.13	20	16	Uniform fibers
	l	0.13	25	16	Non-uniform fibers with nodules

* Uniformity criteria were the fiber abundance and lack of nodules observed in the micrographs shown in Figure 1. CS: chitosan; PVA: poly (vinyl alcohol).

Figure 1 shows the CS/PVA electrospun sample using a 2 wt % CS/8 wt % PVA solution, with the parameters featured in Table 1, varying flow rate, distance, and voltage during the performance of the electrospinning technique, where it is observed that the fibers are non-uniform: some fibers are thinner than others and some nodules can be seen, unlike what is observed under the optimal conditions mentioned above. These experiments were performed on a glass surface.

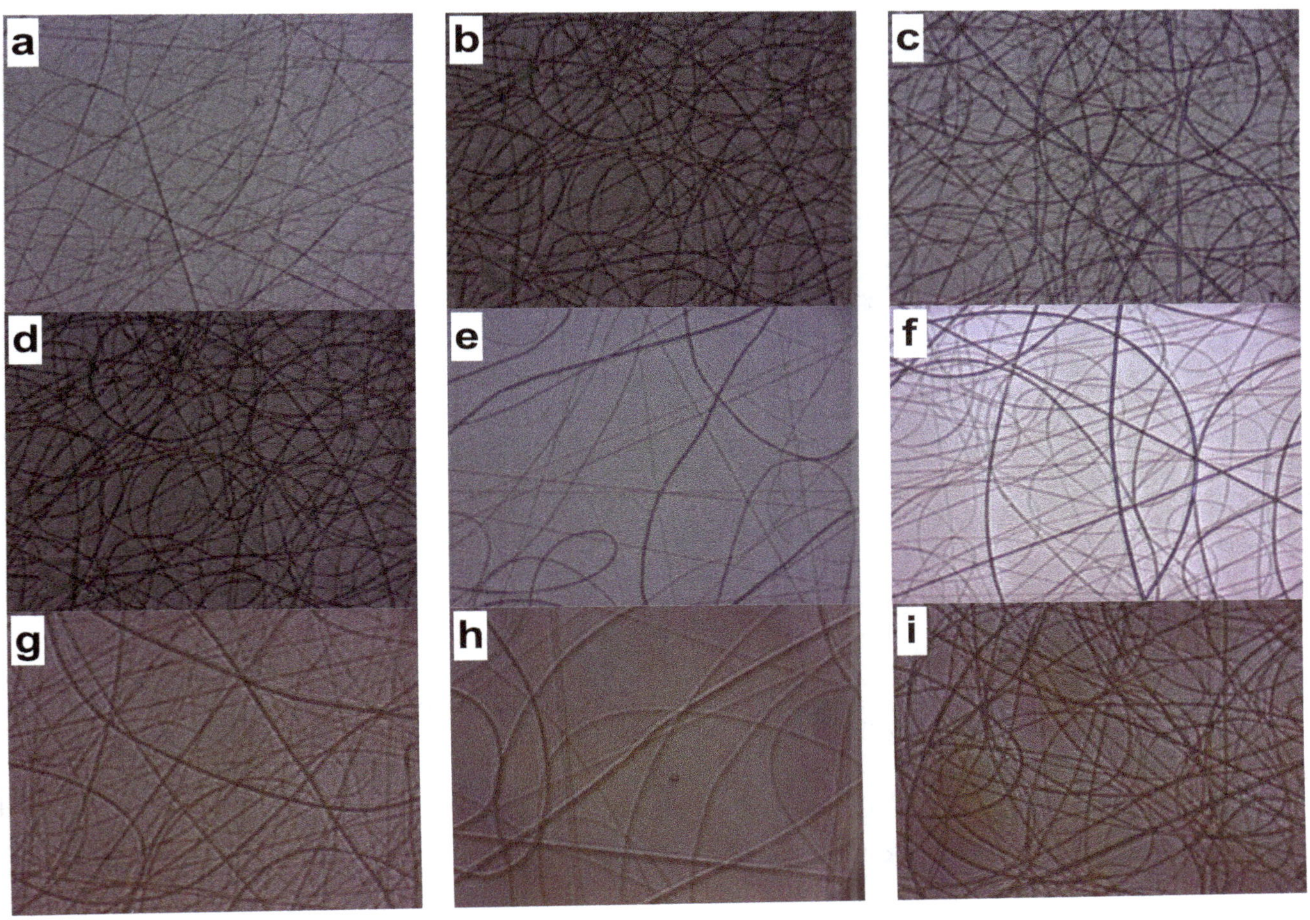

Figure 1. *Cont.*

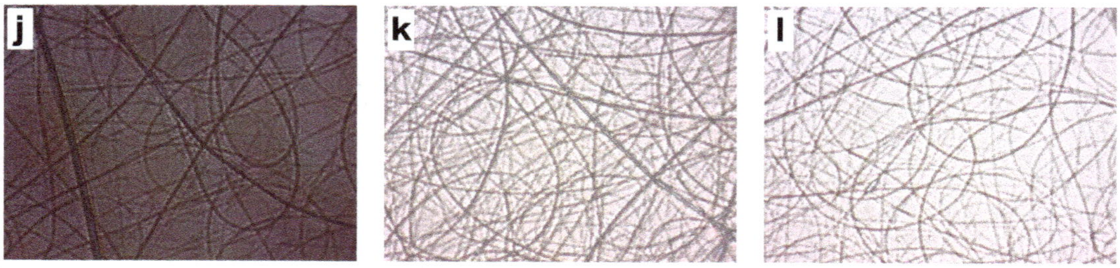

Figure 1. Micrographs of fibers produced by using a 2 wt % CS/8 wt % PVA solution (40× resolution): (**a**) 0.12 mL/h, 15 cm and 14 kV; (**b**) 0.13 mL/h, 15 cm and 14 kV; (**c**) 0.13 mL/h, 15 cm and 15 kV; (**d**) 0.13 mL/h, 15 cm and 14 kV; (**e**) 0.12 mL/h, 20 cm and 14 kV; (**f**) 0.12 mL/h, 25 cm and 14 kV; (**g**) 0.13 mL/h, 20 cm and 16 kV; (**h**) 0.12 mL/h, 25 cm and 16 kV; (**i**) 0.13 mL/h, 20 cm and 14 kV; (**j**) 0.13 mL/h, 25 cm and 14 kV; (**k**) 0.13 mL/h, 20 cm and 16 kV; (**l**) 0.13 mL/h, 25 cm and 16 kV.

Figure 2 shows a comparison of optical images for threedifferent solution compositions with optimal conditions. Non-uniform fibers are observed for 0.5 (a) and 0.1 wt % (b) of CS. For 2 wt % CS (c), uniform fibers were obtained.

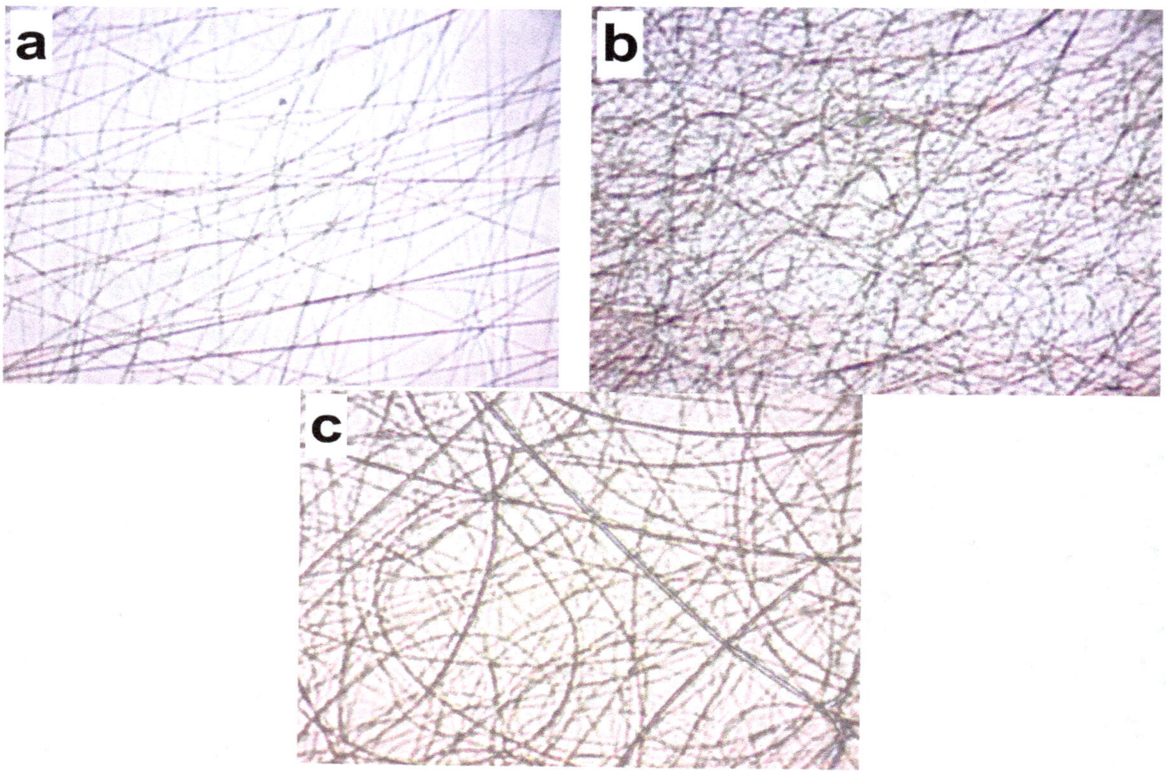

Figure 2. Micrographs of fibers produced with different CS concentrations (wt %): (**a**) 0.5; (**b**) 1; and (**c**) 2 with a voltage, distance, and flow rate of 0.13 mL/h, 20 cm, and 16 kV, respectively (40× resolution).

2.2. Electrospun CS/PVA Samples Collected on Reticulated Vitreous Carbon

Figure 3 shows the images of the electrospun fibers on the reticulated vitreous carbon (RVC) support at different times. The deposition of the fibers was observed during the spinning process, and the surface of the RVC support was modified by showing a white coloration caused by the electrospun fibers.

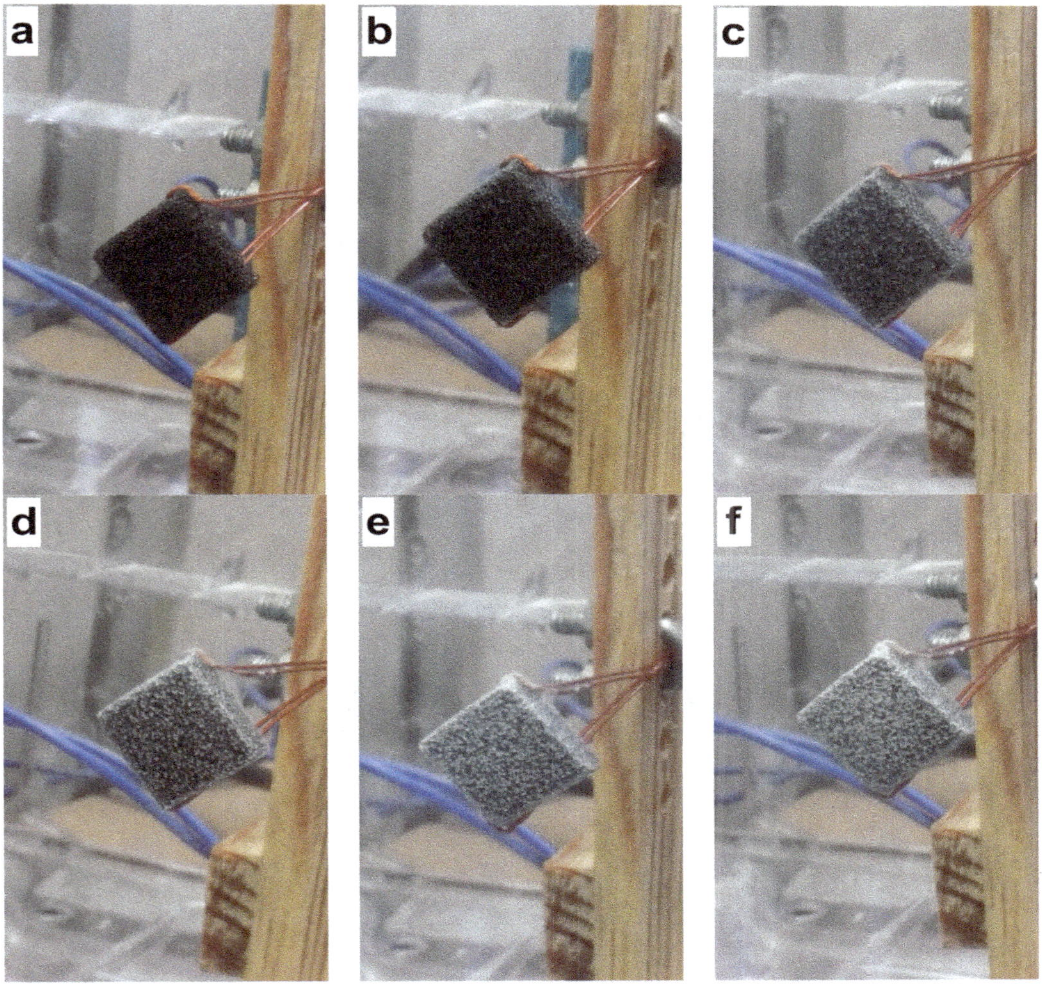

Figure 3. Electrospun fibers on the RVC support: (**a**) 0; (**b**) 15; (**c**) 30; (**d**) 60; (**e**) 90; and (**f**) 120 min.

Zargarian and Haddadi [15] indicated that the electrospinning technique can produce fibers with diameters ranging from 5 nm to 1 μm under a high-voltage electrostatic field that runs between a metal capillary syringe and an electrically grounded collector. Fibers made of polymer blends can be produced by this technique. In their study, the authors fabricated nanofibrous scaffolds of poly (ε-caprolactone)/hydroxyapatite/chitosan–PVA in order to investigate the influence of several parameters that affect the process and morphology such as concentration, solvent composition, voltage and tip-to-collector distance (TCD). In this work, the surface morphology of the nanofibrous material was studied using SEM micrographs. The distribution of fiber diameter frequencies was plotted to facilitate the optimization process. The voltage played an important role, while the TCD was largely responsible for the determination of the average diameter of the fibers.

2.3. Morphology of the RVC Support and CS/PVA Fibers on RVC

As for RVC, it is a macroporous polymeric carbon form as it can be seen in Figure 4. This material is obtained, for example, by carbonization of an expanded polymer or a vacuum removed material [16,17]. This material is a solid foam constituted by an open cell network composed of vitreous carbon, a material with high electrical and thermal conductivity. The adjective reticulated means "constructed, arranged or marked as a network or part of a network". RVC has a remarkably high vacuum volume and surface area. Its rigid structure has low density and withstands low temperatures. Figure 4 shows a SEM micrograph of unmodified RVC.

Figure 5 shows the micrographs of the electrospun fibers on RVC at different times with 2 wt % CS/8 wt % PVA, with a voltage, distance, and flow rate of 0.13 mL/h, 20 cm, and 16 kV, respectively. It is observed that the deposited fibers increased as the electrospinning time increased too; very few defects and a relatively uniform morphology can be observed.

Figure 4. SEM micrograph of reticulated vitreous carbon used as support material.

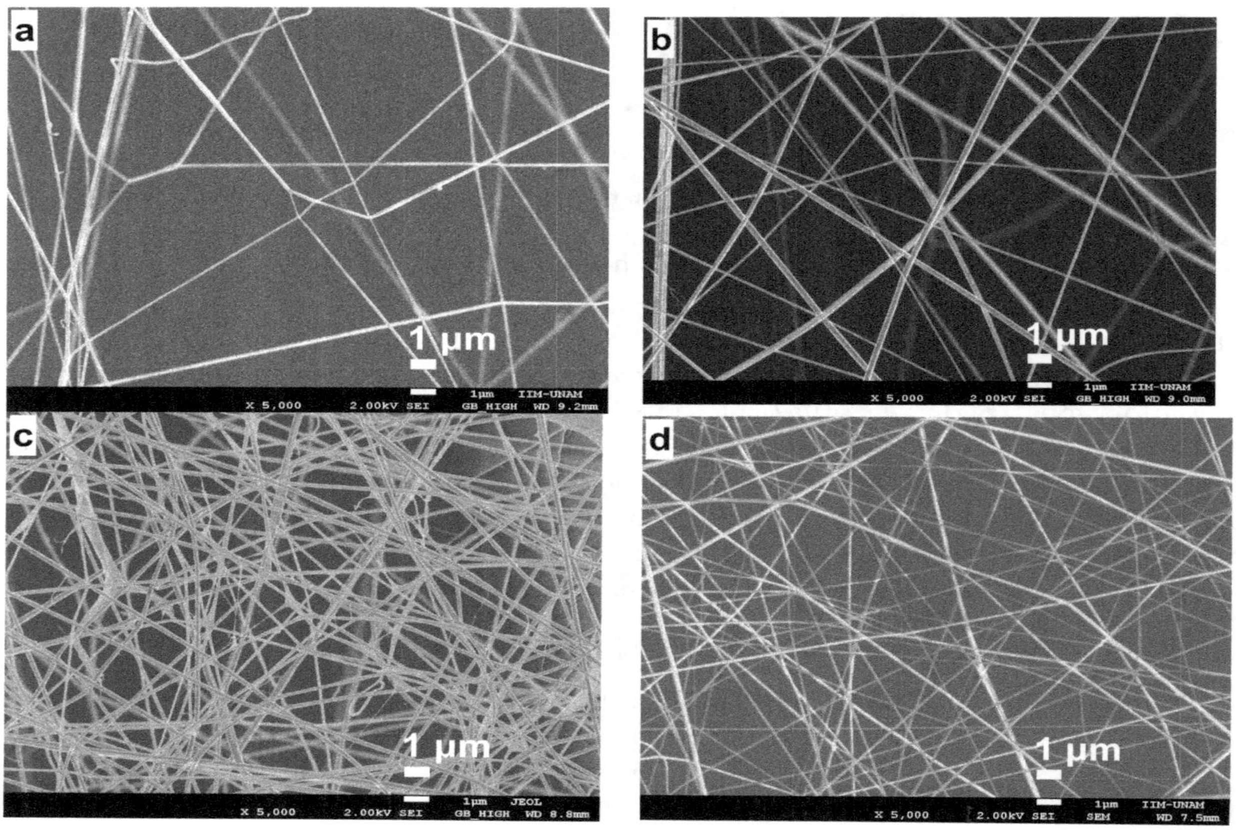

Figure 5. *Cont.*

Figure 5. SEM characterization of electrospun nanofibers on a RVC support at different times: (**a**) 15; (**b**) 30; (**c**) 60; (**d**) 90; and (**e**) 120 min.

For the diameter frequency histograms, about 100 electrospun fibers were analyzed with particle counting. Figure 6 presents the histograms of the nanofibers analyzed above, where diameters from 90 to 390 nm were observed.

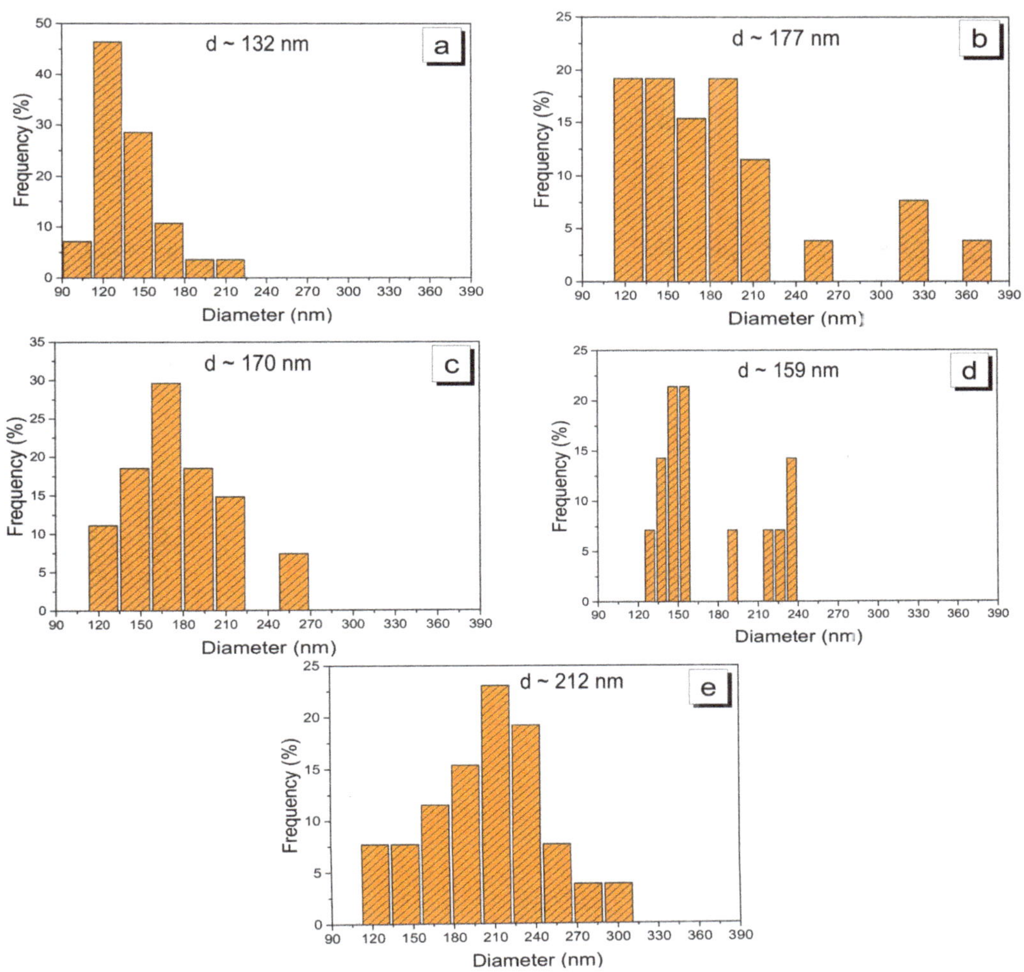

Figure 6. Histograms of nanofibers on the RVC support: (**a**) 15; (**b**) 30; (**c**) 60; (**d**) 90; and (**e**) 120 min.

Table 2 shows the average fiber diameters at electrospinning times of 15 min, 0.5, 1, 1.5 and 2 h. Diameters from 132 to 212 nm were observed with a standard deviation from ±6 to ±20 nm.

Table 2. Diameters of the electrospun nanofibers.

Parameter	15 min	30 min	1 h	1.5 h	2 h
Diameter (nm)	132	177	170	159	212
Standard deviation (nm)	±20	±9	±7	±11	±6

2.4. Stabilization of CS/PVA Nanofibers

Figure 7 shows micrographs of CS/PVA nanofibers deposited on RVC at an electrospinning time of 90 min before ((a) and (c)) and after ((b) and (d)) the ethanolic NaOH solution treatment used to stabilize the nanofibers. This treatment resulted in a fiber diameter decrease, which is in good agreement with results reported in the literature [18].

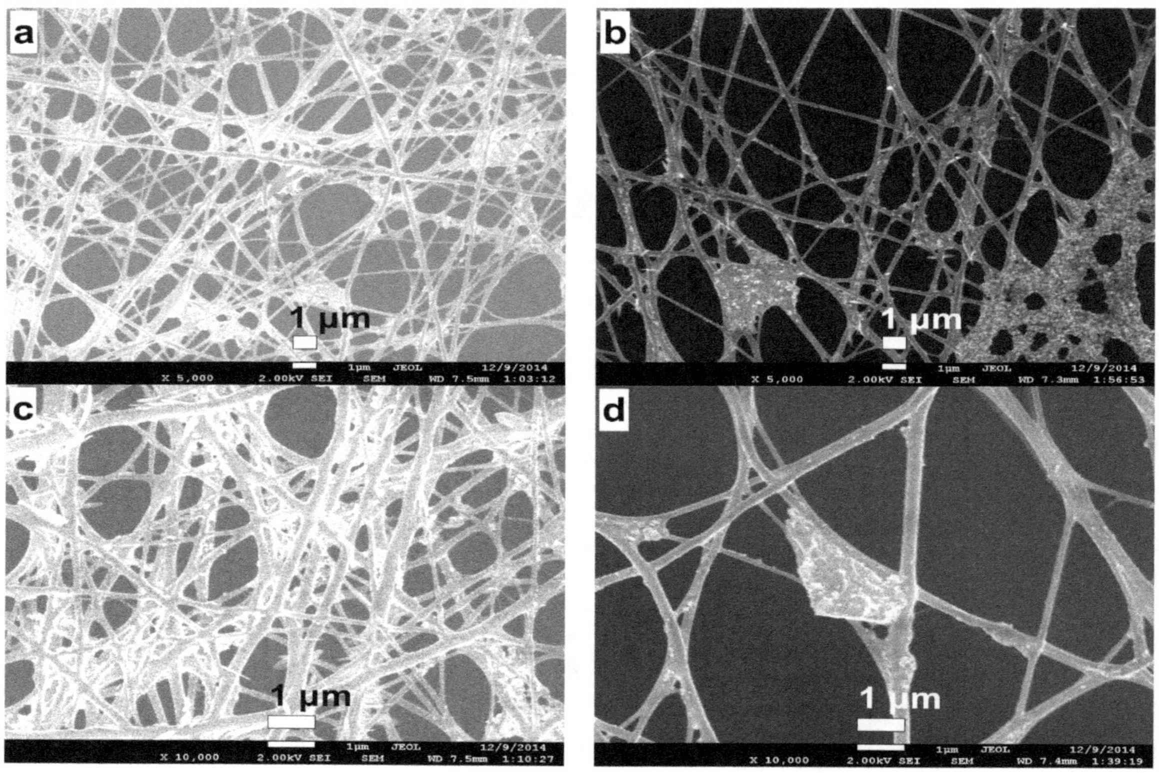

Figure 7. SEM micrographs of CS/PVA electrospun nanofibers deposited on RVC for 90 min before (**a**,**c**) and after (**b**,**d**) ethanolic NaOH solution treatment.

Table 3 presents the diameter of the electrospun fibers for 90 min before and after the alkaline treatment, which produced a diameter decrease of 44 nm.

Table 3. Diameters of the electrospun fibers before and after the alkaline treatment.

Parameter	Before	After
Diameter (nm)	159 ± 11	115 ± 9

2.5. Solution Viscosity

The viscosity of the CS/PVA solution decreased with time (weeks) from 10,500 to 410 cP, which was due to the hydrolysis of chitosan, as stated elsewhere [19,20]. Based on this fact, the electrospun nanofibers were produced by using a recently prepared solution. The results are shown in Figure 8.

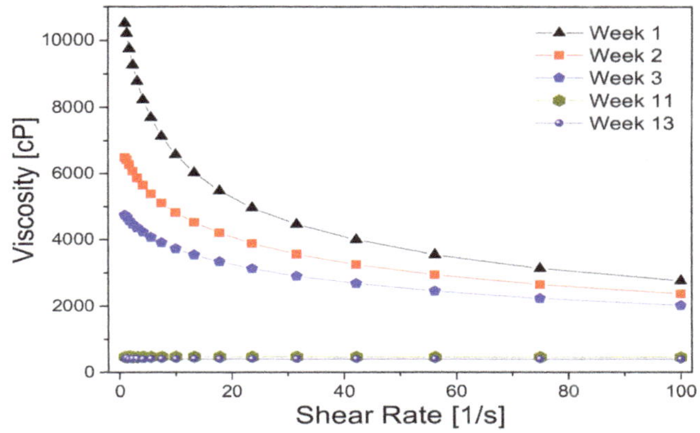

Figure 8. Viscosity variation of the 2 wt % CS/8 wt % PVA solution.

Figure 9 shows the micrographs obtained from the electrospun process on RVC, where cross ((a) and (b)) and topview ((c) and (d)) sections were taken for the SEM analysis. It is observed that the fibers are only found on the RVC surface, confirming the fact that they did not penetrate into the pores of the RVC support.

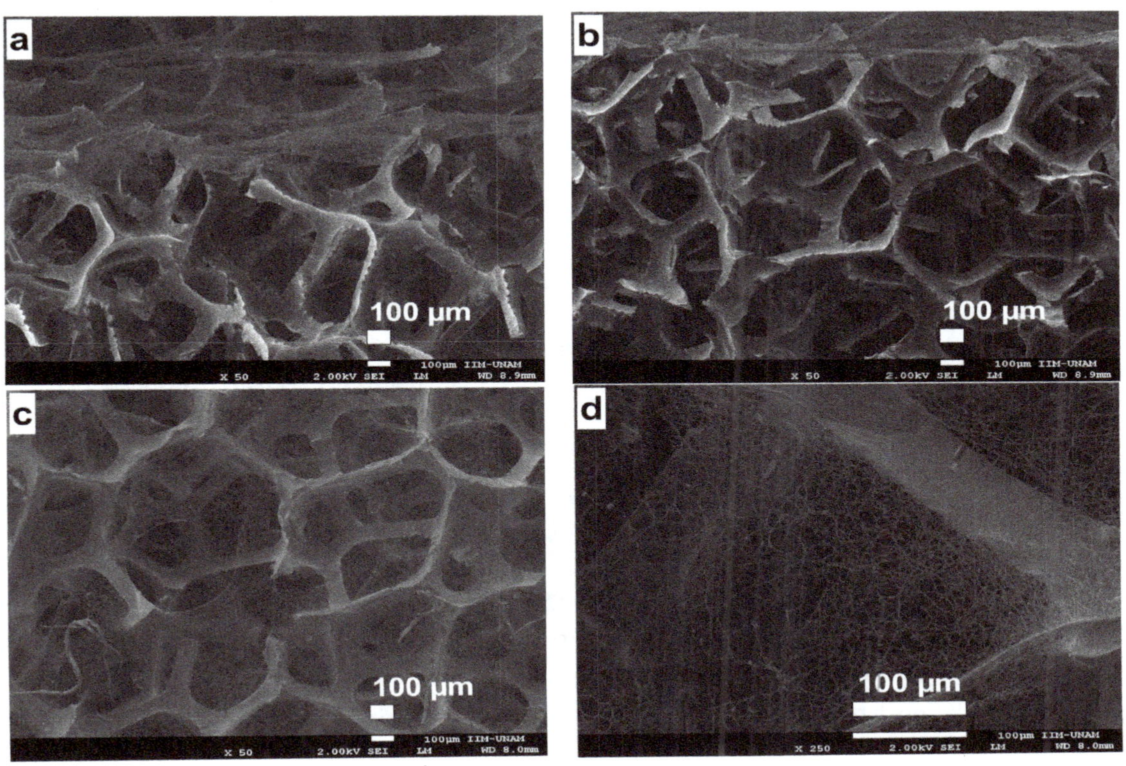

Figure 9. SEM micrographs of electrospun nanofibers deposited on RVC (90 min): (**a**,**b**) cross sections; (**c**,**d**) top-view sections.

3. Materials and Methods

3.1. Materials

CS from crab shells with deacetylation degree of 85% (lot number 91k1265) was used as received from Sigma Aldrich. Medium molecular weight PVA was purchased from Sigma Aldrich and used as received. Acetic acid (99.7%) was acquired from J.T Baker. RVC 60 ppi (normal pores per lineal inch) was obtained from Electrosynthesis, Co. Inc., Lancaster, NY, USA.

3.2. Preparation of the CS/PVA Solution

The electrospinning solutions were prepared by dissolving a maximum of 3 wt % of CS in acetic acid at 2 wt %. This solution was magnetically stirred until complete dissolution at room temperature. PVA powder was dissolved in deionized water at 8 wt % near its boiling point to facilitate complete dissolution. Subsequently, both solutions were homogenized by magnetic stirring.

In order to study the component ratio effect on the electrospinning process, the CS concentration was varied using 0.5, 1, 2, and 3 wt % while the PVA concentration remained at 8 wt % for each prepared solution.

3.3. Electrospinning Process

Electrospinning solutions were pumped using a 5 mL commercial plastic syringe and a Programmable Syringe Pump NE-1600. The studied flow rates were adjusted to 0.10, 0.12, 0.13, 0.15, and 0.20 mL/h at three different working voltages: 14, 15, and 16 kV, controlled by a high DC (Direct Current) voltage supply (Glassman High Voltage EH60). A glass plate or RVC was used as a collector, and the tip-to-collector distanceswere set at 15, 20, and 25 cm. All electrospinning procedures were carried out at room temperature (i.e., at around 23 °C). As the solution viscosity is a very important parameter, all the electrospinning experiments were carried out with recently prepared CS/PVA solutions (no more than 1 week old).

3.4. Alkaline Treatment with NaOH

The as-obtained electrospun nanofibers were immersed overnight in a saturated NaOH ethanolic solution in order to analyze the alkaline treatment effect. The treated samples were then rinsed with water and dried in air to be observed by SEM.

3.5. Viscometric Studies

The solution viscosity was measured in a rheometer (MCR Physica 301, Anton Paar) with cone and plate geometry, varying the shear rate between 1 and 100 rad/s. The measurements were carried out after 1, 2, 3, 11, and 13 weeks of their preparation. The storage temperature was 25 °C.

4. Conclusions

In the present study, CS/PVA nanofibers were manufactured by the electrospinning technique. The use of RVC allowed the formation of homogeneous fibers, compared to the fibers collected on a glass plate, with diameters ranging from 132 to 212 nm, making RVC a suitable substrate to obtain thin fibers. However, the deposit of fibers occurred only on the substrate surface. After testing CS concentrations of 0.5, 1, 2, and 3 wt %, it was found that the best solution to produce fibers was that containing 2 wt % CS and 8 wt % PVA. After evaluating electrospinning parameters such as flow rate, tip-to-collector distance, and voltage on either glass or RVC substrates, the optimum parameters for producing CS/PVA electrospun nanofibers were a voltage, distance, and flow rate of 16 kV, 20 cm, and 0.13 mL/h, respectively. , which allowed the manufacture of uniform fibers that were observed by SEM. The use of RVC improved the formation of homogeneous fibers compared to the fibers collected on the glass plate, with diameters ranging from 132 to 212 nm. In order to obtain chemically stable CS/PVA nanofibers, it was necessary to treat the nanofibers with an ethanolic NaOH solution.

Author Contributions: D.I.S.-A. performed the experimental tests. J.G.-P. designed and supervised the experimental part aimed at obtaining the polymeric materials. He also was in charge of the manuscript revision process. U.P.-G. carried out the discussion of the work overall results. A.M.-C. supervised the electrospinning experiments. R.D.M.-O. ran and interpreted the SEM experiments and results, respectively. R.V.-G. provided the infrastructure to perform the electrospinning experiments.

Acknowledgments: Diana Isela Sanchez-Alvarado acknowledges the scholarship granted by CONACYT-Mexico (269921/218106).

References

1. Fong, H.; Reneker, D.H. *Structure Formation in Polymeric Fibers*; Hanser Gardner Publishers: Munich, Germany, 2001; pp. 225–246.
2. Huang, Z.; Zhang, Y.; Kotaki, M.; Ramakrishna, S. A review on polymer nanofibers by electrospinning and their applications in nanocomposites. *Compos. Sci. Technol.* **2003**, *63*, 2223–2253. [CrossRef]
3. Gibson, P.; Gibson, H.; Rivin, D. Transport properties of porous membranes based on electrospun nanofibers. *Colloids Surf. A* **2001**, *187–188*, 469–481. [CrossRef]
4. Desai, K.; Kit, K.; Li, J.; Zivanovic, S. Morphological and surface properties of electrospun chitosan nanofibers. *Biomacromolecules* **2008**, *9*, 1000–1006. [CrossRef] [PubMed]
5. Geng, X.; Kwon, O.; Jang, J. Electrospinning of chitosan dissolved in concentrated acetic acid solution. *Biomaterials* **2005**, *26*, 5427–5432. [CrossRef] [PubMed]
6. Homayoni, H.; Hosseini Ravandi, S.A.; Valizadeh, M. Electrospinning of chitosan nanofibers: Processing optimization. *Carbohydr. Polym.* **2009**, *77*, 656–661. [CrossRef]
7. Li, L.; Hsieh, Y.-L. Chitosan bicomponent nanofibers and nanoporous fibers. *Carbohydr. Res.* **2006**, *341*, 374–381. [CrossRef] [PubMed]
8. Kumar, K.; Tripathi, B.P.; Shahi, V.K. Crosslinked chitosan/polyvinyl alcohol blend beads for removal and recovery of Cd(II) from wastewater. *J. Hazard. Mater.* **2009**, *172*, 1041–1048. [CrossRef] [PubMed]
9. Kim, J.K.; Kim, J.Y.; Lee, Y.M.; Kim, K.Y. Properties and swelling characteristics of cross-linked poly (vinyl alcohol)/chitosan blend membrane. *J. Appl. Polym. Sci.* **1992**, *45*, 1711–1717. [CrossRef]
10. Habiba, U.; Siddique, T.A.; Talebian, S.; Lee, J.J.L.; Salleh, A.; Ang, B.C.; Afifi, A.M. Effect of deacetylation on property of electrospun chitosan/PVA nanofibrous membrane and removal of methyl orange, Fe(III) and Cr(VI) ions. *Carbohydr. Polym.* **2017**, *177*, 32–39. [CrossRef] [PubMed]
11. Jawalkar, S.; Raju, K.; Halligudi, S.; Sairam, M.; Aminabhavi, T. Molecular modeling simulation to predict compatibility of poly (vinyl alcohol) and chitosan blends: A comparison with experiments. *J. Phys. Chem. B* **2007**, *111*, 2431–2439. [CrossRef] [PubMed]
12. Kumar, N.; Prabhakar, M.; Prasad, V.; Rao, M.; Reddy, A.; Rao, C.; Subha, M. Compatibility studies of chitosan/PVA blend in 2% aqueous acetic acid solution at 30 °C. *Carbohydr. Polym.* **2010**, *82*, 251–255. [CrossRef]
13. Lewandowska, K. Viscometric Studies in dilute solution mixtures of chitosan and microcrystalline chitosan with poly (vinyl alcohol). *J. Solut. Chem.* **2013**, *42*, 1654–1662. [CrossRef] [PubMed]
14. Paipitak, K.; Pornpra, T.; Mongkontalang, P.; Techitdheera, W.; Pecharapa, W. Characterization of PVA-chitosan Nanofibers prepared by electrospinning. *Procedia Eng.* **2011**, *8*, 101–105. [CrossRef]
15. Zargarian, S.S.H.; Haddadi-Asl, V. A nanofibrous composite scaffold of PCL/hydroxyapatite-chitosan/PVA prepared by electrospinning. *Iran. Polym. J.* **2010**, *19*, 457–468.
16. Kizling, K.; Dzwonek, M.; Olszewski, B.; Bącal, P.; Tymecki, L.; Więckowska, A.; Stolarczyk, K.; Bilewicz, R. Reticulated vitreous carbon as a scaffold for enzymatic fuel cell designing. *Biosens. Bioelectron.* **2017**, *95*, 1–7. [CrossRef] [PubMed]
17. Walsh, F.C.; Arenas, L.F.; Ponce de León, C.; Reade, G.W.; Whyte, I.; Mellor, B.G. The continued development of reticulated vitreous carbon as a versatile electrode material: Structure, properties and applications. *Electrochim. Acta* **2016**, *215*, 566–591. [CrossRef]
18. Huang, X.-J.; Ge, D.; Xu, Z.-K. Preparation and characterization of stable chitosan nanofibrous membrane for lipase immobilization. *Eur. Polym. J.* **2007**, *43*, 3710–3718. [CrossRef]
19. Skaugrud, O. Chitosan-New biopolymer for cosmetics and drugs. *Drug Cosmet. Ind.* **1991**, *148*, 24–29.
20. Ravi Kumar, M.N.V. A review of chitin and chitosan applications. *React. Funct. Polym.* **2000**, *46*, 1–27. [CrossRef]

15

Impact of pH Modification on Protein Polymerization and Structure–Function Relationships in Potato Protein and Wheat Gluten Composites

Faraz Muneer [1], Eva Johansson [1], Mikael S. Hedenqvist [2], Tomás S. Plivelic [3] and Ramune Kuktaite [1,*]

1 Department of Plant Breeding, Swedish University of Agricultural Sciences, Box 101, SE-23053 Alnarp, Sweden; faraz.muneer@slu.se (F.M.); eva.johansson@slu.se (E.J.)
2 KTH Royal Institute of Technology, School of Engineering Sciences in Chemistry, Biotechnology and Health, Fibre and Polymer Technology, SE-10044 Stockholm, Sweden; mikaelhe@kth.se
3 MAX-IV Laboratory, Lund University, Box 118, SE-22100 Lund, Sweden; tomas.plivelic@maxiv.lu.se
* Correspondence: ramune.kuktaite@slu.se

Abstract: Wheat gluten (WG) and potato protein (PP) were modified to a basic pH by NaOH to impact macromolecular and structural properties. Films were processed by compression molding (at 130 and 150 °C) of WG, PP, their chemically modified versions (MWG, MPP) and of their blends in different ratios to study the impact of chemical modification on structure, processing and tensile properties. The modification changed the molecular and secondary structure of both protein powders, through unfolding and re-polymerization, resulting in less cross-linked proteins. The β-sheet formation due to NaOH modification increased for WG and decreased for PP. Processing resulted in cross-linking of the proteins, shown by a decrease in extractability; to a higher degree for WG than for PP, despite higher β-sheet content in PP. Compression molding of MPP resulted in an increase in protein cross-linking and improved maximum stress and extensibility as compared to PP at 130 °C. The highest degree of cross-linking with improved maximum stress and extensibility was found for WG/MPP blends compared to WG/PP and MWG/MPP at 130 °C. To conclude, chemical modification of PP changed the protein structures produced under harsh industrial conditions and made the protein more reactive and attractive for use in bio-based materials processing, no such positive gains were seen for WG.

Keywords: wheat gluten; potato protein; chemical pre-treatment; structural profile; tensile properties

---

1. Introduction

Wheat gluten (WG) and potato protein (PP) are industrial side-streams originating from wheat and potato processing into ethanol and potato starch, respectively. Both protein-rich by-products are currently mostly utilized in baking (e.g., WG) and animal feed industries (PP). In addition, both of these proteins have attractive properties as raw materials for non-food applications such as in bio-based plastics [1,2]. During the production processes of WG and PP, high drying temperature is utilized, and for PP, an acid mediated coagulation step is also included. Both procedures are known to denature the chemical and structure-determining bonds in the protein and may also result in ionization of amino, carboxyl and phenolic groups [3] and reduced protein functionality [2,4,5]. In addition, for PP being treated at a low pH ≈ 4 during industrial processing, causing definite changes in chemical and physical properties to occur. The chemical processes resulting in intra- and intermolecular changes of the protein limit their processing window, thereby decreasing their suitability for processing into bio-based materials [2,6]. Similarly for WG, the relatively high temperature during drying negatively impacts the proteins and results in aggregated gluten with limited functionality [5]. An alternative and added-value opportunity is to use chemical tools to retain the WG and PP processing ability, both in their present or modified form, for benefit of industry, the consumer and society, thereby contributing to the circular bio-economy. Manipulating the chemical nature of industrial WG and PP would create new possibilities and contribute to the development of alternative uses and an added-value of these proteins for different applications. Development of new bio-based materials from WG and PP proteins is one such promising application.

Un-plasticized WG and PP are difficult to process as their glass transition temperature (T_g) is located close to their thermal degradation temperature [2,7–9]. Thus, easy processing requires the addition of chemical agents or polyol-based plasticizers that depress T_g and widen the processing window [2,4,10,11]. In general, proteins perform better in terms of their structure–functional properties when they are processed at basic pH [12,13]. The basic pH is known to denature and unfold the proteins and thereby expose sulfhydryl and hydrophobic protein sections, which open-up for new protein interactions when processed [14].

Chemical additives, such as NaOH, NH_4OH or urea, create basic conditions for the protein, resulting in changes of secondary and supramolecular protein structures which correlate with improved functional properties of processed materials [7,9,10,12,13,15–17]. However, previous studies have mainly focused on additives or chemical modifiers being added directly to the protein where the blend was immediately processed, thereby leaving a short reaction time for additives to interact with the protein. Although, pre-treatment of the proteins with chemical modifiers in order to modify the protein and later use it for processing has not been evaluated before. This has left a gap for further investigations on the impact of such chemical pre-treatment on protein structure and also functional properties of the final product. In this study, we pre-treated the WG and PP with NaOH in a solution to achieve a basic pH and to promote protein unfolding and induce reactivity. The pH modification was aimed to promote new protein–protein interactions, that could positively impact the mechanical performance of the protein films and composites. The objective was also to understand the effect of the chemical modification on protein molecular and secondary structures, processing behavior and mechanical performance of composites with varying WG and PP ratios. Increased understanding of protein–protein interactions and chemistry, as well as structure-function relationships of proteins in a blend can offer novel opportunities to create versatile and attractive performance of films and composites.

2. Results and Discussion

2.1. Protein Extractability in pH-Modified Protein Powders, Films and Composites

In this study, SE-HPLC was used to assess protein polymerization behavior in the pH modified protein powders and high temperature processed films (Figure 1). The modified potato protein (MPP) and modified wheat gluten (MWG) powders showed greater amounts of extractable polymeric proteins (HMw and LMw) compared to the non-modified samples, PP and WG (Figure 1a–d). This was particularly evident for the HMw proteins extracted during 2Ex and 3Ex extractions (Figure 1b,d). The results indicated, that the alkaline pH induced unfolding in the studied protein and made the proteins more extractable.

For protein films and composites processed at two temperatures, 130 and 150 °C, variation in protein extractability was observed. The processing of the protein powders into films at high temperature induced a higher degree of protein cross-linking and promoted the formation of larger protein aggregates and polymers, thereby decreasing their extractability during SE-HPLC analysis as has previously been reported in processed protein based materials [2,4,12,16,18–20]. In addition, possible interactions of the proteins studied with residual starches (wheat starch from wheat gluten fraction and potato starch from potato protein fraction) could have potentially occurred through non-covalent reactions such as, hydrogen and van der Waals-bonding between the protein and reactive hydroxyl groups from starch. Although those interactions were minor and did not substantially impact protein cross-linking.

For the pressed materials in this study an increase in protein extractability was seen for protein materials pressed at 150 °C, as compared to those pressed at 130 °C (Figure 1e). This increase in extractability in films pressed at the higher temperature (150 °C), was mainly due to an increase in LMw proteins being extracted as a consequence of protein degradation (except, for MWG, in which increase in HMw proteins at higher temperature was also observed) and lack of protein–protein interactions. In addition, proteins from films produced from the chemically modified proteins were in general more easily extractable, with the exception of MPP pressed at 130 °C (compare MPP with PP pressed at 130 °C; Figure 1e).

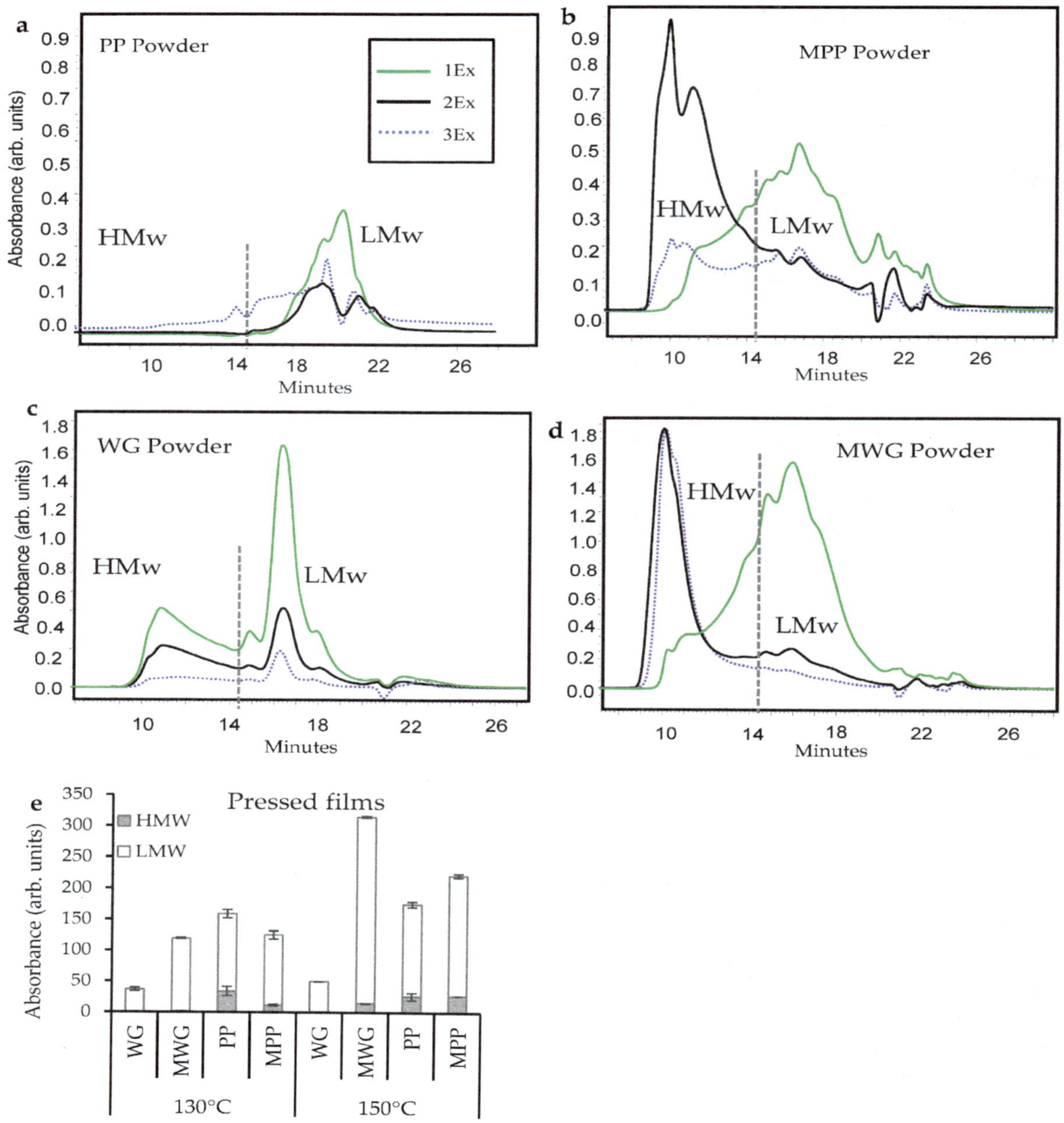

Figure 1. Representative chromatograms of protein molecular size distribution of potato protein powder (PP) (**a**), modified potato protein powder (MPP) (**b**), wheat gluten (WG) (**c**) and modified wheat gluten (MWG) powder (**d**), and the total protein extractability of films from the 3 extractions of non-modified and pH-modified proteins pressed at 130 °C and 150 °C (**e**).

The protein extractability results clearly suggests that the basic pH-modification of protein powders contributed to a protein network unfolding, decrease in protein–protein interactions and de-polymerization of protein aggregates being present in the non-modified protein powder. The unfolding, decrease of protein–protein interactions and de-polymerization of the proteins after modification resulted in an increased extractability of the proteins. The fact that the extractability increased primarily as HMw proteins were extracted during 2Ex and 3Ex extraction steps (where sonication energy is used), suggests that primarily protein aggregates/polymers were affected by the pH adjustment (chemical modification). However, the majority of the protein extraction taking place at 2Ex and 3Ex indicated the presence of some of the cross-links, although with lower degree of intermolecular disulphide bonds, while hydrogen bonds and non-covalent protein interactions might still be in place [21,22].

Compression molding of the powder into films resulted in excessive cross-linking and re-formation of intermolecular disulphide bonds for all samples, as this has been previously reported for wheat gluten and potato protein materials [2,11,16,19,23]. In particular, most of the HMw proteins in all the hot pressed samples ended up in forming large polymers, which were not extractable in the films after pressing. It is also known from previous studies that compression molding of films under alkaline conditions leads to increased protein–protein interactions with the formation of disulphide cross-links in WG, hydrogen bonding in WG and PP and also the formation of isopeptide bonds such as dehydroalanin, lanthionine and lysinoalanine [7,16,24–26]. In addition, alkaline pH and the use of cross-linking enzymes, such as transglutaminase, for protein pre-treatment can further introduce new protein cross-linking opportunities, as has been shown in grass pea flour films [27]. Also, newly sourced transglutaminase species that have been shown to positively impact certain functional behavior of proteins in wheat dough [28,29] should further be explored in protein-based materials.

In this study, the proteins were found to be less polymerized after hot pressing at 150 °C than at 130 °C, most likely due to protein degradation, with HMw proteins breaking into LMw fragments [8,23]. In terms of nano-structure for unmodified WG and PP samples pressed at 130 and 150 °C, few differences due to high temperature were observed in the main scattering broad peak, d_1 and for the scattering distance d_2, studied by SAXS, for both types of proteins (supporting information, Figure S1). For the PP film, the broad peak (d_1) became "sharper", while d_2 intensity decreased with increasing temperature. While for the WG film, at 150 °C the scattering distance, d_2, disappeared, indicated a less complex structural morphology compared to the previously observed structural morphologies in WG films pressed at 130 °C [5]. A possible explanation is a decrease in polymerization and an increase in protein breakdown as shown by SE-HPLC (Figure 1e). Regarding the PP film morphology, intensity of the d_2 peak was higher at 130 °C than at 150 °C, indicating more structural complexity, although without any specific peak ratio [2]. This change in peak position and intensity may be affected by protein polymerization behavior, which was observed at different pressing temperatures. Regarding the MWG and MPP films morphology, the MPP showed a clear peak, d_1, indicating similar structural morphology observed previously for the PP films pressed at 130 °C [2] (supporting information, Figure S2). While MWG showed d_1 and d_2, scattering distances, that can be referred to some undefined, though poorly hierarchically arranged structural entities (as compared to previously observed hexagonal arrangement of WG proteins). Thus, modification may not have considerably changed the nano-structure of MPP, while MWG nanostructure was not improved.

Previous study of WG has indicated that a less polymerized structure in the powder used for processing contributes to a higher degree of re-polymerization, chemical flexibility for proteins to interact and cross-link during processing, and the availability of chemically reactive sites for disulphide bond formation [5]. Although, the described effect could not be seen for the WG proteins in this study, indicating the alkali modification either leads to irreversible changes in the protein conformation (negative impact) or to chemical reactions not involving disulphide bond formation.

2.2. *Protein Extractability in Pressed Composites*

Among the pressed composites the highest protein extractability was clearly seen for the MWG/MPP samples at 150 °C, almost independent of the ratio of the two modified proteins within the composite (Figure 2). These results corresponded well with the high protein extractability found for the modified proteins pressed individually at 150 °C (compare Figure 1e), although protein extractability was even higher in the composites than in the pressed films of individual proteins. Thus, mixing of the two modified proteins did not contribute to increased protein polymerization as compared to protein polymerization for the modified proteins separately. Similarly, a high proportion of MWG (50 and 70%) in the composites pressed at 130 °C resulted in increased protein extractability as compared to the films from each of the individually modified proteins (Figure 2a compared with Figure 1). The extractability was also increased with the increase in temperature from 130 to 150 °C and with higher MWG content in the blend (Figure 2). An increased protein extractability with increase in MWG content (at 130 °C) could be due to the fact that MWG has a more complex protein molecular structure and chemistry, compared to PP, and was more denatured and depolymerized in the pH-modification process compared to PP (see MWG and MPP powder profiles; Figure 1b,d). Therefore, an MWG increase in the MWG/MPP blends (i.e., the 70/30 sample) resulted in a decrease in polymerization and lack of disulphide cross-link formation compared to 30/70 and 50/50 samples. However, using a 150 °C pressing temperature reduced the protein polymerization (increased protein extractability) for all MWG/MPP compositions compared to the individual modified proteins (compare Figures 1 and 2b). This increase in protein extractability at higher temperature may be due to the lack of disulphide and irreversible cross-links between the protein chains and also due to breakdown of HMw protein into LMw fragments.

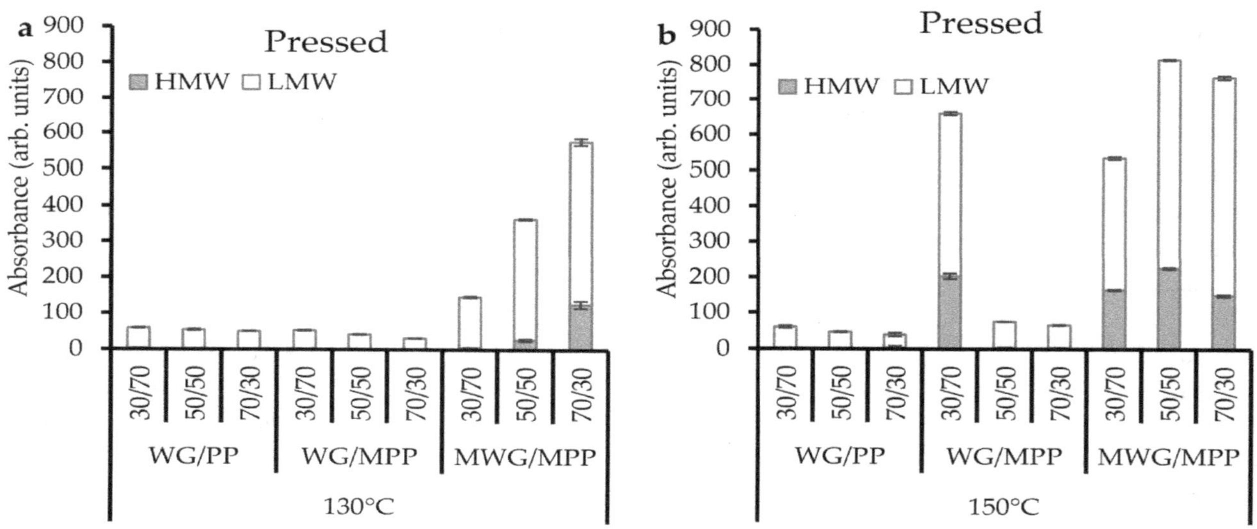

Figure 2. Total protein extractability of WG/PP, WG/MPP and MWG/MPP samples pressed at 130 °C (**a**) and 150 °C (**b**), as studied by SE-HPLC.

For the composites based on non-modified proteins (WG/PP), such an increase in protein extractability (as seen for modified samples) was not seen (compare Figures 1 and 2); instead a slight decrease in protein extractability was observed, especially at high WG ratios (70/30). The non-modified (as received) WG and PP have a low extractability due to the formation of HMw protein aggregates during their industrial processing [2,5]. Therefore, their further processing at high temperatures favored the formation of non-reducible protein cross-links and also incorporation of the LMw fragments in larger protein networks, thus reducing their extractability [8,22,30]. This suggests that composites of non-modified proteins produced in this study did show reduced protein extractability as reported for individual protein based materials produced previously [2,19].

As for the WG (non-modified) composites with different ratios of MPP, the protein extractability was in a similar range to that of WG/PP films at both pressing temperatures (compare Figures 1 and 2) except for WG/MPP 30/70 at 150 °C with surprisingly higher amounts of extractable proteins. Hence, higher pressing temperature (as 150 °C) increased protein polymerization and decreased extractability when $\leq$ 50 *wt.*% of MPP was used. No such effect of temperature and amount used of MPP was seen in the samples pressed at 130 °C, suggesting that high temperature may be one of the main driving factors in determining the protein polymerization.

Interestingly, the lowest protein extractability (highest degree of protein polymerization) was found for the 70/30 WG/MPP composite film at 130 °C, and for the 70/30 WG/PP film at 150 °C (Figure 2). Thus, to increase protein polymerization in WG based films, modified potato protein (MPP) is beneficial at 130 °C, while non-modified PP is beneficial to have at higher temperature. Increased protein polymerization in 70/30 WG/MPP could be explained by a higher degree of protein cross-linking of WG at 130 °C as has been reported in previous studies [5,19,31]. However, an increased protein polymerization in WG/PP 70/30 sample at 150 °C suggests formation of new protein–protein interactions in a composite, because the extractability of these blends decreased slightly more when compared to both non-modified proteins pressed at 150 °C.

2.3. *Effect of Chemical Modification on Secondary Structure of Proteins*

As shown in previous studies [5,10,12,31–33], the non-modified WG powder showed relatively low amounts of β-sheets, as indicated by a flat shoulder in the amide I region (1620–1625 cm^{-1}), while peaks at 1660, 1650 and 1641 cm^{-1} verified the presence of α-helices, α-helices and random coils, and unordered structures, respectively (Figure 3a). Chemical modification of the WG protein (MWG sample) resulted in a slight decrease in the intensity of peaks in the region 1640–1660 cm^{-1}, indicating that some of either α-helices, α-helices and random coils, or unordered structures became involved in formation of strong hydrogen bonded β-sheets (Figure 3a). Thus, FT-IR data indicate formation of β-sheets at the expense of either α-helices, α-helices and random coils or unordered structures, may indicate development of novel interactions (presumably via hydrogen bonds, protein-protein and peptide–protein interactions [34]), although not verified by SE-HPLC results (Figure 1).

The FT-IR spectra of PP showed a very distinct and high intensity peak at 1622 cm^{-1} verifying the presence of strongly hydrogen-bonded β-sheets. The formation of these protein structures are known to originate from the harsh treatment of the PP during industrial extraction with the use of high temperature and acidic processing conditions promoting a high degree of cross-linking [2]. The high degree of cross-linking in PP was also supported by the SE-HPLC data, showing low extractability of PP powder (Figure 1a). Furthermore, results of the SE-HPLC were supported by FT-IR data for the MPP powder sample, in general showing a lower amount of strongly hydrogen-bonded β-sheets (less aggregated structure) as reported in previous studies [4,34,35].

The pressed WG/PP blended films resulted in FT-IR spectra with a clear and high intensity peak at 1620 cm^{-1} and also several well defined peaks in the 1640–1660 and 1622 cm^{-1} region (Figure 3b,c), thereby showing structural features resembling both single powders used in the blend (Figure 3a) and indicating presence of strongly bonded β-sheet interactions, α-helices, α-helices and random coils, and unordered structures [2,5]. Including MPP in the blends (WG/MPP) for compression molding resulted in FT-IR spectra with reduced intensity and a shift to lower frequency of the major β-sheet peak (especially at 150 °C when the shoulder also became broader), and a decrease in intensity of the peaks in the region of 1640–1660 cm^{-1} (Figure 3b,c). In general protein aggregation has been correlated to an increased amount of strongly hydrogen-bonded β-sheets contributing to ordered conformation and improved tensile strength [2,4,7,10,12,31–34,36–38]. However, the lowest extractability of the proteins (indicating protein polymerization) as resolved by SE-HPLC results (Figure 2) was found for WG/MPP 70/30 at 130 °C and WG/PP 70/30 at 150 °C, not corresponding with blends showing the most intense peak for β-sheets by FT-IR (Figure 3b,c). These results suggest that the protein polymerization behavior and relationships with structural properties seems to be more complicated for compression molded composites than for individual proteins, especially while modifying of the chemical structures of the proteins used in the blends.

FT-IR spectra of MWG/MPP composites showed more differentiation between the ratios of proteins, where the strongly hydrogen-bonded β-sheets related structural peak was more pronounced at higher PP ratio and lower temperature (supporting information, Figure S3), corresponding to lower protein extractability by SE-HPLC (Figure 2). This might indicate that differences in protein polymerization were too limited among WG/PP and WG/MPP blended compression molded films to be differentiated by FT-IR in most cases, and basically only differences between separate blended protein components were seen. Thus, only for differences in protein polymerization behavior of MWG/MPP blends were large enough to be differentiated by FT-IR as also indicated by SE-HPLC, and corresponded to the characteristic relationship of SE-HPLC, FT-IR and functional properties data [12,31].

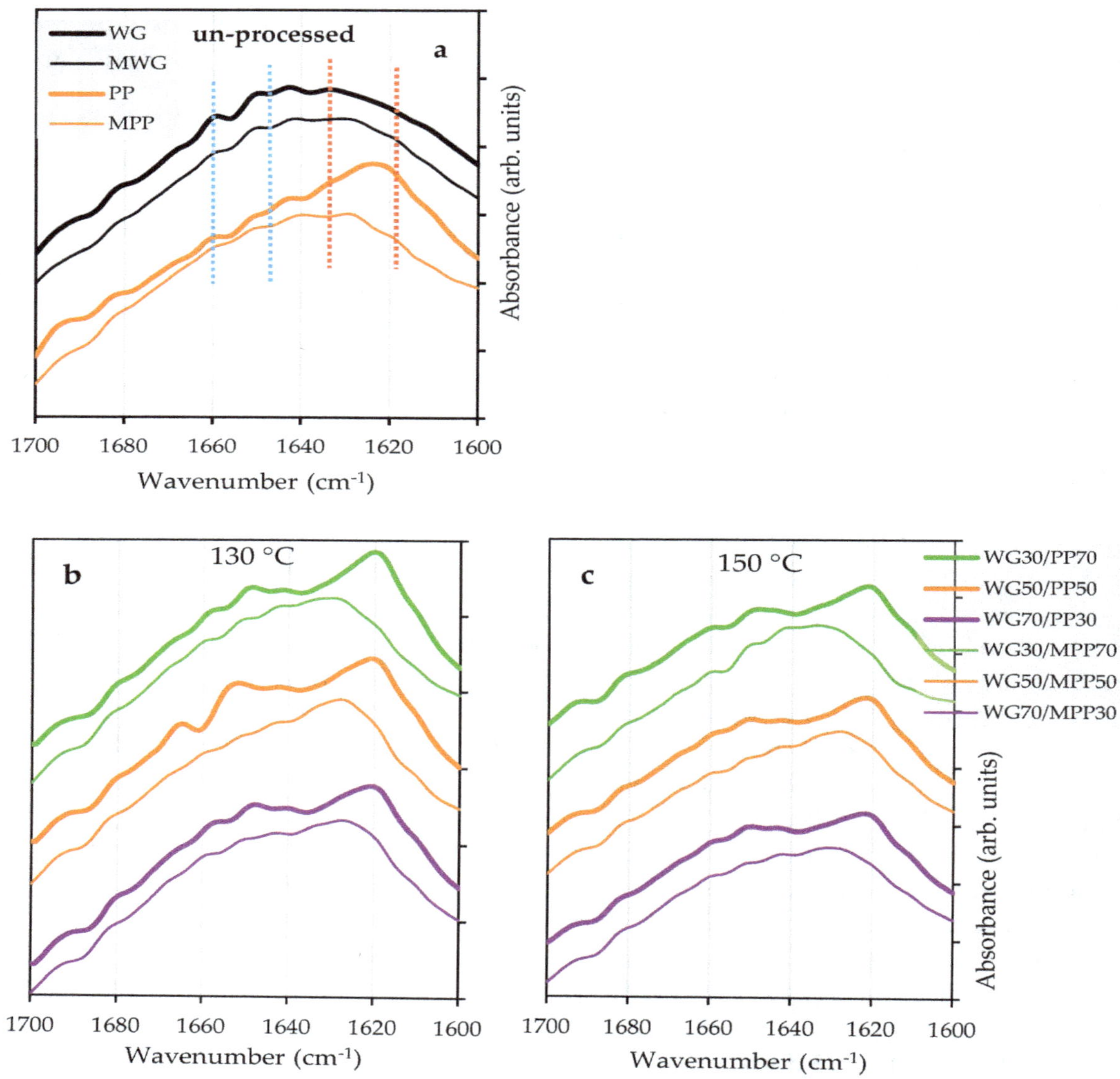

Figure 3. FT-IR spectra of individual protein powders and pressed films at 130 and 150 °C. WG, MWG, PP and MPP powders (**a**), pressed samples of WG/MPP blends along with controls (**b**,**c**).

2.4. *Effect of Protein Modification on Mechanical Performance of the Pressed Composites*

Compression molding of WG, MWG, PP, MPP and their blends, yielded samples of different colors as can be seen in Figure 4. High WG and MWG content generally resulted in a sample with light brown color (Figure 4, MWG100 and WG/MPP 70/30), while high PP and MPP content generally resulted in samples with dark brown (WG/PP 30/70) or almost black in color (MPP100), which can influence the composites' attractiveness and end-use. It is however an easy task to incorporate a pigment or dye to obtain a desired color (excluding white color). WG and PP samples showed opposite tensile performance, WG having lower strength and E-modulus but higher extensibility (elongation at break) than PP. Thus, comparing these two protein types, the tensile strength correlated positively with protein extractability. This observation was different to many previous studies, where a higher degree of cross-linking yielded higher strengths [2,5,19,22,39,40]. Here, the WG proteins most likely have a higher ability to cross-link (due to their more complex molecular structure characteristics and specific polymerization properties [8,22]) during pressing than the PP as SE-HPLC results indicate, although FT-IR shows a higher degree of β-sheets in the pressed PP which might explain their higher maximum stress and E-modulus [2].

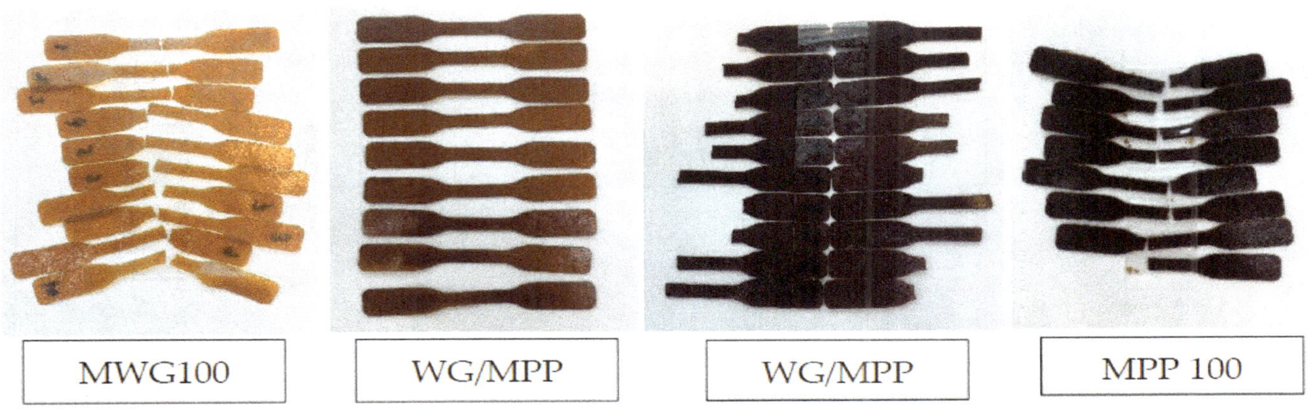

Figure 4. Representation of the range of colors of films and composites MWG100, WG/MPP 70/30, WG/MPP 30/70 and MPP100 pressed at 130 °C.

Modification of WG (MWG) resulted in lower maximum stress and E-modulus and reduced extensibility compared to WG for both evaluated pressing temperatures (Figure 5). Modification of the PP (MPP), resulted in a decrease in E-modulus for samples pressed at both temperatures and maximum stress decreased at 150 °C while the extensibility increased particularly at 130 °C (Figure 5). SE-HPLC data indicated an increase in cross-linking at 130 °C, although a decrease in β-sheet content was shown by FT-IR, which might explain the improvement in the extensibility and decrease in E-modulus for the MPP as compared to PP. In addition, a previous study reported an increase in extensibility and a decrease in modulus for PP modified with NaOH, when pressed with 30% glycerol [4]. Thus, modification of PP prior to processing (e.g., hot pressing) leads to improved tensile properties but also contributes to opportunities for using lower processing temperatures than was used in previous studies ($\geq$ 150 °C) [2], thereby saving energy input during the production process.

The pressed composites of WG and PP resulted in tensile properties generally as expected from the ratios of the individually pressed proteins (Figure 5). In addition, pressed WG/MPP samples showed tensile properties generally between those of the individually pressed protein samples, although minor differences were seen among the samples. Furthermore, as also indicated by comparing FT-IR and SE-HPLC results, differences in the protein cross-linking obtained by SE-HPLC could not explain the differences in tensile properties between the WG/PP and WG/MPP blends.

The most rubber-like properties were observed for the pressed MWG material (lower stiffness than WG), which also resulted in that the MWG/MPP blends were with the most rubber-like properties (low modulus but still with appreciable extensibility) (Figure 5), thereby correlating with SE-HPLC and FT-IR data. It is noteworthy that the higher pressing temperature (150 °C) leads to lower stiffness, strength and extensibility for the modified materials and their blends (MWG, MPP and MWG/MPP) (Figure 5).

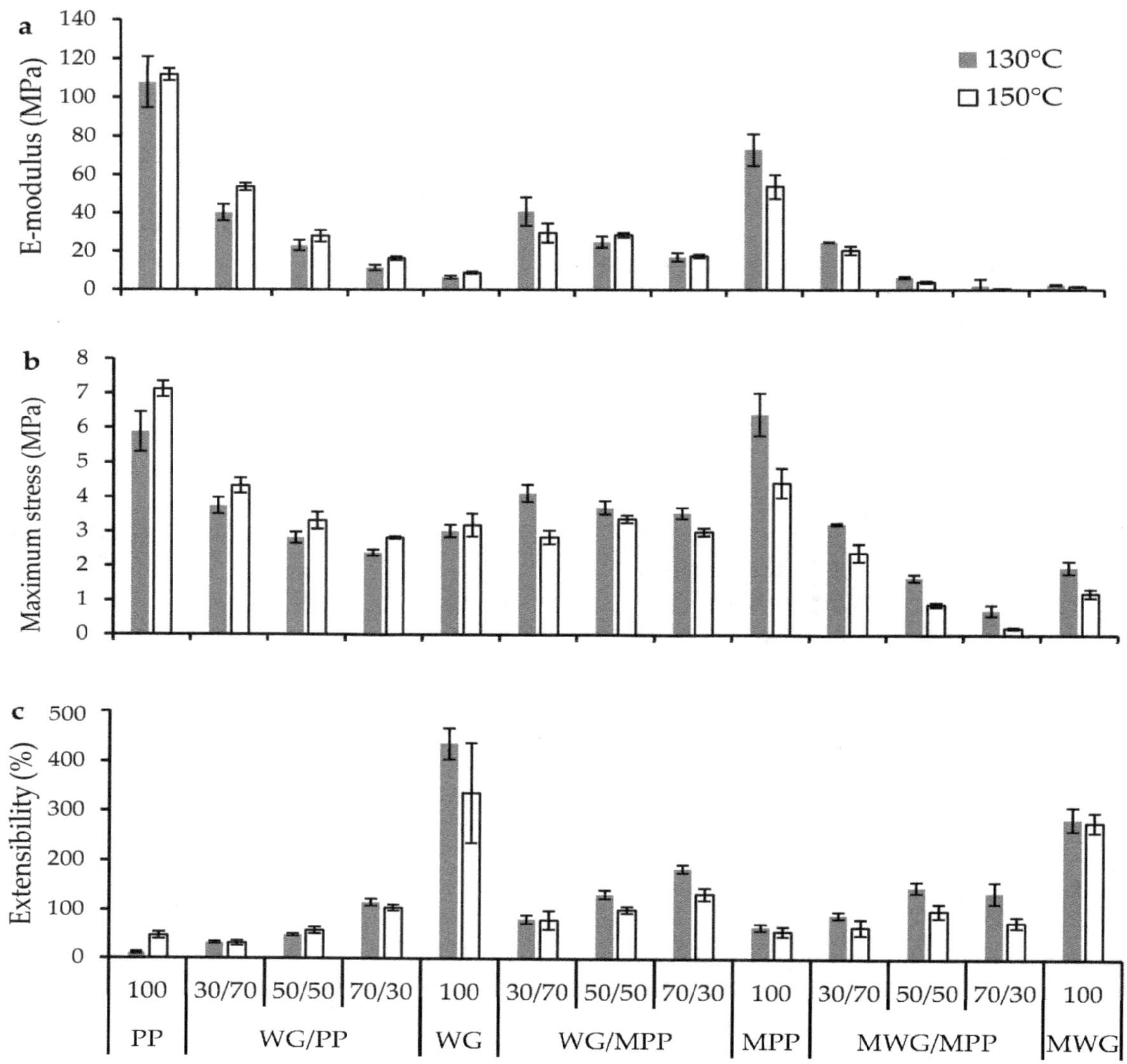

Figure 5. Mechanical properties of individual protein films and WG/PP, WG/MPP and MWG/MPP blends pressed at 130 and 150 °C; (**a**) E-modulus, (**b**) maximum stress and (**c**) extensibility.

3. Materials and Methods

3.1. Wheat Gluten and Potato Proteins

Wheat gluten with protein content of 77.7 *wt.*%, starch 5.8 *wt.*%, moisture 6.9 *wt.*% and fat 1.2 *wt.*%, was purchased from Lantmännan Reppe AB, Lidköping, Sweden. Commercial PP was supplied by Lyckeby Starch AB, Kristianstad, Sweden. The potato protein content was 82.2 *wt.*% (Dumas method, Flash 2000 NC Analyser, Thermo Scientific, USA, NX 6.25) and moisture content of 8.1% (dry basis, dried at 105 °C for 3 h). Glycerol (purity 99.5 *wt.*%, 0.5 *wt.*% water) was supplied by Karlshamns Tefac AB, Karlshamn, Sweden.

3.2. Protein Modification

Potato protein powder, 50 g was slowly dispersed in 600 mL distilled water under stirring. The suspension was adjusted to pH 10 by addition of a 5 M NaOH solution, followed by stirring for 30 min at 75 °C (± 3 °C). The PP was well dispersed in these conditions forming a dark brown suspension. The PP suspension was then lyophilized and ground to powder using an IKA A10 grinder (IKA, Germany) and was designated as modified PP (MPP) in this study. WG was modified in the same way as the PP above and was designated as modified WG (MWG). Protein blends were prepared by dispersing various ratios of WG and PP in water followed by the pH modification step as described for the separate protein sources. Thereafter, MWG/MPP blend were lyophilized and ground to obtain powder for further sample processing.

3.3. Sample Preparation and Compression Molding

Powders from individually modified PP and WG proteins were mixed with glycerol, (70 *wt.*% protein and 30 *wt.*% glycerol) prior to compression molding and later pressed into films. Each type of protein was pressed into two films, one at 130 °C and another one at 150 °C. For blends, different ratios of WG/PP and WG/MPP (30/70, 50/50 and 70/30) were manually mixed with glycerol, (70 *wt.*% protein and 30 *wt.*% glycerol) prior to compression molding. The sample compositions are presented in Table 1. The prepared individual protein films and blends were then placed in a 0.5 mm thick aluminum frame with a 100 × 100 mm opening to control the size and thickness of the film. The film and blends were then molded for 5 min at 200 bar in a hydraulic press (Polystat 400s, Servitech, Germany) between pre-heated aluminum plates with poly(ethylene terephthalate) release films. From each of the blends two films were compression molded, one at 130 °C and another one at 150 °C. Pressed films were removed from the hot press and left to cool between two room temperature aluminum plates.

Table 1. Compositions of the blends of WG, PP, MWG and MPP.

Type	Abbreviation	WG (*wt.*%)	PP (*wt.*%)	MWG (*wt.*%)	MPP (*wt.*%)
Controls	WG/PP 30/70	30	70	—	—
	WG/PP 50/50	50	50	—	—
	WG/PP 70/30	70	30	—	—
Only PP Modified	WG/MPP 30/70	30	—	—	70
	WG/MPP 50/50	50	—	—	50
	WG/MPP 70/30	70	—	—	30
MWG/MPP Modified in Composite	MWG/MPP 30/70	—	—	30	70
	MWG/MPP 50/50	—	—	50	50
	MWG/MPP 70/30	—	—	70	30

3.4. SE-HPLC to Assess Protein Polymerization in Processed Composites

The soluble amount of protein and protein size distribution of the protein powders, compression molded samples and composites were examined in this study. For this analysis, films were cut into approximately 0.2 mm pieces using a scalpel. Thereafter, 16.5 mg (± 0.05 mg) of each sample was added to 1.4 mL buffer solution (0.5% SDS, 0.05 M NaH_2PO_4, pH 6.9) in a Eppendorf tube (1.5 mL). To obtain the first extraction (1Ex) tubes were vortexed for 10 s (Whirli Vib 2, Labassco, Sweden) and then shaken for 5 min (IKA-VIBRAX VXR, IKA, Germany) at 2000 rpm. The tubes were then centrifuged (Legend Micro 17, Sorvall, Germany) for 30 min at 12,500 rpm and the supernatant was collected in HPLC vials. For the 2nd extraction (2Ex), extraction buffer (1.4 mL) was added to the pellet from 1Ex and then sonicated for 30 s at an amplitude of 5 microns using a Sanyo Soniprep 150 Ultrasonic Disintegrator (Tamro, UK) and thereafter centrifuged for 30 min and the supernatant collected in HPLC vials for analysis. The third extraction (3Ex) was similar to 2Ex, the pellet of 2Ex was used with sonication intervals of 30 + 60 + 60 s, to avoid overheating, samples were left to cool at room temperature between each sonication interval. All three extractions (1Ex, 2Ex and 3Ex) were analyzed with a Waters 2690 Separation Module connected to a Waters 996 Photodiode Array Detector (Waters, Millford, MA, USA).

An SE-HPLC column (Biosep-SEC-S 4000, Phenomenex, Torrance, CA, USA) was used for protein size distribution determination. For each sample, 20 µL was injected onto the column at an isocratic flow of 0.2 mL/min (50% acetonitrile, 0.1% TFA; 50% H_2O, 0.1% TFA). Chromatograms were obtained at 210 nm and integrated using Empower Pro software (Waters, USA). The chromatograms were divided into two sections depending on the elution time of the proteins and total area was calculated. The high molecular weight (HMw) proteins were eluted between 7–14.5 min and low molecular weight (LMw) between 14.5–28 min. HMw are referred to as polymeric proteins, and LMw as smaller molecular size proteins.

3.5. *Tensile Testing*

For mechanical testing the films were conditioned for at least 48 h at 23 °C and 50% relative humidity and then cut into dumbbell shaped samples (ISO 37 type 3, Elastocon, Sweden) and thereafter tested in the same conditions. Prior to testing the thickness of all the samples was measured at 5 different points (Mitutoyo IDC 112B) in the test area and the average was used for calculations. A clamp separation distance of 40 mm, crosshead speed of 10 mm/min and a 500 N load cell was used to test all samples on an Instron 5566 universal testing machine with a Bluehill software (Instron AB, Danderyd, Sweden).

3.6. *Fourier Transform Infrared Spectroscopy*

All the samples were dried for at least 72 h in a desiccator over silica gel prior to the analysis. FT-IR spectroscopy was carried out on all films and powders using a Spectrum 2000 FT-IR spectrometer (Perkin-Elmer inc., Norwalk, CN, USA) equipped with single reflection ATR (Golden Gate, Speac Ltd., USA). Data was Fourier self-de-convoluted using the Spectrum software with γ=2 and a smoothing factor of 70%.

4. Conclusions

Chemical modification of plant proteins, through heating at a basic pH, results in a protein with a decreased amount of intermolecular disulphide bonds and thereby a less cross-linked structure, as shown here for both WG and PP. Cross-linking of the proteins in the starting powder, when producing bio-based materials, largely impacts the properties of compression molded materials both in terms of opportunities for the proteins to further cross-link and form new bonds and structures during processing. From our previous study, mild protein extraction of WG has been shown to contribute with less cross-linked protein powder which when hot compression molded into films resulted in improved tensile properties (E-modulus and maximum stress) [5]. However, a less cross-linked structure in the protein starting material does not necessarily contribute to improved properties of processed bio-based materials. Here, a decrease in cross-linking of proteins by chemical modification, improved cross-linking (during temperature pressing) and tensile properties (E-modulus and maximum stress) of the material of PP but not for WG. Thus, using basic pH for modification of WG must contribute to irreversible changes in the protein conformation/structure and/or to chemical reactions not involving disulphide bond formation, which reduces the opportunities for the proteins to re-cross-link when processed. For PP, instead the chemical modification through basic pH contributed to the breaking of disulphide bonds which were created during industrial production of the PP starting material (the powder).

A major advantage with the chemical modification of PP was that it enabled the production of films with improved properties at a pressing temperature as low as 130 °C, thereby contributing to a lower environmental foot-print due to a reduction of energy use.

Thus, chemical modification of proteins for bio-based material production on an industrial scale may potentially help to widen the functional properties and improve the socio-economic value of the product (especially in the case of PP). Figure 6 provides a schematic summary of the important findings of this study.

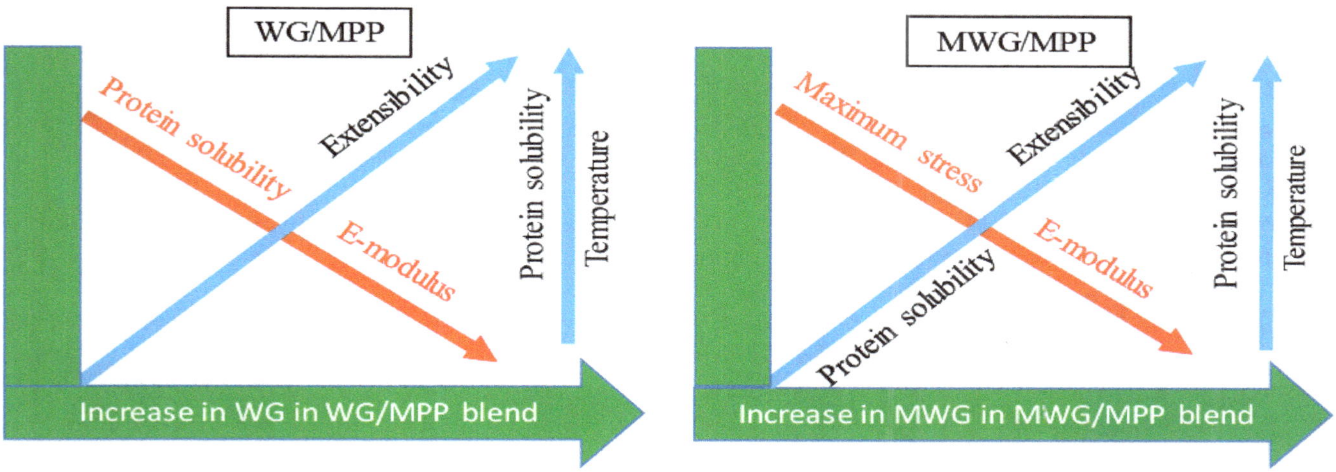

Figure 6. Schematic diagram of summarized effects of temperature and composition of the blend on protein polymerization and mechanical properties.

Author Contributions: R.K. conceived and designed the experiment together with F.M., F.M. performed the experiments and wrote the manuscript with contribution from the co-authors. M.S.H. performed FT-IR measurements and contributed with critical reading of the manuscript. R.K. contributed with performance of some SAXS analysis, coordinated SAXS experiments and obtained a beamtime for SAXS analysis, while T.S.P. contributed with interpretation of SAXS results. E.J. and R.K. contributed with obtaining of partial funding for this study and with critical commenting and writing of this manuscript.

Acknowledgments: We acknowledge Maria Luisa-Prieto Linde for her kind assistance in the laboratory, Lyckeby Starch AB for their kind provision of the potato protein and William Newson for his kind editing of the manuscript and English.

Abbreviations

Ex	Extraction
FT-IR	Fourier Transform Infrared Spectroscopy
HMw	High molecular weight proteins
LMw	Low molecular weight proteins
MPP	Modified potato protein
MWG	Modified wheat gluten
PP	Potato protein
SE-HPLC	Size Exclusion High Performance Liquid Chromatography
T_g	Glass transition temperature
WG	Wheat gluten

References

1. Kuktaite, R.; Newson, W.R.; Rasheed, F.; Plivelic, T.S.; Hedenqvist, M.S.; Gällstedt, M.; Johansson, E. Monitoring Nanostructure Dynamics and Polymerization in Glycerol Plasticized Wheat Gliadin and Glutenin Films: Relation to Mechanical Properties. *ACS Sustain. Chem. Eng.* **2016**, *4*, 2998–3007. [CrossRef]
2. Newson, W.R.; Rasheed, F.; Kuktaite, R.; Hedenqvist, M.S.; Gällstedt, M.; Plivelic, T.S.; Johansson, E. Commercial potato protein concentrate as a novel source for thermoformed bio-based plastic films with unusual polymerisation and tensile properties. *RSC Adv.* **2015**, *5*, 32217–32226. [CrossRef]
3. Neurath, H.; Greenstein, J.P.; Putnam, F.W.; Erickson, J.A. The chemistry of protein denaturation. *Chem. Rev.* **1944**, *34*, 157–265. [CrossRef]
4. Du, Y.; Chen, F.; Zhang, Y.; Rempel, C.; Thompson, M.R.; Liu, Q. Potato protein isolate-based biopolymers. *J. Appl. Polym. Sci.* **2015**, 132. [CrossRef]
5. Rasheed, F.; Hedenqvist, M.S.; Kuktaite, R.; Plivelic, T.S.; Gällstedt, M.; Johansson, E. Mild gluten separation–A non-destructive approach to fine tune structure and mechanical behavior of wheat gluten films. *Ind. Crops Prod.* **2015**, *73*, 90–98. [CrossRef]

6. Van Koningsveld, G.A.; Gruppen, H.; de Jongh, H.H.J.; Wijngaards, G.; van Boekel, M.A.J.S.; Walstra, P.; Voragen, A.G.J. Effects of pH and Heat Treatments on the Structure and Solubility of Potato Proteins in Different Preparations. *J. Agric. Food Chem.* **2001**, *49*, 4889–4897. [CrossRef] [PubMed]
7. Ullsten, N.H.; Cho, S.-W.; Spencer, G.; Gallstedt, M.; Johansson, E.; Hedenqvist, M.S. Properties of Extruded Vital Wheat Gluten Sheets with Sodium Hydroxide and Salicylic Acid. *Biomacromolecules* **2009**, *10*, 479–488. [CrossRef] [PubMed]
8. Pommet, M.; Morel, M.-H.; Redl, A.; Guilbert, S. Aggregation and degradation of plasticized wheat gluten during thermo-mechanical treatments, as monitored by rheological and biochemical changes. *Polymer* **2004**, *45*, 6853–6860. [CrossRef]
9. Olabarrieta, I.; Cho, S.W.; Gällstedt, M.; Sarasua, J.R.; Johansson, E.; Hedenqvist, M.S. Aging properties of films of plasticized vital wheat gluten cast from acidic and basic solutions. *Biomacromolecules* **2006**, *7*, 1657–1664. [CrossRef]
10. Ture, H.; Gallstedt, M.; Kuktaite, R.; Johansson, E.; Hedenqvist, M.S. Protein network structure and properties of wheat gluten extrudates using a novel solvent-free approach with urea as a combined denaturant and plasticiser. *Soft Matter* **2011**, *7*, 9416–9423. [CrossRef]
11. Rasheed, F.; Newson, W.R.; Plivelic, T.S.; Kuktaite, R.; Hedenqvist, M.S.; Gällstedt, M.; Johansson, E. Macromolecular changes and nano-structural arrangements in gliadin and glutenin films upon chemical modification: Relation to functionality. *Int. J. Biol. Macromol.* **2015**, *79*, 151–159. [CrossRef] [PubMed]
12. Kuktaite, R.; Plivelic, T.S.; Cerenius, Y.; Hedenqvist, M.S.; Gallstedt, M.; Marttila, S.; Ignell, R.; Popineau, Y.; Tranquet, O.; Shewry, P.R.; et al. Structure and Morphology of Wheat Gluten Films: From Polymeric Protein Aggregates toward Superstructure Arrangements. *Biomacromolecules* **2011**, *12*, 1438–1448. [CrossRef] [PubMed]
13. Ullsten, N.H.; Gällstedt, M.; Spencer, G.M.; Johansson, E.; Marttila, S.; Ignell, R.; Hedenqvist, M.S. Extruded High Quality Materials From Wheat Gluten. *Polym. Renew. Resour.* **2010**, *1*, 173–186. [CrossRef]
14. Gennadios, A.; Brandenburg, A.H.; Weller, C.L.; Testin, R.F. Effect of pH on properties of wheat gluten and soy protein isolate films. *J. Agric. Food Chem.* **1993**, *41*, 1835–1839. [CrossRef]
15. Kuktaite, R.; Türe, H.; Hedenqvist, M.S.; Gällstedt, M.; Plivelic, T.S. Gluten Biopolymer and Nanoclay-Derived Structures in Wheat Gluten–Urea–Clay Composites: Relation to Barrier and Mechanical Properties. *ACS Sustain. Chem. Eng.* **2014**, *2*, 1439–1445. [CrossRef]
16. Newson, W.R.; Kuktaite, R.; Hedenqvist, M.S.; Gällstedt, M.; Johansson, E. Effect of Additives on the Tensile Performance and Protein Solubility of Industrial Oilseed Residual Based Plastics. *J. Agric. Food Chem.* **2014**, *62*, 6707–6715. [CrossRef]
17. Kowalczyk, D.; Baraniak, B. Effects of plasticizers, pH and heating of film-forming solution on the properties of pea protein isolate films. *J. Food Eng.* **2011**, *105*, 295–305. [CrossRef]
18. Domenek, S.; Morel, M.-H.; Bonicel, J.; Guilbert, S. Polymerization kinetics of wheat gluten upon thermosetting. A mechanistic model. *J. Agric. Food Chem.* **2002**, *50*, 5947–5954. [CrossRef]
19. Gällstedt, M.; Mattozzi, A.; Johansson, E.; Hedenqvist, M.S. Transport and Tensile Properties of Compression-Molded Wheat Gluten Films. *Biomacromolecules* **2004**, *5*, 2020–2028. [CrossRef]
20. Muneer, F.; Johansson, E.; Hedenqvist, M.S.; Plivelic, T.S.; Markedal, K.E.; Petersen, I.L.; Sørensen, J.C.; Kuktaite, R. The impact of newly produced protein and dietary fiber rich fractions of yellow pea (Pisum sativum L.) on the structure and mechanical properties of pasta-like sheets. *Food Res. Int.* **2018**, *106*, 607–618. [CrossRef]
21. Gontard, N.; Guilbert, S.; Cuq, J.-L. Water and Glycerol as Plasticizers Affect Mechanical and Water Vapor Barrier Properties of an Edible Wheat Gluten Film. *J. Food Sci.* **1993**, *58*, 206–211. [CrossRef]
22. Johansson, E.; Malik, A.H.; Hussain, A.; Rasheed, F.; Newson, W.R.; Plivelic, T.; Hedenqvist, M.S.; Gällstedt, M.; Kuktaite, R. Wheat Gluten Polymer Structures: The Impact of Genotype, Environment, and Processing on Their Functionality in Various Applications. *Cereal Chem. J.* **2013**, *90*, 367–376. [CrossRef]
23. Newson, W.R.; Kuktaite, R.; Hedenqvist, M.; Gällstedt, M.; Johansson, E. Oilseed Meal Based Plastics from Plasticized, Hot Pressed Crambe abyssinica and Brassica carinata Residuals. *J. Am. Oil Chem. Soc.* **2013**, *90*, 1229–1237. [CrossRef]
24. Gerrard, J.A. Protein–protein crosslinking in food: Methods, consequences, applications. *Trends Food Sci. Technol.* **2002**, *13*, 391–399. [CrossRef]
25. Lagrain, B.; Goderis, B.; Brijs, K.; Delcour, J.A. Molecular basis of processing wheat gluten toward biobased materials. *Biomacromolecules* **2010**, *11*, 533–541. [CrossRef] [PubMed]

26. Rombouts, I.; Lagrain, B.; Delcour, J.A.; Türe, H.; Hedenqvist, M.S.; Johansson, E.; Kuktaite, R. Crosslinks in wheat gluten films with hexagonal close-packed protein structures. *Ind. Crops Prod.* **2013**, *51*, 229–235. [CrossRef]
27. Giosafatto, C.; Al-Asmar, A.; D'Angelo, A.; Roviello, V.; Esposito, M.; Mariniello, L. Preparation and Characterization of Bioplastics from Grass Pea Flour Cast in the Presence of Microbial Transglutaminase. *Coatings* **2018**, *8*, 435. [CrossRef]
28. Ceresino, E.B.; de Melo, R.R.; Kuktaite, R.; Hedenqvist, M.S.; Zucchi, T.D.; Johansson, E.; Sato, H.H. Transglutaminase from newly isolated Streptomyces sp. CBMAI 1617: Production optimization, characterization and evaluation in wheat protein and dough systems. *Food Chem.* **2018**, *241*, 403–410. [CrossRef]
29. Ceresino, E.B.; Kuktaite, R.; Sato, H.H.; Hedenqvist, M.S.; Johansson, E. Impact of gluten separation process and transglutaminase source on gluten based dough properties. *Food Hydrocolloid.* **2019**, *87*, 661–669. [CrossRef]
30. Morel, M.H.; Redl, A.; Guilbert, S. Mechanism of heat and shear mediated aggregation of wheat gluten protein upon mixing. *Biomacromolecules* **2002**, *3*, 488–497. [CrossRef]
31. Rasheed, F.; Newson, W.R.; Plivelic, T.S.; Kuktaite, R.; Hedenqvist, M.S.; Gallstedt, M.; Johansson, E. Structural architecture and solubility of native and modified gliadin and glutenin proteins: Non-crystalline molecular and atomic organization. *RSC Adv.* **2014**, *4*, 2051–2060. [CrossRef]
32. Blomfeldt, T.O.J.; Kuktaite, R.; Johansson, E.; Hedenqvist, M.S. Mechanical Properties and Network Structure of Wheat Gluten Foams. *Biomacromolecules* **2011**, *12*, 1707–1715. [CrossRef] [PubMed]
33. Muneer, F.; Andersson, M.; Koch, K.; Menzel, C.; Hedenqvist, M.S.; Gallstedt, M.; Plivelic, T.S.; Kuktaite, R. Nanostructural Morphology of Plasticized Wheat Gluten and Modified Potato Starch Composites: Relationship to Mechanical and Barrier Properties. *Biomacromolecules* **2015**, *16*, 695–705. [CrossRef] [PubMed]
34. Nowick, J.S. Exploring β-sheet structure and interactions with chemical model systems. *Acc. Chem. Res.* **2008**, *41*, 1319–1330. [CrossRef] [PubMed]
35. Jackson, M.; Mantsch, H.H. The Use and Misuse of FTIR Spectroscopy in the Determination of Protein Structure. *Critical Rev. Biochem. Mol. Biol.* **1995**, *30*, 95–120. [CrossRef] [PubMed]
36. Cho, S.W.; Gällstedt, M.; Johansson, E.; Hedenqvist, M.S. Injection-molded nanocomposites and materials based on wheat gluten. *Int. J. Biol. Macromol.* **2011**, *48*, 146–152. [CrossRef] [PubMed]
37. Blomfeldt, T.O.; Kuktaite, R.; Plivelic, T.S.; Rasheed, F.; Johansson, E.; Hedenqvist, M.S. Novel freeze-dried foams from glutenin-and gliadin-rich fractions. *RSC Adv.* **2012**, *2*, 6617–6627. [CrossRef]
38. Muneer, F.; Andersson, M.; Koch, K.; Hedenqvist, M.S.; Gällstedt, M.; Plivelic, T.S.; Menzel, C.; Rhazi, L.; Kuktaite, R. Innovative Gliadin/Glutenin and Modified Potato Starch Green Composites: Chemistry, Structure, and Functionality Induced by Processing. *ACS Sustain. Chem. Eng.* **2016**, *4*, 6332–6343. [CrossRef]
39. Sun, S.; Song, Y.; Zheng, Q. Thermo-molded wheat gluten plastics plasticized with glycerol: Effect of molding temperature. *Food Hydrocolloid.* **2008**, *22*, 1006–1013. [CrossRef]
40. Muneer, F.; Johansson, E.; Hedenqvist, M.S.; Gällstedt, M.; Newson, W.R. Preparation, Properties, Protein Cross-Linking and Biodegradability of Plasticizer-Solvent Free Hemp Fibre Reinforced Wheat Gluten, Glutenin, and Gliadin Composites. *BioResources* **2014**, *9*, 5246–5261. [CrossRef]

Permissions

All chapters in this book were first published by MDPI; hereby published with permission under the Creative Commons Attribution License or equivalent. Every chapter published in this book has been scrutinized by our experts. Their significance has been extensively debated. The topics covered herein carry significant findings which will fuel the growth of the discipline. They may even be implemented as practical applications or may be referred to as a beginning point for another development.

The contributors of this book come from diverse backgrounds, making this book a truly international effort. This book will bring forth new frontiers with its revolutionizing research information and detailed analysis of the nascent developments around the world.

We would like to thank all the contributing authors for lending their expertise to make the book truly unique. They have played a crucial role in the development of this book. Without their invaluable contributions this book wouldn't have been possible. They have made vital efforts to compile up to date information on the varied aspects of this subject to make this book a valuable addition to the collection of many professionals and students.

This book was conceptualized with the vision of imparting up-to-date information and advanced data in this field. To ensure the same, a matchless editorial board was set up. Every individual on the board went through rigorous rounds of assessment to prove their worth. After which they invested a large part of their time researching and compiling the most relevant data for our readers.

The editorial board has been involved in producing this book since its inception. They have spent rigorous hours researching and exploring the diverse topics which have resulted in the successful publishing of this book. They have passed on their knowledge of decades through this book. To expedite this challenging task, the publisher supported the team at every step. A small team of assistant editors was also appointed to further simplify the editing procedure and attain best results for the readers.

Apart from the editorial board, the designing team has also invested a significant amount of their time in understanding the subject and creating the most relevant covers. They scrutinized every image to scout for the most suitable representation of the subject and create an appropriate cover for the book.

The publishing team has been an ardent support to the editorial, designing and production team. Their endless efforts to recruit the best for this project, has resulted in the accomplishment of this book. They are a veteran in the field of academics and their pool of knowledge is as vast as their experience in printing. Their expertise and guidance has proved useful at every step. Their uncompromising quality standards have made this book an exceptional effort. Their encouragement from time to time has been an inspiration for everyone.

The publisher and the editorial board hope that this book will prove to be a valuable piece of knowledge for researchers, students, practitioners and scholars across the globe.

List of Contributors

Kostiantyn Protchenko, Fares Zayoud, Marek Urbański and Elżbieta Szmigiera
Faculty of Civil Engineering, Warsaw University of Technology, 00-637 Warsaw, Poland

Bobo Zhang, Chongxing Huang, Hui Zhao, Jian Wang, Cheng Yin, Lingyun Zhang and Yuan Zhao
College of Light Industry & Food Engineering, Guangxi University, Nanning 530004, China

J. Vincent Edwards, Krystal Fontenot, Alfred D. French and Brian D. Condon
Southern Regional Research Center, USDA, New Orleans, LA 70124, USA

Falk Liebner and Nicole Doyle nee Pircher
University of Natural Resources and Life Sciences Vienna, Konrad-Lorenz-Straße 24, A-3430 Tulln an der Donau, Austria

Na Liang
Key Laboratory of Photochemical Biomaterials and Energy Storage Materials, Heilongjiang Province, College of Chemistry & Chemical Engineering, Harbin Normal University, Harbin 150025, China

Shaoping Sun, Xianfeng Gong and Qiang Li
Key Laboratory of Chemical Engineering Process & Technology for High-Efficiency Conversion, College of Heilongjiang Province; School of Chemistry and Material Science, Heilongjiang University, Harbin 150080, China

Pengfei Yan
Key Laboratory of Chemical Engineering Process & Technology for High-Efficiency Conversion, College of Heilongjiang Province; School of Chemistry and Material Science, Heilongjiang University, Harbin 150080, China
Key Laboratory of Functional Inorganic Material Chemistry, Heilongjiang University, Harbin 150080, China

Fude Cui
School of Pharmacy, Shenyang Pharmaceutical University, Shenyang 110016, China

Maria Cristina Righetti, Norma Mallegni, Carlo Andrea Massa and Simona Bronco
CNR-IPCF, National Research Council—Institute for Chemical and Physical Processes, Via Moruzzi 1, 56124 Pisa, Italy

Patrizia Cinelli and Andrea Lazzeri
CNR-IPCF, National Research Council—Institute for Chemical and Physical Processes, Via Moruzzi 1, 56124 Pisa, Italy
Department of Civil and Industrial Engineering, University of Pisa, Largo Lucio Lazzarino 1, 56122 Pisa, Italy
National Inter University Consortium of Materials Science and Technology (INSTM), Via Giusti 9, 50121 Florence, Italy

Andreas Stäbler
Fraunhofer Institute for Process Engineering and Packaging IVV, Giggenhauser Straße, 35, 85354 Freising, Germany

Mihaela Tanase-Opedal and Gary Chinga-Carrasco
RISE PFI, Høgskoleringen 6b, 7491 Trondheim, Norway

Eduardo Espinosa and Alejandro Rodríguez
Chemical Engineering Department, Faculty of Science, Universidad de Córdoba, Building Marie-Curie, Campus de Rabanales, 14014 Córdoba, Spain

Maria-Beatrice Coltelli, Vito Gigante, Laura Aliotta and Luca Panariello
Department of Civil and Industrial Engineering, University of Pisa, Via Diotisalvi 2, 56122 Pisa, Italy
National Inter University Consortium of Materials Science and Technology (INSTM), Via Giusti 9, 50121 Florence, Italy

Pierfrancesco Morganti
Skin Pharmacology and Dermatology Unit, Campania University "Luigi Vanvitelli", 80100 Naples, Italy
MAVI SUD, Aprilia (LT), 04011 Aprilia, Italy
Department of Experimental Medicine, University of Campania "Luigi Vanvitelli", 80138 Naples, Italy

Su-Kyoung Baek, Sujin Kim and Kyung Bin Song
Department of Food Science and Technology, Chungnam National University, Daejeon 34134, Korea

Serena Danti
Department of Civil and Industrial Engineering, University of Pisa, 56122 Pisa, Italy
Consorzio Interuniversitario Nazionale per la Scienza e Tecnologia dei Materiali (INSTM), 50121 Florence, Italy

Luisa Trombi and Bahareh Azimi
Consorzio Interuniversitario Nazionale per la Scienza e Tecnologia dei Materiali (INSTM), 50121 Florence, Italy

Alessandra Fusco and Giovanna Donnarumma
Consorzio Interuniversitario Nazionale per la Scienza e Tecnologia dei Materiali (INSTM), 50121 Florence, Italy
Department of Experimental Medicine, University of Campania "Luigi Vanvitelli", 80138 Naples, Italy

Evaldas Balčiūnas, Nadežda Dreižė, Monika Grubliauskaitė, Silvija Urnikytė, Egidijus Šimoliūnas, Virginija Bukelskienė, Mindaugas Valius and Daiva Baltriukienė
Institute of Biochemistry, Life Sciences Centre, Vilnius University, 10257 Vilnius, Lithuania

Sara J. Baldock and John G. Hardy
Department of Chemistry, Lancaster University, Lancaster LA1 4YB, UK
Materials Science Institute, Lancaster University, Lancaster LA1 4YB, UK

Sanna Siljander, Pasi Keinänen, Anna Räty, Karthik Ram Ramakrishnan, Jyrki Vuorinen and Mikko Kanerva
Laboratory of Materials Science, Tampere University of Technology, FI-33720 Tampere, Finland

Sampo Tuukkanen
BioMediTech, Tampere University of Technology, FI-33720 Tampere, Finland

Vesa Kunnari and Ali Harlin
VTT Research Center, FI-02044 Espoo, Finland

Bei Liu, Hui Liang, Yue Liang, Zhibin Jia and Shuxin Shi
School of Materials Science and Engineering, Tianjin University of Technology, Tianjin 300384, China

Yun Zhao, Wei Li and Minfang Chen
School of Materials Science and Engineering, Tianjin University of Technology, Tianjin 300384, China
Key Laboratory of Display Materials and Photoelectric Device (Ministry of Education), Tianjin University of Technology, Tianjin 300384, China

Hongwei Bi
Tianjin Sannie Bioengineering Technology Co., Ltd., Tianjin 300384, China

Jinjun Yang
School of Environmental Science and Safety Engineering, Tianjin University of Technology, Tianjin 300384, China

Ricardo J. B. Pinto, Daniela Bispo, Carla Vilela, Sónia A. O. Santos and Carmen S. R. Freire
Department of Chemistry, CICECO—Aveiro Institute of Materials, University of Aveiro, 3810-193 Aveiro, Portugal

Alexandre M. P. Botas and Rute A. S. Ferreira
Phantom-G, Department of Physics, CICECO—Aveiro Institute of Materials, University of Aveiro, 3810-193 Aveiro, Portugal

Ana C. Menezes, Fábio Campos and Helena Oliveira
Department of Biology & CESAM, University of Aveiro, 3810-193 Aveiro, Portugal

Maria H. Abreu
ALGAplus—Prod. e Comerc. De Algas e Seus Derivados, Lda., 3830-196 Ílhavo, Portugal

Diana Isela Sanchez-Alvarado, Ulises Páramo-García and Reinaldo David Martínez-Orozco
Tecnológico Nacional de México/Instituto Tecnológico de Cd. Madero, Centro de Investigación en Petroquímica, Prol. Bahía de Aldhair y Av. De las Bahías, Parque de la Pequeña y Mediana Industria, Altamira, Tamaulipas 89600, Mexico

Javier Guzmán-Pantoja
Instituto Mexicano del Petróleo, Gerencia de Transformación de Hidrocarburos, Eje Central Lázaro Cárdenas No. 152, Col. San Bartolo Atepehuacán, 07730 Ciudad de México, Mexico

Alfredo Maciel-Cerda and Ricardo Vera-Graziano
Instituto de Investigaciones en Materiales, Universidad Nacional Autónoma de México, Circuito Exterior S/N, Circuito de la Investigación Científica, Ciudad Universitaria, 04510 Ciudad de México, Mexico

Faraz Muneer, Eva Johansson and Ramune Kuktaite
Department of Plant Breeding, Swedish University of Agricultural Sciences, SE-23053 Alnarp, Sweden

Mikael S. Hedenqvist
KTH Royal Institute of Technology, School of Engineering Sciences in Chemistry, Biotechnology and Health, Fibre and Polymer Technology, SE-10044 Stockholm, Sweden

Tomás S. Plivelic
MAX-IV Laboratory, Lund University, SE-22100 Lund, Sweden

Index